Sleep Loss and Obesity

Priyattam J. Shiromani
Tamas Horvath • Susan Redline
Eve Van Cauter
Editors

Sleep Loss and Obesity

Intersecting Epidemics

Springer

Editors
Priyattam J. Shiromani
Department of Psychiatry
and Behavioral Science
Ralph H. Johnson VA and Medical
University of South Carolina
Charleston, SC, USA

Susan Redline
Department of Medicine
Harvard Medical School
Brigham and Women's Hospital
and Beth Israel Deaconess
Medical Center
Boston, MA, USA

Tamas Horvath
Section of Comparative Medicine
Yale University School of Medicine
New Haven, CT, USA

Eve Van Cauter
Department of Medicine
University of Chicago
Chicago, IL, USA

ISBN 978-1-4614-3491-7 e-ISBN 978-1-4614-3492-4
DOI 10.1007/978-1-4614-3492-4
Springer New York Dordrecht Heidelberg London

Library of Congress Control Number: 2012937867

Printed on acid-free paper

Springer is part of Springer Science+Business Media (www.springer.com)

Preface

Obesity levels worldwide have almost doubled since 1980. Currently, it is estimated that 1.46 billion adults are overweight (body mass index greater than 25 kg/m^2) and 502 million are clinically obese (BMI>30 kg/m^2) [1]. These estimates include children where the percentage of children who are overweight or obese has more than doubled since 1974 [2]. If these trends continue then it is estimated that by 2020 three of four Americans and seven of ten people in the UK will be overweight or obese [3]. Obesity is a problem in both high-income countries as well as in many low income countries. Obesity has serious adverse consequences as it is an established risk factor for diseases such as type-2 diabetes, cardiovascular diseases, acid reflux, sleep apnea, kidney failure and many cancers [3]. In America one out of every eight deaths is caused by an illness directly related to being overweight or obese [4]. In the USA alone where current estimates show that two out of three people are overweight or obese [5], total healthcare costs attributable to obesity and overweight are projected to account for 16–18% of total US health-care expenditure by 2030 [3].

The increase in obesity is viewed to be primarily driven by changes in the global food system that makes it easy for individuals to consume energy rich foods and beverages [2]. However, emerging evidence over the last 10 years from neurobiological and clinical studies is pointing to a hitherto unknown variable—inadequate sleep, as a powerful driver of energy balance. It may appear odd to consider a good night's sleep as a weight control method, but the evidence is very strong that energy metabolism, sleep and the cellular processes that control them are inextricably intertwined. Disrupt any one of these and a vicious cycle may be triggered that ultimately leads to positive energy balance, and possible weight gain.

We have invited experts in the field to review and discuss the implication of this new evidence. The collection of reviews in this book makes the argument that poor or inadequate sleep is a trigger for increased energy metabolism. This is supported by well controlled, experimentally rigorous clinical studies in otherwise healthy normal weight subjects that have found changes in glucose regulation and appetite after just a few days of partial sleep restriction (reviewed in Chap. 10). In children and adolescents poor or inadequate sleep increases the risk for obesity (reviewed in Chap. 7). Epidemiological data also link insufficient sleep to hypertension and to direct measures of clinical cardiovascular disease and cardiovascular mortality (Chap. 12).

Disruption of the normal cycle of sleep, such as in night-shift work, contributes to weight gain and obesity (reviewed in Chap. 8).

The clinical evidence is supported by data from neurobiological studies that some of the brain neurons regulating feeding and energy metabolism also regulate sleep and waking (Chap. 3). In turn, these neurons interact with the biological clock that controls the timing of sleep and feeding (Chaps. 1 and 2). These interacting physiological mechanisms provide a biological basis for the observations that disruption of feeding or sleep schedules adversely impacts weight gain. Overweight and obese individuals are likely to have obstructive sleep apnea, which might be undiagnosed (Chap. 9) and clinical guidelines for evaluating and treating obstructive sleep apnea are provided in Chap. 13. In obstructive sleep apnea fat accumulation in the upper airway obstructs normal breathing during sleep, resulting in frequent arousals. Repeated arousals are likely to increase energy metabolism and may alter eating behaviors, thereby perpetuating a dangerous cycle. Obesity risk is also high in other sleep disorders such as insomnia and restless legs syndrome (Chap. 11). Once a person gains weight it may be difficult to lose it not only because consuming palatable food activates reward pathways in the brain (Chap. 4), but also because of the intriguing possibility that our brains are wired to overeat (Chap. 5). Furthermore, eating a high-fat diet can produce structural changes in neurons related to energy metabolism which can then also make it difficult to lose weight (Chaps. 2 and 5).

Given the close relationship between obesity and sleep, the genetic basis for obesity is reviewed (Chap. 6). In the section on treatment strategies Kabra et al. (Chap. 14) review pharmacological approaches to manage weight and Perna et al. (Chap. 15) detail current surgical choices for treating obesity.

Considering that lack of sleep has an insidious effect on weight control and health, we emphasize that it is important to include adequate sleep along with well-balanced nutritious foods and plenty of exercise as part of a healthy life style. And, yes, turn off the smart phone before going to sleep.

Finally, we would like to thank Richard Lansing at Springer for recognizing the need for this book, Joni Fraser for her enormous help and assistance with the publication process, and Reka Sasi for coordinating the printing process.

Charleston, SC, USA — Priyattam J. Shiromani
New Haven, CT, USA — Tamas Horvath
Boston, MA, USA — Susan Redline
Chicago, IL, USA — Eve Van Cauter

References

1. Finucane MM, Stevens GA, Cowan MJ, Danaei G, Lin JK, Paciorek CJ. National, regional, and global trends in body-mass index since 1980: systematic analysis of health examination surveys and epidemiological studies with 960 country-years and 9.1 million participants. Lancet 2011;377:557–567.
2. Swinburn BA , Gary Sacks, Kevin D Hall, Klim McPherson, Diane T Finegood, Marjory L Moodie et al. The global obesity pandemic: Shaped by global drivers and local environments. Lancet 2011;378:804–814.
3. Wang YC, McPherson K, Marsh T, Gortmaker SL, Brown M. Health and economic burden of the projected obesity trends in the USA and the UK. Lancet 2011;378:815–825.
4. Richard H. Carmona (Surgeon General U.S. Public Health Service Acting Assistant Secretary for Health Department of Health and Human Services). "The Obesity Crisis in America", Testimony Before the Subcommittee on Education Reform Committee on Education and the Workforce, United States House of Representatives; July 16, 2003.
5. Flegal KM, Carroll MD, Kit BK, Ogden CL. Prevalence of obesity and trends in the distribution of body mass index among US adults, 1999–2010. JAMA 2012;E1–6 [published online January 17, 2012].

Contents

Contributors

Joseph Bass Department of Medicine, Division of Endocrinology, Feinberg School of Medicine, Northwestern University, Chicago, IL, USA

Guglielmo Beccuti Department of Medicine, University of Chicago, Chicago, IL, USA

Thomas Karl Byrne Department of Surgery, Medical University of South Carolina, Charleston, SC, USA

Thomas Curie Center for Integrative Genomics, University of Lausanne, Lausanne-Dorigny, Switzerland

Ralph J. DiLeone Department of Psychiatry and Neurobiology, Yale School of Medicine, New Haven, CT, USA

Lawrence J. Epstein Division of Sleep Medicine, Brigham and Women's Hospital, Harvard Medical School, Brighton, MA, USA

Paul Franken Center for Integrative Genomics, University of Lausanne, Lausanne-Dorigny, Switzerland

Douglas J. Guarnieri Department of Psychiatry, Yale University, New Haven, CT, USA

Chantelle N. Hart Psychiatry & Human Behavior/Weight Control & Diabetes Research Center, Alpert Medical School of Brown University/ The Miriam Hospital, Providence, RI, USA

Susanna Hofmann Cincinnati Diabetes and Obesity Centre, Division of Internal Medicine, University of Cincinnati, Cincinnati, OH, USA

Kelly A. Holes-Lewis Department of Psychiatry and Behavioral Sciences, Medical University of South Carolina, Charleston, SC, USA

Hee-Kyung Hong Department of Medicine, Feinberg School of Medicine, Northwestern University, Chicago, IL, USA

Tamas Horvath Section of Comparative Medicine, Yale University School of Medicine, New Haven, CT, USA

Wenyu Huang Department of Medicine, Division of Endocrinology, Feinberg School of Medicine, Northwestern University, Chicago, IL, USA

Elissa Jelalian Psychiatry & Human Behavior/Bradley Hasbro Children's Research Center, Alpert Medical School of Brown University/The Miriam Hospital, Providence, RI, USA

Dhiraj G. Kabra Cincinnati Diabetes and Obesity Centre, Division of Internal Medicine, University of Cincinnati, Cincinnati, OH, USA

Uma D. Kabra Cincinnati Diabetes and Obesity Centre, Division of Internal Medicine, University of Cincinnati, Cincinnati, OH, USA

Roda Rani P. Konadhode Department of Psychiatry and Behavioral Science, Medical University of South Carolina, Charleston, SC, USA

Vidya Krishnan Division of Pulmonary, Critical Care and Sleep Medicine, MetroHealth Medical Center, Case Western Reserve University, Cleveland, OH, USA

Elizabeth S. Kuhl Psychiatry & Human Behavior/Weight Control & Diabetes Research Center, Alpert Medical School of Brown University/ The Miriam Hospital, Providence, RI, USA

Biliana Marcheva Department of Medicine, Division of Endocrinology, Feinberg School of Medicine, Northwestern University, Chicago, IL, USA

Rachel R. Markwald Department of Integrative Physiology, University of Colorado at Boulder, Boulder, CO, USA

Martin G. Myers Jr. Departments of Internal Medicine, University of Michigan, Ann Arbor, MI, USA

Department of Molecular and Integrative Physiology, University of Michigan, Ann Arbor, MI, USA

Nandakumar S. Narayanan Department of Neurobiology, Yale Medical School, New Haven, CT, USA

Silvana Pannain Department of Medicine, University of Chicago, Chicago, IL, USA

Sanjay R. Patel Division of Sleep Medicine, Brigham and Women's Hospital, Boston, MA, USA

Dheeraj Pelluru Department of Psychiatry and Behavioral Science, Medical University of South Carolina, Charleston, SC, USA

Mark J. Perna Department of Surgery, Medical University of South Carolina, Charleston, SC, USA

Chitharanja C. Pullattrana Department of Surgery, Medical University of South Carolina, Charleston, SC, USA

Kathryn M. Ramsey Division of Endocrinology, Department of Medicine, Feinberg School of Medicine, Northwestern University, Chicago, IL, USA

Susan Redline Department of Medicine, Harvard Medical School, Brigham and Women's Hospital and Beth Israel Deaconess Medical Center, Boston, MA, USA

Priyattam J. Shiromani Department of Psychiatry and Behavioral Science, Ralph H. Johnson VA and Medical University of South Carolina, Charleston, SC, USA

Matthias H. Tschöp Cincinnati Diabetes and Obesity Centre (CDOC), Division of Internal Medicine, University of Cincinnati, Cincinnati, OH, USA

Eve Van Cauter Department of Medicine, University of Chicago, Chicago, IL, USA

Kenneth P. Wright Jr. Department of Integrative Physiology, University of Colorado at Boulder, Boulder, CO, USA

Circadian Clock Genes and the Regulation of Sleep

1

Thomas Curie and Paul Franken

Abstract

Sleep and waking are controlled by opposing interactions between circadian and homeostatic processes. A circadian process generated by the suprachiasmatic nucleus determines when sleep should occur, while a homeostatic process keeps track of time spent awake and asleep and signals sleep need or sleep propensity. Recent evidence indicates that these two processes employ many of the same set of genes. Herein, we review the basic concepts of the circadian and homeostatic regulation of sleep, and then outline the molecular components of circadian clock. We then discuss the evidence demonstrating a role of clock genes in sleep homeostasis in flies, mice, and humans. We conclude by suggesting that clock genes might be crucial for integrating homeostatic need, not only that of sleep but also of food intake and energy metabolism.

The study of circadian rhythms and of sleep regulation have long been, and to some extent still are, considered as two separate fields of research. Whereas circadian researchers prefer to view sleep as one of several outputs of the circadian clock, sleep researchers have long believed that the circadian clock exerts a modulatory role on sleep but does not impinge on core sleep features such as the homeostatic control of sleep. The latter notion is based mainly on the observation that destruction of the suprachiasmatic nuclei (SCN), which harbor the central circadian pacemaker, results in loss of all circadian organization of overt behaviors while leaving the homeostatic sleep response to enforced wakefulness intact. Meanwhile, it has become clear that sleep and waking are controlled by the constant and fine-tuned opposing interaction between circadian and homeostatic processes. This integration of the two processes seems to have a molecular correlate because the genes thought to be exclusively involved in setting up the negative feedback loops driving circadian rhythms, referred to as clock genes, are also involved in the homeostatic regulation of sleep. This dual role of clock genes is the emphasis of this chapter.

The outline of this chapter is as follows: after introducing the basic concepts of the circadian and homeostatic regulation of sleep, we only briefly review the molecular components of self-sustained

T. Curie, PhD • P. Franken, PhD (✉)
Center for Integrative Genomics, University of Lausanne, Genopode Building, Quartier Sorge, Lausanne-Dorigny 1015, Switzerland
e-mail: Thomas.Curie@unil.ch; paul.franken@unil.ch

P.J. Shiromani et al. (eds.), *Sleep Loss and Obesity: Intersecting Epidemics*,
DOI 10.1007/978-1-4614-3492-4_1,

cellular circadian clocks as this has been the topic of many excellent earlier reviews. We then discuss the evidence demonstrating a role of clock genes in sleep homeostasis in flies, mice, and humans. We conclude by suggesting that clock genes might be crucial for integrating homeostatic need, not only that of sleep but also of food intake and energy metabolism.

Sleep Homeostasis and Circadian Rhythms: Concepts and Definitions

Although sleep is a behavior, sleep research in mammals, including humans, relies heavily on the recording of electrophysiological potentials measured from the scalp (in humans) or the surface of the cerebral cortex (in animals), both referred to as the electro-encephalogram (EEG). In combination with muscle tone (electro-myogram or EMG) and eye movements (electro-oculogram or EOG), three major behavioral states can reliably be assessed: wakefulness, rapid eye movement (REM) sleep, and non-REM (NREM) sleep. Analysis of these three states gives information on the amount, distribution, and quality of sleep indexed as sleep efficiency or sleep consolidation. Besides its utility to discriminate among behavioral states, the EEG signal contains a wealth of information concerning the processes governing neuronal activity in the structures implicated in generating the EEG. The electrophysiological correlates of sleep have been thoroughly examined and studies in both mice and humans demonstrated that amplitude and frequency of the oscillations comprising the EEG are strongly determined by genetic factors [1]. One widely used EEG component is slow wave activity or EEG delta power which quantifies the prevalence and amplitude of the delta oscillations (1–4 Hz) characteristic of deep NREM sleep. EEG delta power is considered a reliable physiological marker of homeostatic sleep need because its level during NREM sleep changes in function of the sleep–wake history such that it monotonically increases as time spent awake progresses and decreases with increasing time spent in NREM sleep [2, 3]. Although other aspects of mammalian sleep, such as the duration of NREM sleep and REM sleep, are homeostatically regulated as well, their dynamics are less predictable and depend on various other factors including circadian factors. Leading hypotheses on sleep function have their foundation in the cellular mechanisms underlying the sleep–wake-driven waxing and waning of EEG delta power [4, 5]. Also, the dynamics of this homeostatic process are genetically determined [6–9] demonstrating that the regulation of homeostatic sleep need has a molecular substrate [10–12].

Circadian rhythms are biological rhythms with a period of about 24 h (with "circa" referring to "about" and "dies" to "day") and have been observed in most life forms studied from bacteria, fungi, plants, to animals. In animals, locomotor activity has long been the main state variable to study circadian rhythmicity while in humans, physiological measures such as core body temperature and plasma melatonin levels serve to index the state of the clock characterized by the phase of peak or trough levels, amplitude, and period of the rhythm. Circadian rhythms are generated endogenously and, because the period deviates from 24 h, can only remain synchronized to the environmental light–dark cycle through daily phase resetting of the rhythm. Circadian rhythms are thought to have evolved as an adaption to the changes in environment accompanying the earth's rotation around its axis. An endogenous circadian clock allows the organism to anticipate these changes and organize, in time, biochemical and physiological processes thereby increasing efficient use of resources. Most physiological processes and behaviors like body temperature control, melatonin secretion, feeding, and the sleep–wake distribution are under circadian control.

A master clock has been identified in the brain of mammals and it is located in the SCN [13] resting on top of the optic chiasm. Lesioning and transplant studies in animals conclusively demonstrated that the SCN is both necessary and sufficient to endow the animal with a circadian organization of overt behavior and physiology [14]. Light information gathered in retinal ganglion cells reaches the SCN through a monosynaptic pathway, the retinal-hypothalamic tract (RHT). Through this pathway light information can alter the phase of circadian rhythm generated in the SCN thus maintaining correct phase alignment between the external light–dark cycle and internal time. Also in the brain of insects such as the fruit fly, *Drosophila melanogaster*, dedicated neurons have been identified needed for circadian organization of overt behaviors and physiology and for entrainment to light–dark cycles [15].

Opponent Processes Regulate Sleep

By far the most influential conceptual model of the regulation of sleep is Alexander Borbely's "two-process model" [2, 3]. This model stipulates that the interaction between two independent processes governs sleep–wake regulation: a circadian process generated by the SCN that sets the time of day at which sleep should preferably occur, and a homeostatic process that keeps track of time spent awake and asleep and which signals sleep need or sleep propensity. The dynamics of the latter process is directly derived from the sleep–wake-dependent changes in EEG delta power measured during NREM sleep. In the two-process model, the homeostatic sleep need increases during wakefulness until an upper threshold is reached (in humans after ca. 16 h of wakefulness) after which sleep is initiated. During sleep, homeostatic sleep need decreases until a lower threshold is reached (after ca. 8 h) at which time sleep ends. By modulating this lower (and upper) thresholds in a circadian manner, the model can reliably predict the nonmonotonic relationship between the time of sleep onset and subsequent sleep duration [3, 16].

The general concepts of the model are now supported by a wealth of observations, although it also has become evident that the interaction between the two processes extends well beyond the two time points of the start and end of the sleep period. In a so-called forced desynchrony protocol in which subjects were forced to adhere to an 28 h day which forces the sleep–wake cycle to desynchronize from the endogenous circadian rhythm, it was demonstrated that the circadian system generates a sleep–wake propensity rhythm that is timed to oppose homeostatic changes in sleep need [17]. This enables us to stay awake and alert throughout the day despite an accumulating need for sleep, and to remain asleep throughout the night despite a decreasing sleep need [18, 19]. These observations suggest that at any given time of the day, sleep propensity depends on the momentary balance between the two processes. When this balance is disturbed, for instance by misalignment between the two such as experienced during jet lag and shift work, then there is an immediate decrement in sleep quality and daytime functioning [17, 20, 21].

In the mouse and in humans, changes in EEG delta power are so predictable that its time course can be mathematically calculated in detail solely based on the sleep–wake distribution both under baseline conditions and under conditions of sleep deprivation [7, 22, 23] The close approximation of predicted and actual measures of EEG delta power supports the view that most of its variance reflects a sleep–wake-driven homeostatic process. Studies in animals rendered arrhythmic by lesioning the SCN or through a shift of the light–dark cycle show that the sleep deprivation-induced increase in EEG delta power is not affected and thus does not depend on a functional circadian system [24, 25]. Moreover, the above-mentioned forced desynchrony protocol revealed that the circadian contribution to the time course of EEG delta power during sleep is small [18]. These observations contributed to the notion that sleep homeostasis and circadian rhythm generation are separate processes and that EEG delta power is a robust marker of homeostatic sleep need.

Nevertheless, several examples indicate that the separation between the two processes is not so clear and that cross-talk exists. For instance, sleep deprivation has been demonstrated to phase shift the circadian clock in hamsters under constant dark conditions [26]. Moreover, firing rates of SCN neurons were found to be lower during NREM sleep and negatively correlated with the levels of EEG delta power attained during this state during baseline and sleep deprivation conditions [27, 28]. Also, in humans, the amplitude of the observed circadian variation in several physiological and behavioral variables depends on sleep homeostatic processes [19]. Finally, circadian clock genes, besides their established role in generating circadian rhythms, seem also to play a role in sleep homeostasis, suggesting that the molecular circuitry used to set internal time-of-day, could be also utilized to track and anticipate sleep need (see "Circadian Rhythms as an Emerging Property of Transcriptional Feedback Loops" section).

Circadian Rhythms as an Emerging Property of Transcriptional Feedback Loops

The molecular correlates underlying the sleep homeostat are unknown. On the other hand, the specific genes and how they interact to generate circadian rhythms in mammals have been worked out in detail over the last 20 years [29]. In the fruit fly, mutagenesis screen yielded the earliest evidence for a molecular substrate of circadian rhythm generation [30]. A similar approach in the mouse led in 1994 to the identification and later to the cloning of the gene Circadian Locomotor Output Cycles Kaput (CLOCK) as a pivotal or "core" circadian clock gene [31, 32].

At the molecular level, circadian rhythms have been explained by an oscillatory network of several transcriptional regulators engaged in transcriptional feedback loops [33]. The core of this self-sustained oscillation consists of positive and negative elements. In mammals, the core positive elements are three basic helix–loop–helix (bHLH) Per-Arnt-Sim (PAS)-domain-containing transcription factors: CLOCK, NPAS2 (neuronal PAS-domain protein 2), and BMAL1 (brain and muscle Arnt-Like 1). BMAL1 can partner with either CLOCK or NPAS2 to form transcriptionally active heterodimers. Both dimers can drive the transcription of *Period* (*Per1*, *Per2*, *Per3*) and *Cryptochrome* (*Cry1*, *Cry2*) genes. After translation and reentry into the cell nucleus, PER and CRY protein complexes can suppress CLOCK:BMAL1 and NPAS2:BMAL1-mediated transcription thereby forming the negative elements in the feedback loop. After this inhibition is removed through post-translational protein modifications including phosphorylation, acetylation, or protein degradation, a new cycle can be initiated (Fig. 1.1). This network is far more complex and additional feedback loops between these core clock genes, interactions at the level of transcription and translocation back into the nucleus with numerous modulators and post-transcriptional and -translational modifications, add further complexity and stability to this oscillatory network [34–36]. The importance of these clock genes in circadian rhythm generation has been demonstrated by constructing mice carrying targeted disruptions for one or a combination of these genes. Thus, *Bmal1* knock-out mice and *Clock-Npas2*, *Cry1-Cry2,* and *Per1-Per2* double knock-out mice, lack circadian organization of overt behavioral rhythms when placed under constant dark conditions [29].

Clock genes are not expressed only in the SCN, they are expressed in most tissues peripheral to the SCN (including the brain) and capable of generating cell-autonomous and self-sustained oscillations in gene expression. This allows for the study of circadian rhythms in tissue explants, cell cultures, and in blood and buccal samples and changes of mRNA levels of clock genes have become as widely used as markers of circadian state as locomotor activity [29, 37–41]. Because cells throughout the body can generate circadian rhythms, the view of the role of the SCN has changed; instead of being the only central clock driving oscillations in the periphery, the SCN is now seen as an orchestrator keeping phase throughout the body so that the various organs are appropriately synchronized to the external light–dark cycle [40, 42].

Although incredible advances have been made in unraveling the molecular circadian clock, still much is to be learned. Besides transcription, post-translation changes such as protein phosphorylation and degradation are critical in obtaining periods in the range of 24 h. Circadian rhythms in protein level of some clock genes have been observed in the absence of rhythms in their transcript [43, 44] and *in vitro* results demonstrate that the phosphorylation state of the cyanobacteria clock gene *KaiC* is sufficient to drive circadian

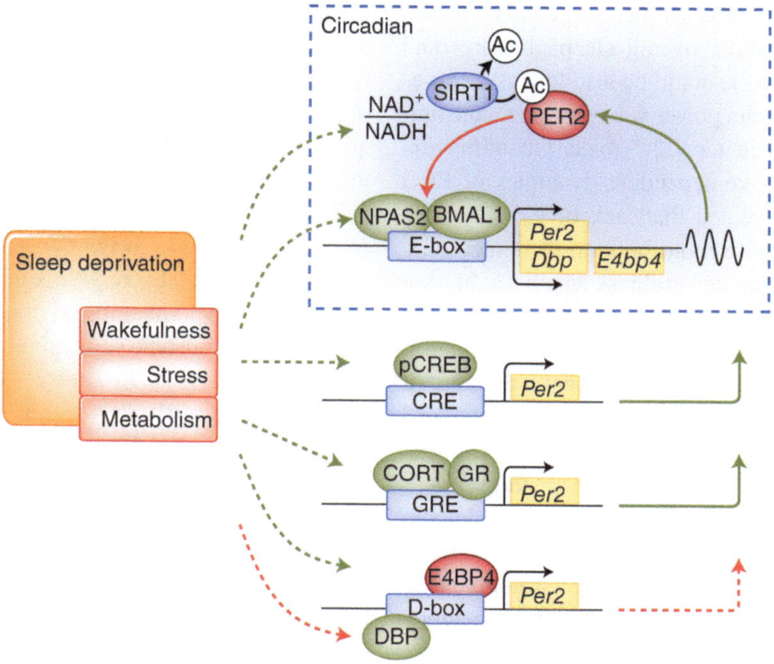

Fig. 1.1 Schematic summary of the molecular mechanisms by which sleep deprivation could alter clock genes and especially *Per2* expression in the forebrain. A negative feedback loop underlies circadian rhythm generation. The positive arm of this feedback loop consists of NPAS2 (or CLOCK)::BMAL1 dimers that activate the transcription of *Per* and *Cry* (only *Per2* shown) through their E-box enhancers. PER and CRY proteins then increase (*green arrow*) and, upon nuclear entry, associate with the NPAS2::BMAL1 complex to inhibit their own transcriptional activation thus providing negative feedback (*red arrow*). PER and CRY proteins then decrease resulting in a removal of this inhibition allowing the cycle to restart. As a result NPAS2::BMAL1 target genes, such as *Per2*, *Dbp*, and *E4bp4*, cycle with a near 24 h period in many tissues and have been widely utilized as a state variable of the circadian clock. Besides this circadian organization, *Per2* expression is regulated by many other factors through which sleep deprivation might act: for example, the cAMP-response element (CRE), glucocorticoid response element (GRE), and D-box enhancers through which phosphorylated CRE binding protein (pCREB), ligand-bound glucocorticoid receptors (GR), and DBP can induce *Per2* expression, respectively, while E4BP4, also acting on the D-box can repress *Per2* expression. Sleep deprivation, besides extending waking, also activates the HPA (hypothalamic–pituitary–adrenal) axis and metabolism, evidenced by its induction of corticosterone (CORT), increases CREB phosphorylation, and increases in *Npas2* and *E4bp4* expression (*green dashed arrows*), and decreases *Dbp* expression (*red dashed arrow*). In addition, SD-induced changes in metabolic state as evidenced by changes in NAD+ can directly affect NPAS2-mediated transcriptional activation and the SIRT1-mediated deacetylation (Ac) of PER2 (see also Chap. 2)

rhythms [45]. Recent results in red blood cells (that do not have a cell nucleus) show that non-transcriptional events are sufficient to sustain cellular circadian rhythms [46, 47].

Clock Genes and Sleep Homeostasis

Sleep regulation was examined in mice lacking both *Cry1* and *Cry2* circadian genes (*Cry1,2*$^{-/-}$ double knock-out mice) to further demonstrate the independence of circadian and sleep homeostatic processes [48]. *Cry1,2*$^{-/-}$ mice lack a functioning circadian clock and are behaviorally arrhythmic when kept under constant dark conditions [49, 50]. *Cry1,2*$^{-/-}$ mice did, however, not have a sleep phenotype which could be expected of an arrhythmic animal such as a noncircadian distribution of sleep and wakefulness with fewer consolidated bouts. Instead, these mice spent more time in NREM sleep, and NREM sleep was more consolidated (i.e., longer, uninterrupted NREM sleep episodes) and EEG delta power was higher. With the increased sleep time and increased sleep consolidation, overall sleep pressure (and EEG delta power) should be also low as a consequence. EEG delta power was, however, constitutively elevated in *Cry1,2*$^{-/-}$ mice. Quantification of the sleep–wake-dependent dynamics of EEG delta power revealed that lack of *Cry1,2* genes resulted in a more rapid buildup of homeostatic sleep need during wakefulness which might have contributed to the higher levels in EEG delta power compared to wild-type mice. Thus, sleep in *Cry1,2*$^{-/-}$ mice has all the characteristics of sleep in sleep-deprived wild-type mice suggesting that a fundamental aspect of sleep homeostasis was altered. This homeostatic phenotype was not observed in *Cry1*$^{-/-}$ and *Cry2*$^{-/-}$ single knock-out mice [51], confirming functional redundancy between the two CRY1 and CRY2 proteins as has been observed for circadian rhythm generation [49, 50].

In other mouse lines carrying targeted disruptions for one or two related circadian genes, aberrant sleep homeostatic phenotypes were also observed. Mice homozygous for a *Bmal1* deletion showed increases in total sleep time, sleep fragmentation, and EEG delta power under baseline conditions, and an attenuated compensatory response to sleep deprivation [52]. In *Clock* mutant mice, decreases in NREM sleep time and consolidation were reported under baseline conditions together with a reduced compensatory rebound in REM sleep after sleep deprivation [53]. Similarly, mice lacking the clock-controlled gene *Dbp* (albumin D-binding protein) also showed altered homeostatic regulation of REM sleep [54]. Knock-out mice for *Npas2* were found to sleep less in the latter half of the baseline dark period, a time-of-day when sleep need is high and wild-type mice showed a consolidated period of sleep (i.e., nap), conceivably to discharge accumulated sleep pressure [55]. After further raising sleep pressure experimentally by means of an 8-h sleep deprivation, *Npas2*$^{-/-}$ knock-out mice were, like in baseline, incapable of initiating the appropriate compensatory behavior (i.e., sleep) during the circadian phase when mice are usually awake (i.e., the dark period) [55]. Besides the blunted homeostatic response in NREM sleep time after sleep deprivation, *Npas2*$^{-/-}$ mice could not adjust to a novel feeding schedule in which food availability was restricted to the "wrong" time-of-day (i.e., the light phase) and, as a consequence, would have starved if not taking care of [56]. This underscores the importance of NPAS2 in sensing "homeostatic drive," both for sleep and food. In the same way, *Clock* mutant mice, besides having severely disrupted circadian rhythms, are hyperphagic and obese and develop a metabolic syndrome indicative of an impaired energy balance [57]. In *Per1,2*$^{-/-}$ double knock-out mice (*Per1*$^{-/-}$, *Per2*$^{Brdm/Brdm}$), EEG delta power seemed enhanced to a greater extent after sleep deprivation compared to wild-type mice [58], suggesting that also *Per1* and *Per2* genes are functionally implicated in sleep homeostasis. Evidence for circadian genes being implicated in the homeostatic regulation of sleep also exist in the fruit fly, *D. melanogaster* [59–62]. This was most dramatically illustrated in flies lacking a functional *Cycle* gene, a *Bmal1* homolog [60]. These flies have an exaggerated homeostatic response to sleep deprivation and die after sleep deprivations longer than 10 h. The effect of the mutation of *Cycle* gene on sleep homeostasis was found to be sexually dimorphic and also to shorten life-span [61]. More recently, evidence of a noncircadian involvement of circadian genes in sleep–wake regulation was obtained also in humans [8], pointing to a possible evolutionary conserved pathway.

The lack of functional circadian genes in the different transgenic mouse models affects the expression level of the other elements of the circadian molecular feedback loop described above. For instance, the deletion of both *Cry1* and *Cry2* genes disinhibits CLOCK:BMAL1 and NPAS2:BMAL1-mediated transcriptional activation and, as a consequence, *Per1* and *Per2* transcripts are increased in *Cry1,2*$^{-/-}$ double knock-out mice both in liver and in brain [48, 63, 64]. Similarly, *Per1* and *Per2* expression is decreased in *Clock* mutant mice [65, 66]. The homeostatic sleep phenotype observed in these transgenic mice seems to correlate with *Per1* and *Per2* gene expression; in *Cry1,2*$^{-/-}$ double knock-out mice, both homeostatic sleep need and *Per1* and *Per2* gene expression in the brain seem upregulated [48], while the opposite is observed for *Clock* mutant mice [53].

If indeed sleep need covaries with *Per1* and *Per2* expression, then sleep deprivation should increase *Per1* and *Per2* gene expression in intact, wild-type mice. Cortical expression of both *Per1* and *Per2* genes increased after enforced wakefulness [11, 12, 48, 55, 67] and was found to monotonically increase in function of the duration of the time that mice were kept awake [68]. *In situ* hybridization studies revealed that in the brain and especially in the cerebral cortex, the thalamic nuclei, and the cerebellum, *Per1* and *Per2* expression was increased after sleep deprivation [51, 68]. Importantly, *Per1* and *Per2* expression reverted to control levels within 2 h of recovery sleep confirming a close relationship with sleep need [48, 68]. Also in wild-type animals that are not sleep-deprived, levels of *Per1* and *Per2* expression are high at times when EEG delta power is high. *Per1* and *Per2* expression seems, however, to consistently follow NREMs propensity only in the forebrain leaving expression in the SCN unaffected [56, 69–71]. In the cerebral cortex of both nocturnal (mice) and diurnal species (ground squirrels), *Per1* and *Per2* gene expression is maximal in conjunction with the major waking episode, while in the SCN their expression peaks in the middle of the (subjective) light period [70, 72]. Similarly, under conditions where the circadian phase (restricted feeding or methamphetamine administration) or the circadian distribution ("circadian splitting") of locomotor activity is altered, *Per1* and *Per2* expression in the cerebral cortex follows the altered behavioral rhythm of wakefulness, again leaving *Per1* and *Per2* expression in the SCN unaffected [56, 69–71]. Consistent with these observations, *Per2* mRNA and protein levels in the SCN are not affected by sleep deprivation [118]. Taken together, these observations suggest that the mRNA level of the genes *Per1* and *Per2* in the forebrain covary with homeostatic sleep need.

Thus, in contrast to their role in the SCN, circadian genes seem not to be a component of a self-sustained circadian oscillator in the forebrain. Instead, their expression illustrated here by *Per1* and *Per2* gene expression seems to follow the sleep–wake cycle. An implicit assumption is that the sleep–wake-dependent changes in cortical *Per1* and *Per2* expression are mediated by CLOCK:BMAL1- and/or NPAS2:BMAL1-induced transcription. The altered homeostatic response in mice with targeted disruptions of clock genes supports this assumption. Moreover, the observation in *Npas2*$^{-/-}$ knock-out mice that *Per2* gene expression in the cerebral cortex no longer follows the circadian sleep–wake distribution [73], together with the fact that *Per2* gene expression is reduced twofold after sleep deprivation compared to wild-type mice [55], underscores that at least part of the sleep–wake dependent changes in circadian gene expression underlie the same circuitry governing circadian-dependent changes in gene expression. This suggests that NPAS2 and not CLOCK is critical in coupling overt behavior with *Per2* gene expression in the forebrain. The neuroanatomical distribution of NPAS2 is particularly suitable to fulfill such a function since it is abundantly expressed in the brain with dense expression in thalami nuclei and the cerebral cortex and with no noticeable expression in the SCN [74]. Apart from a blunted homeostatic response in NREM sleep time and in *Per2* gene expression after sleep deprivation, the EEG activity of thalamacortical origin (such as EEG spindles and delta activity) is altered in *Npas2*$^{-/-}$ knock-out mice [55].

The circadian expression of *Per* and *Cry* genes are regulated by CLOCK:BMAL1 and NPAS2:BMAL1 through binding on E-boxes sequences [29]. Nevertheless, other *cis*-acting elements, such as cAMP-response (CRE) [75–77], DBP-binding (D-box) elements [78, 79], and glucocorticoids response elements (GRE) [44, 80] are known to be also involved in the regulation of the *Per1* and *Per2* expression (Fig. 1.1). Since CREB signaling also increases with prolonged wakefulness [81, 82], this pathway might have contributed to the increase in *Per1* and *Per2* expression after sleep deprivation. Moreover, in a recent study, we demonstrated that the SD-induced increase in the *Per1* and *-3* gene expression, but not that of *Per2*, was to a large extent regulated by the SD-associated increase in the secretion of the glucocorticoid, corticosterone (Fig. 1.1) [12].

There are other examples of molecules known to be affected by sleep and waking that also affect *Per1* or *Per2* expression but their influence in explaining the wake-dependent increase in *Per1* and *Per2* gene expression is less straightforward. For instance, the transcription factor DBP, itself a target of CLOCK- and NPAS2-mediated transcription (Fig. 1.1) [83], promotes *Per1* and *Per2* gene expression by acting on D-boxes on the promoter of *Per1* and *Per2* [78, 79]. *Dbp* expression is, however, known to decrease during sleep deprivation [48, 68]. Similarly, tumor-necrosis factor alpha (TNFα), a cytokine involved in sleep–wake regulation and which increases with sleep loss [5, 84], suppresses *Per1*, *Per2*, and *Per3* gene expression by interfering with E-box-mediated transcription [85]. This seems, however, inconsistent with the widely reported increase in *Per1*, *Per2,* and *Per3* gene expression after sleep deprivation.

In humans, the role of circadian genes in the homeostatic sleep process and circadian process has been investigated by assessing the consequences of natural variation in these genes (i.e., polymorphisms and mutations). A variable number tandem repeat (VNTR) polymorphism in the *Per3* gene associates with diurnal preference and delayed sleep phase syndrome [86]. In a subsequent study, the *Per3* VNTR polymorphism was shown to also affect the electrophysiological and behavioral markers of sleep homeostasis such as EEG delta power and waking performance [8] and in particular executive function in the early morning after a day and a night without sleep [87]. These observations imply that the *PER3* VNTR polymorphism affects the homeostatic aspect of sleep regulation and confirm that the above observations are not restricted to mice and flies but point to a possible conserved molecular pathway important for sleep homeostasis.

In this chapter, the different mutations for the circadian genes and their sleep–wake-dependent change in expression make a strong case for their involvement in sleep homeostasis and demonstrate that circadian and sleep processes of sleep, at a molecular level, cannot easily be separated. The present results might also be of relevance for research on the biology of mood disorders. In this field, there is increasing interest in the relation between clock genes and e.g., seasonal affective disorders [88] or bipolar disorders [89]. The effects of sleep deprivation on clock-gene expression discussed here, combined with the fact that sleep deprivation is a well-known antidepressant [90, 91], add a molecular component to the intricate relationship between mood, circadian rhythms, and sleep (reviewed in [88, 89, 92–95]).

Functional Implications

A tight relation between sleep and metabolism has been extensively documented and has led to hypotheses on sleep function [96–98]. These relationships can be observed at various levels of organization from insufficient sleep as a predisposing factor for metabolic syndrome including type 2 diabetes and obesity [99], to extracellular changes in redox potential and lactate associated with the sleep–wake distribution and sleep deprivation [100, 101]. Increasing evidence also tightly links circadian clock genes to metabolism; for example, *Clock* mutant mice develop metabolic syndrome indicative of an impaired energy balance [57]. Clock genes are also implicated in adapting to impose feeding cycles [102]. Direct molecular evidence comes from the discovery that NPAS2 and CLOCK transcription factors can directly sense cellular energy state and that their transcriptional activity depends on intracellular redox potential [64, 103]. Moreover, NPAS2 also targets metabolic genes like the *Lactate dehydrogenase-A* (*Ldha*) gene necessary for converting pyruvate into lactate, a reaction that depends on the oxidation of the NADH to NAD^+, thereby affecting redox potential [103]. More recent data add to the tight relationship between metabolism and circadian genes. The sirtuin 1 (SIRT1) deacetylase enzyme has been shown to be able to modulate CLOCK-mediated chromatin remodeling and to deacetylate PER2 protein [36, 104]. The activity of this enzyme depends on the reduction of NAD^+ to NADH (Fig. 1.1). Other direct molecular links between metabolism and circadian circuitry have been observed for the peroxisome proliferator-activated receptors

(*Ppars*) the retinoic-related orphan receptor (*Rorα*), *RevErbα*, and *Dbp* [105–113]. Thus, the molecular circadian network can affect and is affected by its intracellular environment from which it cannot be dissociated and metabolism should be regarded as an integral part of the core circadian pacemaker [114–117].

References

1. Franken P, Tafti M. Genetics of sleep and sleep disorders. Front Biosci. 2003;8:e381–97.
2. Borbely AA. A two process model of sleep regulation. Hum Neurobiol. 1982;1(3):195–204.
3. Daan S, Beersma DG, Borbely AA. Timing of human sleep: recovery process gated by a circadian pacemaker. Am J Physiol. 1984;246(2 Pt 2):R161–83.
4. Tononi G, Cirelli C. Sleep function and synaptic homeostasis. Sleep Med Rev. 2006;10(1):49–62.
5. Krueger JM, Rector DM, Roy S, Van Dongen HP, Belenky G, Panksepp J. Sleep as a fundamental property of neuronal assemblies. Nat Rev Neurosci. 2008;9(12):910–9.
6. Andretic R, Franken P, Tafti M. Genetics of sleep. Annu Rev Genet. 2008;42:361–88.
7. Franken P, Chollet D, Tafti M. The homeostatic regulation of sleep need is under genetic control. J Neurosci. 2001;21(8):2610–21.
8. Viola AU, Archer SN, James LM, et al. PER3 polymorphism predicts sleep structure and waking performance. Curr Biol. 2007;17(7):613–8.
9. Retey JV, Adam M, Honegger E, et al. A functional genetic variation of adenosine deaminase affects the duration and intensity of deep sleep in humans. Proc Natl Acad Sci USA. 2005;102(43):15676–81.
10. Cirelli C. The genetic and molecular regulation of sleep: from fruit flies to humans. Nat Rev Neurosci. 2009;10(8):549–60.
11. Maret S, Dorsaz S, Gurcel L, et al. Homer1a is a core brain molecular correlate of sleep loss. Proc Natl Acad Sci USA. 2007;104(50):20090–5.
12. Mongrain V, Hernandez SA, Pradervand S, et al. Separating the contribution of glucocorticoids and wakefulness to the molecular and electrophysiological correlates of sleep homeostasis. Sleep. 2010;33(9):1147–57.
13. Klein DC, Moore RY, Reppert SM. Suprachiasmatic nucleus. The mind's clock. New York: Oxford University Press; 1991.
14. Ralph MR, Foster RG, Davis FC, Menaker M. Transplanted suprachiasmatic nucleus determines circadian period. Science. 1990;247(4945):975–8.
15. Park JH, Helfrich-Forster C, Lee G, Liu L, Rosbash M, Hall JC. Differential regulation of circadian pacemaker output by separate clock genes in *Drosophila*. Proc Natl Acad Sci USA. 2000;97(7):3608–13.
16. Akerstedt T, Gillberg M. The circadian variation of experimentally displaced sleep. Sleep. 1981;4(2):159–69.
17. Dijk DJ, Duffy JF, Czeisler CA. Circadian and sleep/wake dependent aspects of subjective alertness and cognitive performance. J Sleep Res. 1992;1(2):112–7.
18. Dijk DJ, Czeisler CA. Contribution of the circadian pacemaker and the sleep homeostat to sleep propensity, sleep structure, electroencephalographic slow waves, and sleep spindle activity in humans. J Neurosci. 1995;15(5 Pt 1):3526–38.
19. Dijk DJ, Franken P. Interaction of sleep homeostasis and circadian rhythmicity: dependent or independent systems? In: Kryger MH, Roth T, Dement W, editors. Principles and practice of sleep medicine. Saunders: Elsevier; 2005.
20. Sack RL, Auckley D, Auger RR, et al. Circadian rhythm sleep disorders: part I, basic principles, shift work and jet lag disorders. An American academy of sleep medicine review. Sleep. 2007;30(11):1460–83.
21. Lockley SW, Arendt J, Skene DJ. Visual impairment and circadian rhythm disorders. Dialogues Clin Neurosci. 2007;9(3):301–14.
22. Achermann P, Borbely AA. Simulation of human sleep: ultradian dynamics of electroencephalographic slow-wave activity. J Biol Rhythms. 1990;5(2):141–57.
23. Huber R, Deboer T, Tobler I. Effects of sleep deprivation on sleep and sleep EEG in three mouse strains: empirical data and simulations. Brain Res. 2000;857(1–2):8–19.
24. Trachsel L, Edgar DM, Seidel WF, Heller HC, Dement WC. Sleep homeostasis in suprachiasmatic nuclei-lesioned rats: effects of sleep deprivation and triazolam administration. Brain Res. 1992;589(2):253–61.
25. Larkin JE, Yokogawa T, Heller HC, Franken P, Ruby NF. Homeostatic regulation of sleep in arrhythmic Siberian hamsters. Am J Physiol Regul Integr Comp Physiol. 2004;287(1):R104–11.
26. Antle MC, Mistlberger RE. Circadian clock resetting by sleep deprivation without exercise in the Syrian hamster. J Neurosci. 2000;20(24):9326–32.
27. Deboer T, Detari L, Meijer JH. Long term effects of sleep deprivation on the mammalian circadian pacemaker. Sleep. 2007;30(3):257–62.
28. Deboer T, Vansteensel MJ, Detari L, Meijer JH. Sleep states alter activity of suprachiasmatic nucleus neurons. Nat Neurosci. 2003;6(10):1086–90.
29. Takahashi JS, Hong HK, Ko CH, McDearmon EL. The genetics of mammalian circadian order and disorder: implications for physiology and disease. Nat Rev Genet. 2008;9(10):764–75.
30. Konopka RJ, Benzer S. Clock mutants of *Drosophila melanogaster*. Proc Natl Acad Sci USA. 1971;68(9):2112–6.
31. Vitaterna MH, King DP, Chang AM, et al. Mutagenesis and mapping of a mouse gene, Clock, essential for circadian behavior. Science. 1994;264(5159):719–25.

32. King DP, Vitaterna MH, Chang AM, et al. The mouse Clock mutation behaves as an antimorph and maps within the W19H deletion, distal of kit. Genetics. 1997;146(3):1049–60.
33. Lowrey PL, Takahashi JS. Mammalian circadian biology: elucidating genome-wide levels of temporal organization. Annu Rev Genomics Hum Genet. 2004;5:407–41.
34. Grimaldi B, Nakahata Y, Kaluzova M, Masubuchi S, Sassone-Corsi P. Chromatin remodeling, metabolism and circadian clocks: the interplay of CLOCK and SIRT1. Int J Biochem Cell Biol. 2009;41(1):81–6.
35. Sehgal A. Ac-ing the clock. Neuron. 2008;57(1): 8–10.
36. Nakahata Y, Kaluzova M, Grimaldi B, et al. The NAD+-dependent deacetylase SIRT1 modulates CLOCK-mediated chromatin remodeling and circadian control. Cell. 2008;134(2):329–40.
37. Cajochen C, Jud C, Munch M, Kobialka S, Wirz-Justice A, Albrecht U. Evening exposure to blue light stimulates the expression of the clock gene PER2 in humans. Eur J Neurosci. 2006;23(4): 1082–6.
38. Jud C, Chappuis S, Revell VL, et al. Age-dependent alterations in human PER2 levels after early morning blue light exposure. Chronobiol Int. 2009;26(7): 1462–9.
39. Nagoshi E, Saini C, Bauer C, Laroche T, Naef F, Schibler U. Circadian gene expression in individual fibroblasts: cell-autonomous and self-sustained oscillators pass time to daughter cells. Cell. 2004;119(5): 693–705.
40. Yoo SH, Yamazaki S, Lowrey PL, et al. PERIOD2:LUCIFERASE real-time reporting of circadian dynamics reveals persistent circadian oscillations in mouse peripheral tissues. Proc Natl Acad Sci USA. 2004;101(15):5339–46.
41. Cermakian N, Boivin DB. The regulation of central and peripheral circadian clocks in humans. Obes Rev. 2009;10 Suppl 2:25–36.
42. Schibler U. The daily rhythms of genes, cells and organs. Biological clocks and circadian timing in cells. EMBO Rep. 2005;6 Spec No.:S9–13.
43. Fujimoto Y, Yagita K, Okamura H. Does mPER2 protein oscillate without its coding mRNA cycling? Post-transcriptional regulation by cell clock. Genes Cells. 2006;11(5):525–30.
44. Yamamoto Y, Yagita K, Okamura H. Role of cyclic mPer2 expression in the mammalian cellular clock. Mol Cell Biol. 2005;25(5):1912–21.
45. Tomita J, Nakajima M, Kondo T, Iwasaki H. No transcription-translation feedback in circadian rhythm of KaiC phosphorylation. Science. 2005; 307(5707):251–4.
46. O'Neill JS, Reddy AB. Circadian clocks in human red blood cells. Nature. 2011;469(7331):498–503.
47. O'Neill JS, van Ooijen G, Dixon LE, et al. Circadian rhythms persist without transcription in a eukaryote. Nature. 2011;469(7331):554–8.
48. Wisor JP, O'Hara BF, Terao A, et al. A role for cryptochromes in sleep regulation. BMC Neurosci. 2002;3:20.
49. van der Horst GT, Muijtjens M, Kobayashi K, et al. Mammalian Cry1 and Cry2 are essential for maintenance of circadian rhythms. Nature. 1999;398(6728): 627–30.
50. Vitaterna MH, Selby CP, Todo T, et al. Differential regulation of mammalian period genes and circadian rhythmicity by cryptochromes 1 and 2. Proc Natl Acad Sci USA. 1999;96(21):12114–9.
51. Wisor JP, Pasumarthi RK, Gerashchenko D, et al. Sleep deprivation effects on circadian clock gene expression in the cerebral cortex parallel electroencephalographic differences among mouse strains. J Neurosci. 2008;28(28):7193–201.
52. Laposky A, Easton A, Dugovic C, Walisser J, Bradfield C, Turek F. Deletion of the mammalian circadian clock gene BMAL1/Mop3 alters baseline sleep architecture and the response to sleep deprivation. Sleep. 2005;28(4):395–409.
53. Naylor E, Bergmann BM, Krauski K, et al. The circadian clock mutation alters sleep homeostasis in the mouse. J Neurosci. 2000;20(21):8138–43.
54. Franken P, Lopez-Molina L, Marcacci L, Schibler U, Tafti M. The transcription factor DBP affects circadian sleep consolidation and rhythmic EEG activity. J Neurosci. 2000;20(2):617–25.
55. Franken P, Dudley CA, Estill SJ, et al. NPAS2 as a transcriptional regulator of non-rapid eye movement sleep: genotype and sex interactions. Proc Natl Acad Sci USA. 2006;103(18):7118–23.
56. Dudley CA, Erbel-Sieler C, Estill SJ, et al. Altered patterns of sleep and behavioral adaptability in NPAS2-deficient mice. Science. 2003;301(5631): 379–83.
57. Turek FW, Joshu C, Kohsaka A, et al. Obesity and metabolic syndrome in circadian Clock mutant mice. Science. 2005;308(5724):1043–5.
58. Shiromani PJ, Xu M, Winston EM, Shiromani SN, Gerashchenko D, Weaver DR. Sleep rhythmicity and homeostasis in mice with targeted disruption of mPeriod genes. Am J Physiol Regul Integr Comp Physiol. 2004;287(1):R47–57.
59. Hendricks JC, Williams JA, Panckeri K, et al. A non-circadian role for cAMP signaling and CREB activity in Drosophila rest homeostasis. Nat Neurosci. 2001;4(11):1108–15.
60. Shaw PJ, Tononi G, Greenspan RJ, Robinson DF. Stress response genes protect against lethal effects of sleep deprivation in *Drosophila*. Nature. 2002;417(6886):287–91.
61. Hendricks JC, Lu S, Kume K, Yin JC, Yang Z, Sehgal A. Gender dimorphism in the role of cycle (BMAL1) in rest, rest regulation, and longevity in *Drosophila melanogaster*. J Biol Rhythms. 2003;18(1):12–25.
62. Shaw PJ, Franken P. Perchance to dream: solving the mystery of sleep through genetic analysis. J Neurobiol. 2003;54(1):179–202.

63. Yamaguchi S, Mitsui S, Miyake S, et al. The 5′ upstream region of mPer1 gene contains two promoters and is responsible for circadian oscillation. Curr Biol. 2000;10(14):873–6.
64. Rutter J, Reick M, McKnight SL. Metabolism and the control of circadian rhythms. Annu Rev Biochem. 2002;71:307–31.
65. Oishi K, Amagai N, Shirai H, Kadota K, Ohkura N, Ishida N. Genome-wide expression analysis reveals 100 adrenal gland-dependent circadian genes in the mouse liver. DNA Res. 2005;12(3):191–202.
66. Ripperger JA, Shearman LP, Reppert SM, Schibler U. Clock, an essential pacemaker component, controls expression of the circadian transcription factor DBP. Genes Dev. 2000;14(6):679–89.
67. Cirelli C, Tononi G. Locus ceruleus control of state-dependent gene expression. J Neurosci. 2004;24(23):5410–9.
68. Franken P, Thomason R, Heller HC, O'Hara BF. A non-circadian role for clock-genes in sleep homeostasis: a strain comparison. BMC Neurosci. 2007;8:87.
69. Masubuchi S, Honma S, Abe H, et al. Clock genes outside the suprachiasmatic nucleus involved in manifestation of locomotor activity rhythm in rats. Eur J Neurosci. 2000;12(12):4206–14.
70. Abe H, Honma S, Namihira M, Masubuchi S, Honma K. Behavioural rhythm splitting in the CS mouse is related to clock gene expression outside the suprachiasmatic nucleus. Eur J Neurosci. 2001;14(7):1121–8.
71. Wakamatsu H, Yoshinobu Y, Aida R, Moriya T, Akiyama M, Shibata S. Restricted-feeding-induced anticipatory activity rhythm is associated with a phase-shift of the expression of mPer1 and mPer2 mRNA in the cerebral cortex and hippocampus but not in the suprachiasmatic nucleus of mice. Eur J Neurosci. 2001;13(6):1190–6.
72. Mrosovsky N, Edelstein K, Hastings MH, Maywood ES. Cycle of period gene expression in a diurnal mammal (*Spermophilus tridecemlineatus*): implications for nonphotic phase shifting. J Biol Rhythms. 2001;16(5):471–8.
73. Reick M, Garcia JA, Dudley C, McKnight SL. NPAS2: an analog of clock operative in the mammalian forebrain. Science. 2001;293(5529):506–9.
74. Zhou YD, Barnard M, Tian H, et al. Molecular characterization of two mammalian bHLH-PAS domain proteins selectively expressed in the central nervous system. Proc Natl Acad Sci USA. 1997;94(2):713–8.
75. Travnickova-Bendova Z, Cermakian N, Reppert SM, Sassone-Corsi P. Bimodal regulation of mPeriod promoters by CREB-dependent signaling and CLOCK/BMAL1 activity. Proc Natl Acad Sci USA. 2002;99(11):7728–33.
76. Tischkau SA, Mitchell JW, Tyan SH, Buchanan GF, Gillette MU. Ca2+/cAMP response element-binding protein (CREB)-dependent activation of Per1 is required for light-induced signaling in the suprachiasmatic nucleus circadian clock. J Biol Chem. 2003;278(2):718–23.
77. Naruse Y, Oh-hashi K, Iijima N, Naruse M, Yoshioka H, Tanaka M. Circadian and light-induced transcription of clock gene Per1 depends on histone acetylation and deacetylation. Mol Cell Biol. 2004;24(14):6278–87.
78. Yamaguchi S, Mitsui S, Yan L, Yagita K, Miyake S, Okamura H. Role of DBP in the circadian oscillatory mechanism. Mol Cell Biol. 2000;20(13):4773–81.
79. Yamamoto T, Nakahata Y, Soma H, Akashi M, Mamine T, Takumi T. Transcriptional oscillation of canonical clock genes in mouse peripheral tissues. BMC Mol Biol. 2004;5:18.
80. Segall LA, Amir S. Glucocorticoid regulation of clock gene expression in the mammalian limbic forebrain. J Mol Neurosci. 2010;42(2):168–75.
81. Cirelli C, Tononi G. Differential expression of plasticity-related genes in waking and sleep and their regulation by the noradrenergic system. J Neurosci. 2000;20(24):9187–94.
82. Graves LA, Hellman K, Veasey S, Blendy JA, Pack AI, Abel T. Genetic evidence for a role of CREB in sustained cortical arousal. J Neurophysiol. 2003;90(2):1152–9.
83. Stratmann M, Stadler F, Tamanini F, van der Horst GT, Ripperger JA. Flexible phase adjustment of circadian albumin D site-binding protein (DBP) gene expression by CRYPTOCHROME1. Genes Dev. 2010;24(12):1317–28.
84. Churchill L, Rector DM, Yasuda K, et al. Tumor necrosis factor alpha: activity dependent expression and promotion of cortical column sleep in rats. Neuroscience. 2008;156(1):71–80.
85. Cavadini G, Petrzilka S, Kohler P, et al. TNF-alpha suppresses the expression of clock genes by interfering with E-box-mediated transcription. Proc Natl Acad Sci USA. 2007;104(31):12843–8.
86. Archer SN, Robilliard DL, Skene DJ, et al. A length polymorphism in the circadian clock gene Per3 is linked to delayed sleep phase syndrome and extreme diurnal preference. Sleep. 2003;26(4):413–5.
87. Groeger JA, Viola AU, Lo JC, von Schantz M, Archer SN, Dijk DJ. Early morning executive functioning during sleep deprivation is compromised by a PERIOD3 polymorphism. Sleep. 2008;31(8):1159–67.
88. Partonen T, Treutlein J, Alpman A, et al. Three circadian clock genes Per2, Arntl, and Npas2 contribute to winter depression. Ann Med. 2007;39(3):229–38.
89. Roybal K, Theobold D, Graham A, et al. Mania-like behavior induced by disruption of Clock. Proc Natl Acad Sci USA. 2007;104(15):6406–11.
90. Gillin JC, Buchsbaum M, Wu J, Clark C, Bunney Jr W. Sleep deprivation as a model experimental antidepressant treatment: findings from functional brain imaging. Depress Anxiety. 2001;14(1):37–49.

91. Wu JC, Kelsoe JR, Schachat C, et al. Rapid and sustained antidepressant response with sleep deprivation and chronotherapy in bipolar disorder. Biol Psychiatry. 2009;66(3):298–301.
92. Wirz-Justice A. Biological rhythm disturbances in mood disorders. Int Clin Psychopharmacol. 2006;21 Suppl 1:S11–5.
93. Lamont EW, Legault-Coutu D, Cermakian N, Boivin DB. The role of circadian clock genes in mental disorders. Dialogues Clin Neurosci. 2007;9(3): 333–42.
94. McClung CA. Circadian genes, rhythms and the biology of mood disorders. Pharmacol Ther. 2007;114(2):222–32.
95. Hampp G, Ripperger JA, Houben T, et al. Regulation of monoamine oxidase A by circadian-clock components implies clock influence on mood. Curr Biol. 2008;18(9):678–83.
96. Benington JH, Heller HC. Restoration of brain energy metabolism as the function of sleep. Prog Neurobiol. 1995;45(4):347–60.
97. Benington JH, Heller HC. Restoration of brain energy metabolism as the function of sleep. Prog Neurobiol. 1995;45(4):347–60.
98. Maquet P. Sleep function(s) and cerebral metabolism. Behav Brain Res. 1995;69(1–2):75–83.
99. Spiegel K, Tasali E, Leproult R, Van Cauter E. Effects of poor and short sleep on glucose metabolism and obesity risk. Nat Rev Endocrinol. 2009;5(5):253–61.
100. Mottin S, Laporte P, Jouvet M, Cespuglio R. Determination of NADH in the rat brain during sleep-wake states with an optic fibre sensor and time-resolved fluorescence procedures. Neuroscience. 1997;79(3):683–93.
101. Mottin S, Laporte P, Cespuglio R. Inhibition of NADH oxidation by chloramphenicol in the freely moving rat measured by picosecond time-resolved emission spectroscopy. J Neurochem. 2003;84(4):633–42.
102. Challet E, Mendoza J. Metabolic and reward feeding synchronises the rhythmic brain. Cell Tissue Res. 2010;341(1):1–11.
103. Rutter J, Reick M, Wu LC, McKnight SL. Regulation of clock and NPAS2 DNA binding by the redox state of NAD cofactors. Science. 2001;293(5529): 510–4.
104. Asher G, Gatfield D, Stratmann M, et al. SIRT1 regulates circadian clock gene expression through PER2 deacetylation. Cell. 2008;134(2):317–28.
105. Inoue I, Shinoda Y, Ikeda M, et al. CLOCK/BMAL1 is involved in lipid metabolism via transactivation of the peroxisome proliferator-activated receptor (PPAR) response element. J Atheroscler Thromb. 2005;12(3): 169–74.
106. Teboul M, Guillaumond F, Grechez-Cassiau A, Delaunay F. Minireview: the nuclear hormone receptor family round the clock. Mol Endocrinol. 2008;22(12):2573–82.
107. Oishi K, Shirai H, Ishida N. CLOCK is involved in the circadian transactivation of peroxisome-proliferator-activated receptor alpha (PPARalpha) in mice. Biochem J. 2005;386(Pt 3):575–81.
108. Nakamura K, Inoue I, Takahashi S, Komoda T, Katayama S. Cryptochrome and period proteins are regulated by the CLOCK/BMAL1 gene: crosstalk between the PPARs/RXRalpha-regulated and CLOCK/BMAL1-regulated systems. PPAR Res. 2008;2008:348610.
109. Canaple L, Rambaud J, Dkhissi-Benyahya O, et al. Reciprocal regulation of brain and muscle Arnt-like protein 1 and peroxisome proliferator-activated receptor alpha defines a novel positive feedback loop in the rodent liver circadian clock. Mol Endocrinol. 2006;20(8):1715–27.
110. Staels B. When the Clock stops ticking, metabolic syndrome explodes. Nat Med. 2006;12(1):54–5. discussion 55.
111. Mongrain V, Ruan X, Dardente H, Fortier EE, Cermakian N. Clock-dependent and independent transcriptional control of the two isoforms from the mouse Ror γ gene. Genes Cells. 2008;13:1197–210.
112. Duez H, Staels B. The nuclear receptors Rev-erbs and RORs integrate circadian rhythms and metabolism. Diab Vasc Dis Res. 2008;5(2):82–8.
113. Gachon F, Leuenberger N, Claudel T, et al. Proline- and acidic amino acid-rich basic leucine zipper proteins modulate peroxisome proliferator-activated receptor alpha (PPARalpha) activity. Proc Natl Acad Sci USA. 2011;108(12):4794–9.
114. Roenneberg T, Merrow M. Circadian systems and metabolism. J Biol Rhythms. 1999;14(6):449–59.
115. Bass J, Takahashi JS. Circadian rhythms: redox redux. Nature. 2011;469(7331):476–8.
116. Bass J, Takahashi JS. Circadian integration of metabolism and energetics. Science. 2010;330(6009): 1349–54.
117. Tu BP, McKnight SL. Metabolic cycles as an underlying basis of biological oscillations. Nat Rev Mol Cell Biol. 2006;7(9):696–701.
118. Curie T, Mongrain V, Dorsaz S, Mang G, Emmenegger Y, Franken P. PERIOD-2 integrates both homeostatic and time-of-day dependent aspects of sleep-wake regulation. 2012, Submitted.

Clock Genes and Energy Metabolism 2

Hee-Kyung Hong, Wenyu Huang, Kathryn Moynihan Ramsey, Biliana Marcheva, and Joseph Bass

Abstract

Circadian timekeeping is a ubiquitous feature of all eukaryotes and allows appropriate temporal regulation of an organism's physiology, behavior, and metabolism to anticipate and respond to recurrent daily changes in the environment. Animal models provide strong evidence that disruption of circadian pathways are associated with metabolic dysregulation and sleep-related pathologies, while high-fat feeding reveals disrupted circadian rhythms of feeding, activity, and sleep. In humans, short sleep duration is associated with increased risk of metabolic syndrome, including obesity, diabetes, cardiovascular diseases, and cancer. Here, we examine emerging insight into how the circadian clock network influences energy metabolism.

Circadian Clock Networks

The overall goal of the chapter is to gain an appreciation of the feedback relationships between circadian clocks, metabolism, and sleep and to highlight the implications of these relationships for our understanding of circadian disorders of sleep and metabolism (see Fig. 2.1).

H.-K. Hong
Department of Medicine, Feinberg School of Medicine, Northwestern University, 303 East Superior Street, Lurie 7-105, Chicago, IL 60611, USA

W. Huang • K.M. Ramsey • B. Marcheva • J. Bass (✉)
Department of Medicine, Division of Endocrinology, Feinberg School of Medicine, Northwestern University, 303 East Superior Street, Lurie 7-105, Chicago, IL 60611, USA
e-mail: jbass@northwestern.edu

Endogenous Circadian Rhythms and the Master Pacemaker

Virtually, all aspects of human physiology and behavior are orchestrated by an endogenous circadian timing system that allows organisms to respond to predictable 24-h changes in the environment (i.e., the light–dark cycle). In mammals, self-sustained, cell-autonomous circadian clocks exist throughout the body in both the brain and peripheral tissues [1–3]. The "master pacemaker," however, resides within a bilateral structure called the suprachiasmatic nuclei (SCN) in the anterior hypothalamus of the brain, as demonstrated by classical lesioning studies in which rhythmic locomotor activity of SCN-lesioned animals was restored by transplantation of the SCN from wild-time animals [4–8]. The SCN is a heterogeneous structure comprised of approximately 10,000 neurons, each acting as an

P.J. Shiromani et al. (eds.), *Sleep Loss and Obesity: Intersecting Epidemics*,
DOI 10.1007/978-1-4614-3492-4_2, © Springer Science+Business Media, LLC 2012

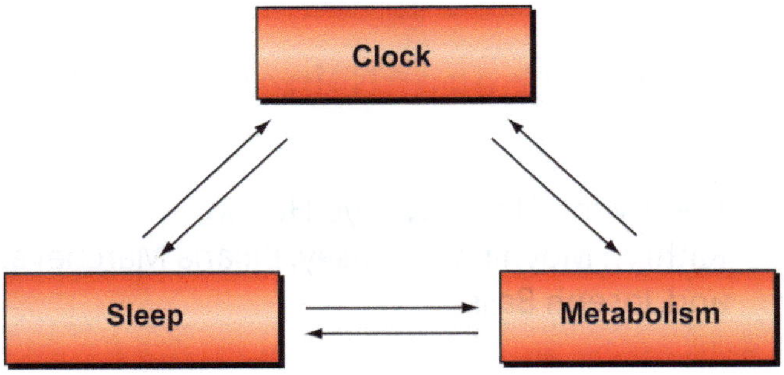

Fig. 2.1 *Interplay between the circadian clock, sleep, and metabolism.* Emerging evidence has revealed complex reciprocal relationships between the circadian clock network, sleep homeostasis, and metabolism, as indicated by the *double arrows*. Disruption of circadian rhythms or sleep results in increased risk of metabolic disorders, including type 2 diabetes. Conversely, changes in metabolism or nutrient availability reciprocally can lead to the disruption of both sleep and circadian homeostasis on both molecular and physiologic levels

autonomous oscillator. Anatomical, electrophysiological, and bioluminescence imaging studies of the SCN from both intact animal models and dispersed neuronal cell cultures suggest that unique neuronal intercellular coupling and network properties enable the SCN to act as the master pacemaker of the organism (reviewed in [9]). The SCN receives environmental light input via the retinohypothalamic tract, which originates in the retina within a small subset of retinal ganglion cells, which express the photopigment melanopsin and integrate photic information (see Fig. 2.2) [11, 12], reviewed in [13]. The SCN mainly projects to the ventral and dorsal subparaventricular zones (vSPZ and dSPZ, respectively) and the dorsomedial nucleus (DMH), which subsequently transmits SCN signals to a variety of nuclei involved in sleep/wake cycles, feeding, thermoregulation, and corticosteroid secretion [14]. In addition, hormonal factors such as transforming growth factor-alpha (TGFα) [15], cardiotropin-like cytokines [16], and prokineticin [17] have also been proposed as signals involved in communication between SCN and other regions of the hypothalamus. Both neuronal and humoral signals originating from the SCN communicate to achieve coherent metabolic, electrical, and secretory rhythms throughout the organism and to coordinate and maintain the relative phasing of diverse physiological processes, including body temperature, heart rate, hormone secretion, glucose and lipid metabolism, cell cycle progression, and feeding behavior (reviewed in [18–21]). In the past 20 years, forward genetic and gene-targeting approaches, as well as the identification of classical spontaneous mutants, have been instrumental in defining the essential set of genes and molecular mechanisms underlying the circadian system in mammals (reviewed in [22, 23]). Further, the recent availability of both global and tissue-specific genetic models has offered unique opportunities to define the role of circadian oscillators in metabolism, energy balance, sleep regulation, and human health. Here, we will focus on the circadian oscillator as a regulator of sleep and metabolism, as the neural circuitry and neurotransmitters involved in regulation of wake, non-REM sleep, and REM sleep is discussed in another chapter (For details, see Chap. 3).

Molecular Basis of Circadian Rhythms: The Core Molecular Clock Mechanism

The core molecular components and the genetics of mammalian circadian system are now well defined [22, 23]. In brief, the molecular basis of circadian timing involves interlocking transcriptional/translational feedback loops, which generate rhythmic expression of a set of core clock genes that encode highly conserved transcription factors and enzymes (see Fig. 2.2) [23, 24]. The positive feedback loop is composed of brain and muscle aryl-hydrocarbon receptor nuclear translocator-like 1/aryl-hydrocarbon receptor

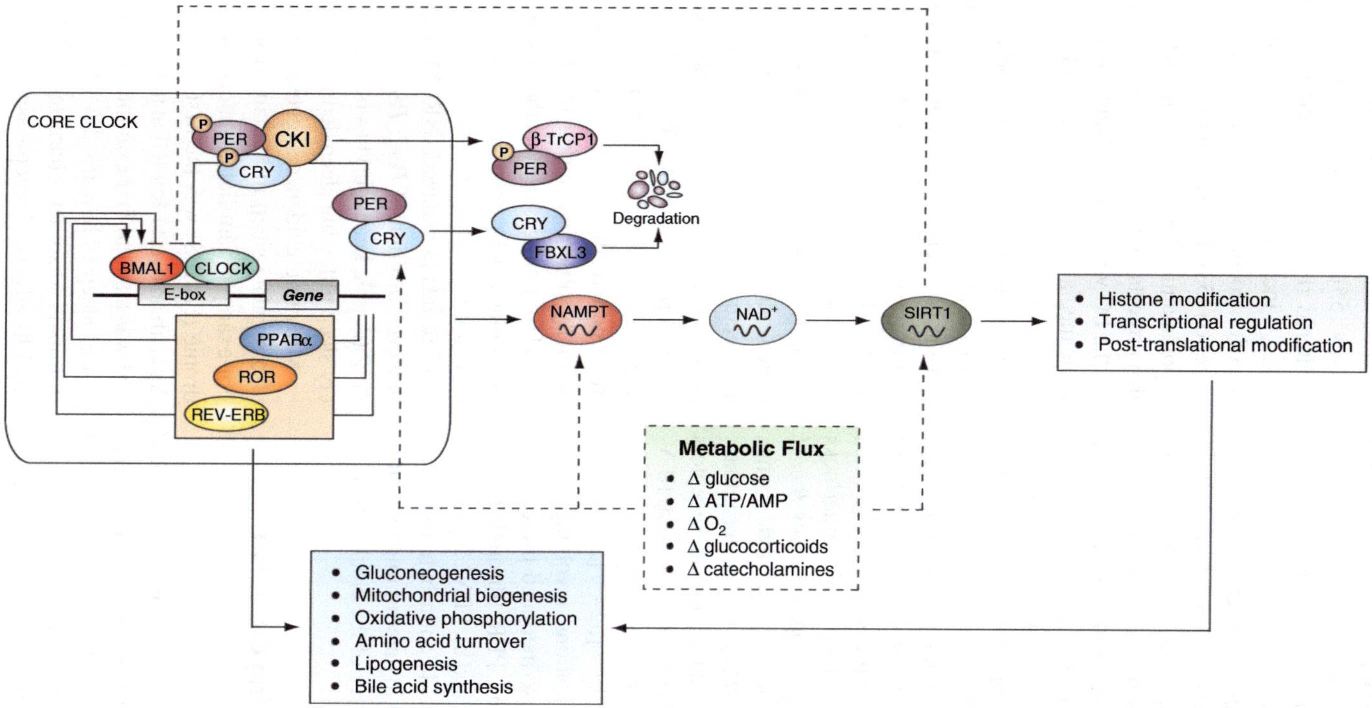

Fig. 2.2 *Direct and indirect outputs of the core clock mechanism.* The core clock consists of a series of transcription/translation feedback loops that synchronize diverse metabolic processes through both direct and indirect outputs, including gluconeogenesis and oxidative metabolism. The clock also receives reciprocal input from nutrient signaling pathways (including SIRT1), which function as rheostats to coordinate metabolic processes with daily cycles of sleep/wakefulness and fasting/feeding. Figure modified with permission from [10]

nuclear translocator-like (BMAL1/ARNTL) and its heterodimeric binding partner circadian locomotor output cycles kaput (CLOCK) or its paralog neuronal PAS domain protein 2 (NPAS2) [25, 26], which transactivate downstream clock and clock-controlled genes (CCGs) by binding to E-box elements that lie within their promoters [27–30]. Among these target genes are *Cryptochrome (Cry1* and *Cry2)* and *Period (Per1, Per2,* and *Per3)*, whose protein products accumulate, heterodimerize with each other in the cytoplasm, and translocate to the nucleus, where they interact with the CLOCK:BMAL1 complex to inhibit their own transcription [31]. As the CRY-PER repressor is degraded and inhibition is relieved, the CLOCK:BMAL1 complex renews its transcriptional cycle, creating a 24-h timekeeping system that ultimately drives oscillations of downstream target gene expression. In addition, a second negative feedback loop, which interlocks with the primary feedback loop, regulates *Bmal1* transcription via retinoic acid-related orphan receptor elements *(ROREs)* in the promoter: retinoic acid receptor-related orphan receptors (RORs: *Rorα, Rorβ, Rorγ*) activate and REV-ERBs *(Rev-erbα, Rev-erbβ)* repress *Bmal1* transcription, respectively [32–37]. Finally, posttranslational modifications involving phosphorylation and E3 ligase-mediated proteasomal degradation of CRY and PER [through Casein Kinase 1 isoforms (α, δ, and ε), FBXL3, *β*-TrCP1, and protein phosphatase 1] contribute to protein turnover and are critical in determining the stability of circadian periodicity of the clock [38–49].

Brain Circuitry Linking Circadian Centers to Sleep

The neural basis for circadian regulation of sleep/wake cycles can be analyzed anatomically by tracing connections between the circadian control centers and the sleep/wake centers (see Fig. 2.3). Efferent projections from the SCN form synapses within neurons in the ventral subparaventricular zone (vSPZ), which in turn project to the DMH. Retrograde tracing studies have demonstrated dense GABAergic connections from the DMH to the VLPO and glutamatergic connections from the DMH to the LHA [51–54]. Functionally, destruction of the vSPZ leads to disrupted circadian rhythms of sleep–wakefulness and locomotion activity [55], while lesioning of the DMH results in altered rhythms of diverse physiological processes, including sleep/wakefulness, feeding behavior, locomotor activity, and glucocorticoid secretion [53]. In addition to communicating with hypothalamic sleep/wake center neurons, the SCN also projects to brain stem arousal regions including the LC via DMH [56] and the VTA via a relay at the medial preoptic nucleus (MPOA) [57].

Genetic deletion of some of the core molecular clock components affects not only the timing of sleep, but also the amount and architecture of sleep and the compensatory response to sleep deprivation (a measure of homeostatic drive). For example, *Clock*$^{\Delta 19/\Delta 19}$ mutant mice have less total sleep time and less recovery REM sleep after sleep deprivation [58]. *Bmal1* knockout mice exhibit disrupted sleep/wake rhythms and altered sleep architecture, including increased total sleep time, sleep fragmentation, and attenuated recovery from sleep deprivation [59]. While NPAS2-deficient mice have a normal sleep pattern during the light (rest) period, they fail to display the brief sleep normally observed during the late dark (active) period [60], suggesting that the forward limb of the molecular clocks may play distinct roles in sleep regulation. In addition, deletion of both *Cry1* and *Cry2* results in enhanced NREM sleep drive [61]; however, deletion of *Per1, Per2,* or both *Per1* and *Per2* (double knockout) has little impact on recovery sleep following sleep deprivation [62, 63].

Mounting evidence suggests that clock genes also regulate sleep in humans (for review, [64]). For example, familial advanced sleep phase syndrome (FASPS) was the first human sleep disorder with an overt Mendelian pattern of inheritance and was also the first example of clock gene control of sleep in humans. FASPS is an autosomal dominant disorder characterized by persistent (3–4 h) advanced sleep onset and early awakening (i.e., "morning larks") [65]. Interestingly, linkage analysis and candidate gene approaches have identified mutations in both *PER2* (S662G) and *CK1δ* (T44A) in two independent FASPS

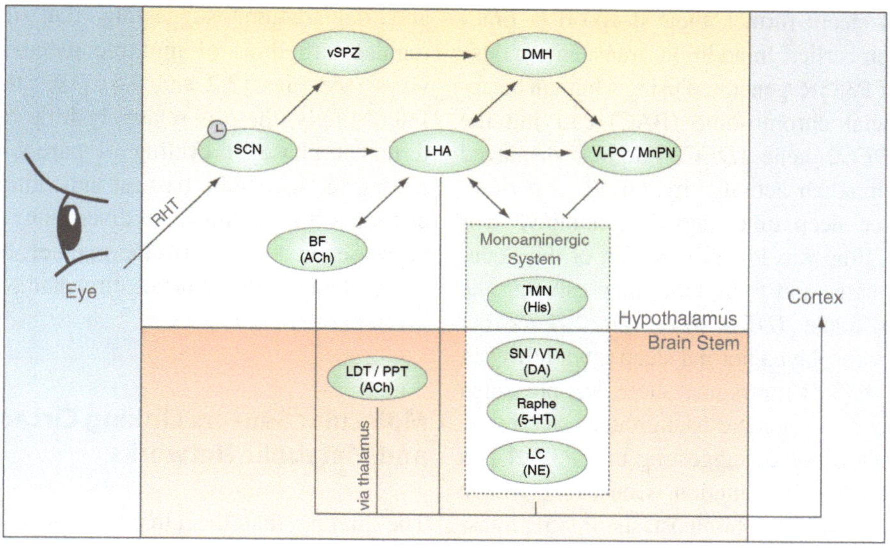

Fig. 2.3 *Map of neural circuits linking SCN and extra-SCN regions important in circadian and energetic control.* The mammalian circadian pacemaker in the hypothalamic suprachiasmatic nucleus (SCN) receives environmental light cycle information through the retinohypothalamic tract (RHT). The SCN projects to the vSPZ, the dorsomedial nucleus (DMH), and lateral hypothalamic area (LHA), the latter acting as a hub to integrate metabolic, circadian, and sleep/wake signals. The LHA possesses reciprocal connections with wake promoting centers including basal forebrain (BF) and the monoaminergic nuclei [tuberomammalary nucleus (TMN); substantia nigra and ventrotegmental area (SN/VTA); dorsal raphe nucleus; locus ceruleus (LC)], in addition to connections with sleep promoting centers including ventrolateral preoptic nucleus (VLPO) and median preoptic nucleus (MnPN). The laterodorsal tegmental nuclei (LDT) and pedunculopontine nuclei (PPT) are also components of the wake promoting centers, which send projection to the cortex via a relay at the thalamus. There are mutually inhibitory communications between VLPO and monoaminergic nuclei, forming a so-called "flip-flop switch" for controlling sleep and wakefulness. Acetocholine (Ach); histamine (His); dopamine (DA); serotonin (5-HT); norepinephrine (NE). *Lines which end with a bar* indicate inhibitory synapses. Figure modified with permission from [50]

pedigrees [48, 66]. Remarkably, both mutations affect the posttranslational modification of PER2 by Casein Kinase 1.

A second sleep disorder, delayed sleep phase syndrome (DSPS), is characterized by a chronic inability to fall asleep or awaken relative to the desired "normal" time of day. The average onset of sleep in DSPS individuals occurs in the early morning (3–6 am), and the average awakening occurs in the late morning to afternoon (11–2 pm) (i.e., "night owls"). While familial cases have been reported, the genetic defect underlying DSPS remains to be identified [67, 68]. However, several association studies of DSPS, including N-24 and extreme diurnal preference, have reported sequence variants of known circadian genes (*CLOCK, CSNK1E, PER2,* and *PER3*) [68–75] reviewed in [64].

In addition to circadian disorders of sleep onset, association studies in humans have reported a link between two known clock genes and sleep homeostasis: *PER3* VNTR (variable-number tandem-repeat) has been implicated with sleep homeostasis and the *TIMELESS* gene has been associated with both depression and sleep disturbances [76, 77]. While animal studies indicate that mutations in clock genes regulate the circadian timing of sleep, only recently clock gene variants have been implicated in sleep homeostasis or architecture in humans. A mutation in a bHLH transcription repressor DEC2, a negative regulator of the circadian clock, was recently identified in a pedigree that segregates a short sleep length phenotype in a dominant fashion [78]. Carriers of the P385R DEC2 mutation slept an average of 6.25 h compared with the noncarriers, who averaged

8.06 h of sleep; further, their sleep-offset times were much earlier. In addition, transgenic mouse models of P385R, generated using a human bacterial artificial chromosome (BAC) carrying the entire *hDEC2* gene (*DEC2-P385R*), displayed normal circadian activity rhythms (i.e., period), but shorter sleep time, and, interestingly, their "activity" time was longer than that of their control littermates and transgenic mice carrying the wild-type allele (*DEC2-WT*). *Dec2* knockout (KO) mice displayed normal sleep time; however, when the P385R transgenic allele was presented in the *Dec2* KO genetic background, sleep time was further reduced, suggesting that P385R is a dominant negative mutation. Consistent with a role of *Dec2* in sleep homeostasis, P385R transgenic mice also exhibited shortening of both NREM and REM sleep during the light phase, fragmented sleep structure, and alterations in sleep rebound following sleep deprivation. Similar rest-sleep disorders were also observed in transgenic flies with inducible expression of *mDec2-WT* and *mDec2-P385R*. Thus, model organisms of human mutations provide an excellent platform to unravel the molecular mechanisms regulating sleep and to explore pharmacological interventions that may lead to therapeutics for sleep-related disorders in humans.

Circadian Regulation of Metabolism

Clock-Driven Rhythms of Metabolic Gene Transcription Networks

Just as the clock produces rhythmic oscillation of core clock genes, the circadian network also drives 24-h transcriptional rhythms of downstream metabolic target genes. Gene microarray studies show that in any given tissue ~10% of all mammalian transcripts exhibit circadian rhythmicity [20, 79–84], a number that will likely increase as further high-resolution transcriptome profiling is performed using more sensitive non-parametric algorithms [81, 85, 86]. Many of these genes encode key transcription factors or enzymes in metabolic processes ranging from glucose and cholesterol metabolism to mitochondrial oxidative phosphorylation and cholesterol synthesis and degradation, suggesting that the clock regulates rhythms of multiple metabolic pathways (see Figs. 2.2 and 2.4) [10, 20, 87–90]. Interestingly, there is relatively little overlap in terms of phase of cycling of particular genes among the metabolic tissues, indicating that the circadian system influences diverse physiological processes in a tissue-specific manner; both local and systemic cues influence circadian output in a given tissue [87, 91].

Molecular Sensors Linking Circadian and Metabolic Networks

The finding that the circadian clock regulates metabolic processes leads to the question of whether there are "molecular sensors" common to both pathways that enable the circadian system to monitor the environment and adjust 24 h fuel utilization cycles according to nutrient availability (reviewed extensively in [92]). In this regard, two families of proteins, the nuclear hormone receptors (NHRs) and the sirtuins (SIRTs), have recently emerged as key metabolic sensors linking the clock and metabolism (see Fig. 2.2). NHRs represent a superfamily of ligand-dependent transcription factors that regulate a diverse range of downstream physiological processes including glucose and lipid metabolism, as well as the core clock itself. More than half of the 49 known NHRs in mice display circadian oscillations in peripheral metabolic tissues such as liver, fat, and skeletal muscle [90], including REV-ERBα, RORα, and PPARα, which also constitute short feedback loops within the core clock mechanism. For example, REV-ERBα regulates hepatic gluconeogenesis and adipocyte differentiation in addition to its role as a negative regulator of *Bmal1* transcription [32, 93–95], while RORα and PPARα regulate lipogenesis and glucose metabolism and are both positive regulators of *Bmal1* transcription [33, 96–98]. Thus, circadian regulation of NHRs likely contributes to the daily oscillation of glucose and lipid metabolism.

Similar to the NHRs, the NAD^+-dependent deacetylase SIRT1 is also a nutrient-responsive transcription factor that comprises a short-feedback loop within the core clock. CLOCK/

BMAL1 directly regulates the expression of the rate-limiting enzyme in NAD+ biosynthesis, nicotinamide phosphoribosyltransferase (NAMPT), in peripheral metabolic tissues [99, 100]. *Nampt* RNA and NAD+ levels are reduced in *Clock* and *Bmal1* mutant mice, whereas they are increased in mice lacking both *Cry1* and *Cry2*, suggesting a direct role for the circadian clock in NAD+ production and downstream NAD+-dependent processes. Further, SIRT1 has been shown to regulate the core clock complex itself by physically interacting with and inhibiting CLOCK/BMAL1 in a NAD+-dependent manner [101, 102]. In addition to regulation of SIRT1 activity by the clock, NAMPT and SIRT1 are also responsive to the nutritional milieu. *Nampt* is upregulated in response to decreased glucose levels in skeletal muscle [103, 104], and fasting and caloric restriction increases SIRT1 activity [105–107], leading to upregulation of downstream metabolic processes including gluconeogenesis and adipocyte differentiation. Because the NHRs and sirtuins are regulated by both the clock and nutrient/hormonal signals, and because these both in turn regulate downstream metabolic processes (in addition to the core clock itself), they constitute a unique node in the integration of temporal and metabolic systems.

Metabolic Consequences of Circadian Disruption in Mice

Genetic mouse models have been instrumental in defining key roles for each of the core clock genes in the generation and maintenance of circadian rhythmicity. While initial assessments of mouse models of circadian disruption revealed changes in circadian phenotypes such as altered locomotor activity rhythms, period length, entrainment to the LD cycle, and sleep (reviewed in [22, 23]), further systematic evaluation of these mouse models revealed that they also display adverse metabolic phenotypes with certain features of the human metabolic syndrome. For instance, mice harboring the dominant negative *Clock*$^{\Delta 19}$ mutation display attenuated diurnal feeding rhythms, hyperphagia, obesity, hyperleptinemia, and hyperglycemic hypoinsulinemia due to impaired insulin secretion and islet proliferation [108, 109]. Tissue-specific *Bmal1* knockout models have elucidated the metabolic consequences of *Bmal1* disruption, as global *Bmal1*$^{-/-}$ mice develop severe myopathy and arthritis, thereby compromising metabolic analyses in these mice [110, 111]. Mice with liver-specific deletion of *Bmal1* exhibit hypoglycemia during the fasting period, loss of rhythmic expression of hepatic glucose regulatory genes, and impaired gluconeogenesis, whereas pancreas-specific *Bmal1* knockout mice display hyperglycemia, impaired glucose tolerance, and decreased insulin secretion owing to secretory defects in islet cells [108, 112]. Notably, pancreas-specific *Bmal1* knockout mice show normal circadian activity, feeding rhythms, and body weight and composition, indicating that the metabolic phenotypes in these mice are specifically due to disruption of the clock network within islets and that *Bmal1* is essential for pancreatic β-cell proliferation and maturation [108]. *Cry* double knockout (*Cry1*$^{-/-}$;*Cry2*$^{-/-}$) mice display increased hepatic gluconeogenesis, due in part to upregulation of cAMP signaling [113]. Interestingly, hepatic overexpression of *Cry1* lowers blood glucose concentrations and improves insulin sensitivity in insulin-resistant *Lepr*$^{db/db}$ mice, suggesting that compounds that enhance cryptochrome activity may provide therapeutic benefit to individuals with type 2 diabetes [113]. In addition, reduced growth and elevated metabolic rate have been reported in another circadian mutant animal model, the Syrian *tau* mutant hamster with a disrupted *CK1ε* gene [114, 115].

The circadian CCG *Nocturnin*, encoding a poly(A) deadenylase, provides another link between the circadian timing system and metabolism of dietary lipids. *Noc*$^{-/-}$ mice are resistant to diet-induced obesity and hepatic steatosis due to defective lipid absorption in the small intestine [116, 117]. The orphan nuclear receptor estrogen-related receptor α (ERRα) has also been implicated as a mediator of clock control of metabolism. ERRα, which exhibits diurnal expression patterns in liver, has been shown to regulate the expression of several genes involved in glycolysis and gluconeogenesis, as well as several core clock genes (for review, [118]).

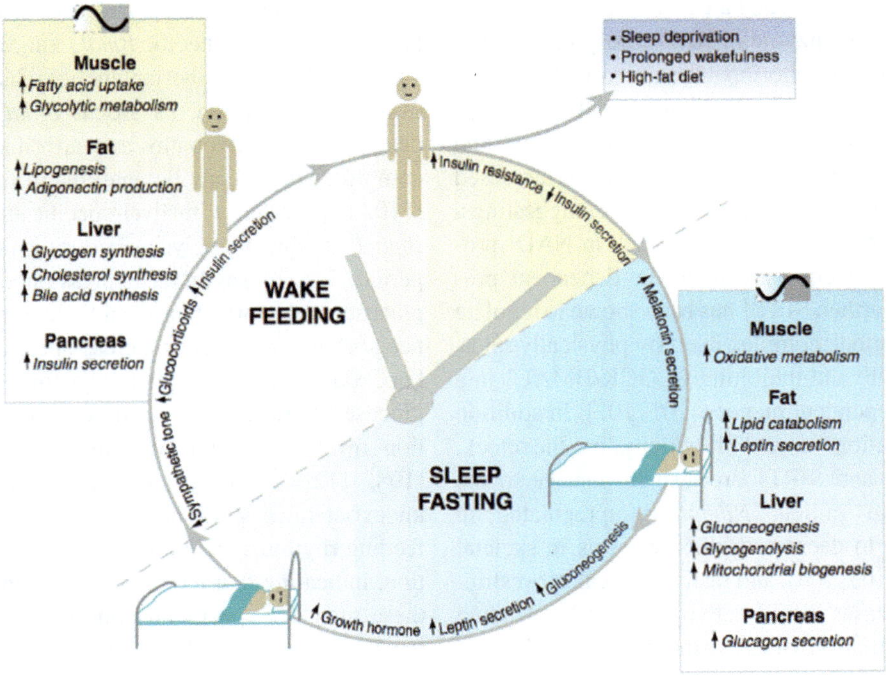

Fig. 2.4 *The clock partitions behavioral and metabolic processes according to the time of day.* The clock coordinates appropriate metabolic responses within peripheral tissues with the light–dark cycle. For example, the liver clock promotes gluconeogenesis and glycogenolysis during the sleep/fasting period, whereas it promotes glycogen and cholesterol synthesis during the wake/feeding period. Proper functioning of peripheral clocks keeps metabolic processes in synchrony with the environment, which is critical for maintaining health of the organism. Different tissues exhibit distinct clock-controlled properties; thus, ablation of the clock in certain tissues will cause opposing effects on metabolic function as uncovered through dynamic challenges at different times in the cycle under different nutrient conditions. Aging, diet, and environmental disruption such as shift work may also affect the integration of circadian and metabolic systems. Reproduced with permission from [10]

Notably, *ERRα* knockout mice are resistant to high-fat diet-induced obesity and display phenotypes correlated with metabolic dysregulation: reduced peripheral fat deposits, hypoglycemia, and altered glucose homeostasis [119]. Locomotor activity rhythms and expression patterns of core clock genes are also altered in *ERRα*-null mice, implicating ERRα as a potential regulator of the circadian clock [119]. Collectively, these genetic studies using animal models support a role for circadian clocks in energy homeostasis.

Circadian Disruption and Metabolic Consequences in Humans

Just as direct genetic evidence has confirmed a role for human clock genes in sleep homeostasis, the circadian network is also critical for regulation of metabolism in humans (see Fig. 2.4). Numerous epidemiological studies have recently demonstrated that internal misalignment of brain and peripheral clocks likely underlies a wide-range of disorders in humans. For instance, night-shift workers, whose activity period is reversed in relation to the day–night cycle, have increased risk of developing colorectal, breast, lymphatic and prostate cancers, as well as gastric ulcers, obesity, diabetes, stroke and coronary heart disease, and metabolic syndrome [120–128]. In addition, shift work, jet lag, and voluntary sleep curtailment are associated with increased hunger, decreased glucose and lipid catabolism, reduced energy expenditure, and alterations in hormonal signals involved in satiety (reviewed [129, 130]). Further, forced circadian misalignment in human subjects induces hypoleptinemia, insulin resistance, inverted cortisol rhythms, increased blood pressure, and impaired sleep efficiency [131].

Several genome-wide association studies (GWAS) have also suggested interconnections between clock gene variations and metabolism in humans: *Bmal1* and *Cry2* were shown to be associated with susceptibility to hypertension, type 2 diabetes and blood glucose concentration, respectively [132, 133], while *Clock* haplotypes were associated with obesity and metabolic syndrome [134]. Single nucleotide polymorphisms in *Clock* are associated with high plasma ghrelin concentration, altered eating behaviors, obesity, short sleep, and evening preference [135, 136]. Interestingly, recent GWAS studies also indicated that melatonin, a circulating hormone whose diurnal expression is tightly controlled by the circadian clock, and its G-protein-coupled receptors MTNR1A and MTNR1B are associated with the development of type 2 diabetes and impaired insulin secretion [132, 137–143]. Melatonin, whose receptors are present in both the SCN and the pancreatic islets, has been shown to modulate a variety of physiological processes, including sleep, glucose metabolism, and insulin secretion (reviewed in [137, 144–149]). Together, these epidemiologic and association studies underscore the clinical importance of interrelationships between melatonin, glucose homeostasis, circadian rhythms, and sleep.

Reciprocal Relations: Metabolism and Sleep Regulate the Clock

Metabolic Alterations Lead to Disrupted Circadian Networks

The aforementioned sections describe the integration of circadian systems, sleep, and metabolism at the neurologic, molecular, and physiologic levels. However, several recent studies have also demonstrated that the reciprocal relationship is also true—alterations in metabolism disrupt circadian rhythms and sleep. For example, feeding mice a high fat alters circadian rhythms by lengthening the locomotor activity period, reducing the amplitude of feeding rhythms, and altering the expression of clock and CCGs [88]. High-fat feeding also disrupts circadian regulation of humoral and CNS metabolic systems by reducing the diurnal rhythm of orexin gene expression and increasing leptin, glucose, insulin, and free fatty acid levels. Finally, high-fat fed mice display increased sleep pressure (i.e., increased NREM time) and difficulties in maintaining wakefulness during the active phase [150]. In addition to the diet-induced changes in circadian rhythms and sleep, genetic mouse models of obesity have demonstrated similar findings with regard to disrupted circadian and sleep–wake rhythms. Leptin-deficient *ob/ob* mice ($Lep^{ob/ob}$), a genetic model of severe obesity and metabolic dysregulation, exhibit disrupted sleep architecture and consolidation, low body temperature, attenuated feeding rhythm behavior, and decreased locomotor activity [151]. Similarly, leptin receptor-deficient mice ($Lepr^{db/db}$) and Zucker obese rats develop obesity, diabetes, and altered sleep regulation, feeding rhythm behavior, and diurnal locomotor activity rhythms [152–154].

Both diet-induced and genetically obese mice exhibit altered circadian locomotor behavior, indicating that the timing of food intake itself may play a significant role in weight gain. Indeed, numerous studies have reported that peripheral clocks are permissive to and can be entrained by temporal changes in food availability. For example, while rhythmicity in the SCN remains phase-locked to the light–dark cycle, liver clocks can be entrained rapidly to changes in feeding schedules [155–157]. In rodents, restricted feeding during the day uncouples circadian liver gene expression from circadian gene expression in the SCN, and SCN-ablated rodents are still able to anticipate mealtime during restricted feeding, as demonstrated by increased locomotor activity, increased body temperature and endocrine changes (reviewed in [158]). Finally, a recent study showed that mice fed a high-fat diet only during the 12-h light phase (the normal "rest" period) gained significantly more weight than mice fed only during the 12-h dark phase (the normal "wake" period) [159], suggesting that temporal changes in food availability impact energy constancy.

Regulation of the Clock by Sleep

Sleep and arousal have also been shown to regulate the circadian clock. Neuronal activity in the SCN increases during REM sleep but decreases during NREM sleep, and a causal relationship exists between the sleep state and SCN neuronal firing [160]. Further, chronic sleep deprivation inhibits SCN neuronal activity, which is most pronounced during NREM and REM sleep [161]. Thus, similar to the interrelationship between circadian rhythms and metabolism, these data suggest the existence of a reciprocal loop between sleep and the circadian network.

The Neuroendocrine System, Energy Balance, and Sleep

Growth Hormone and Sleep

It has been known for decades that secretion of certain hormones, such as growth hormone (GH), is tightly associated with sleep. GH is secreted from the anterior pituitary in a pulsatile manner, and its burst release during slow-wave sleep (SWS) after sleep onset accounts for two-thirds of the daily GH secretion [162]. Sleep deprivation suppresses GH secretion, while GH secretion is restored during recovery sleep [163–165]. Interestingly, sleep-related GH release is also both gender and age dependent. Men have more robust GH secretion after sleep onset than women, whose sleep-related GH release only represents about 50% of daily GH release [166]. Further, during aging, the sleep-associated GH secretion gradually diminishes until age 50 when it practically disappears [167].

GH-releasing hormone (GHRH), somatostatin, and ghrelin (which inhibits and stimulates GH secretion, respectively) may mediate the regulation of GH secretion by sleep [168]. Administration of GHRH antagonists nearly abolishes the GH secretion at sleep onset, suggesting a critical role of GHRH in sleep-related GH secretion [169]. In rats, hypothalamic *Ghrh* mRNA levels peak at the onset of the light (rest) period, then gradually decrease and eventually reach trough levels at night (active period) [170]. On the other hand, GHRH content drops during the light period, indicating its robust release, but accumulates at night likely as a result of either reduced degradation or release [171]. In agreement with these findings, sleep deprivation results in an increase in *Ghrh* mRNA synthesis and a concomitant decrease in GHRH content due to increased release [172].

The relationship between GHRH and sleep appears to be reciprocal. A number of studies have shown that GHRH stimulates NREM sleep in many species including rats, mice, and humans [173, 174]. This phenomenon is not mediated by GH since similar effects of GHRH on sleep were also observed in hypophysectomized animals [174] and in Spontaneous Dwarf rats, which harbor inactivating GH mutations [168]. Consistent with the above findings, disruption of GHRH signaling by various methods, including somatostatin [175], high levels of IGF-1 [176] and GH via their negative feedback [177], GHRH antagonist [178], and GHRH antibody [179], leads to reduced NREM sleep. Moreover, NREM sleep is significantly reduced in dwarf rats and *Lit/Lit* mice, both of which have deficient GHRH signaling [173, 180]. It appears that GHRH signaling synchronizes both sleep and GH secretion and that the connection between the POA sleep center neurons and the GHRH neurons in the arcuate nucleus may underlie this dual function of GHRH.

The HPA Axis and Sleep

In addition to the hypothalamic–somatotropic axis, the hypothalamic–pituitary adrenal axis also interacts with sleep regulation at many levels, as there is a reciprocal interaction between GHRH and corticotropin-releasing hormone (CRH) in sleep control. In humans, NREM sleep and GH release peak during the first half of the night, whereas adrenocorticotropic hormone (ACTH) and cortisol secretion predominate during the second half of the night. GHRH stimulates NREM sleep, GH secretion and reduces cortisol levels [181]. CRH administration decreases GH levels and NREM sleep and increases cortisol levels and awakenings [182].

Hypercortisolism, such as that observed in patients with Cushing's syndrome, is associated with decreased SWS [183, 184]. Similarly, mild elevation in ACTH and cortisol is also present in patients with depression [185], whose sleep pattern is characterized by decreased NREM sleep and disrupted sleep continuity [186, 187]. Additionally, CRH neurons form reciprocal synapses with noradrenergic neurons in the LC, thereby promoting alertness via the wake-promoting center [188]. Studies suggest that the effect of hypercortisolism on sleep is likely mediated by CRH, rather than glucocorticoid hormone itself [189].

Sleep can also modulate the activity of the HPA axis. However, it should be emphasized that the HPA axis is largely controlled by the circadian system, while the effect of sleep on HPA is rather subtle [190]. The peak of the HPA activity is associated with the onset of awakening, the so-called "awakening response." Sleep deprivation leads to an increase in ACTH and cortisol [191]. Interestingly, obstructive sleep apnea (OSA) is characterized by frequent awakenings at night and activation of the HPA axis, the latter of which has been proposed to contribute to the development of metabolic syndrome associated with OSA [191].

Orexin Neurons in the Integration of Circadian Rhythms, Sleep, and Metabolism

Accumulating evidence suggests that orexin neurons in the LHA may serve as a hub for integrating regulatory signals for circadian rhythm, sleep/wake cycles, and energy balance. As discussed in more detail in another chapter (see Chap. 3), orexin neurons communicate with multiple sleep and wake regulating centers including the VLPO, the BF, and the monoaminergic neurons [192]. Moreover, orexin neurons are also under strong circadian control. The SCN projects to orexin neurons in the LHA both directly [54] and indirectly via the DMH [51]. Orexin-1 levels in the CSF display circadian variations, which are absent in SCN-lesioned animals [193, 194].

Recent studies have also demonstrated a key role of orexin neurons in energy homeostasis. Anatomically, orexin neurons in the LHA have dense synapses with multiple neuronal populations implicated in energetics, including reciprocal connections with neurons in the DMH and the NPY/AgRP and POMC/CART neurons in the ARN [50]. Functionally, earlier studies showed that acute ICV injection of orexin leads to increased food intake [195], and orexin antagonists result in hypophagia [196]. However, a recent study demonstrated that mice overexpressing orexin are resistant to diet-induced obesity likely via the orexin 2 receptor and that these effects are dependent on leptin signaling [197], suggesting a role for endogenous orexin signaling in protection against obesity. Another important role of orexin neurons is their involvement in the food anticipatory activity (FAA), which is observed before food presentation when food availability is restricted to certain period of the day. Orexin neuron-ablated mice exhibit normal entrainment of *Per2* expression in the brain and liver by restricted feeding, but fail to demonstrate FAA [198, 199]. Moreover, orexin neurons are important in the maintenance of arousal during fasting, which is essential for alertness and food seeking behavior during food shortage. Consistent with this hypothesis, orexin neuron-ablated mice exhibit loss of activity and arousal in response to fasting [200].

Interestingly, in addition to orexin, orexin neurons may express as-yet unidentified neuropeptides that contribute to wakefulness and energy homeostasis. A recent study showed that orexin neurons, but not the orexin neuropeptide, are critical for the circadian control of REM sleep [201]. In addition, in contrast to orexin neuron-ablated mice, which lack FAA, orexin knockout mice demonstrate normal activity when they are entrained to a restricted feeding schedule [202]. Future studies will further identify the molecular factors within orexin-producing neurons that are involved in FAA.

Conclusions

Rapid strides have been made in recent years to dissect the complex interplay between circadian rhythms, metabolism, and sleep. These three processes are interdependent and closely connected through extensive reciprocal relationships (see Fig. 2.1). Given the concurrent epidemics of sleep loss and obesity, it is more important than ever to gain a detailed understanding of both the molecular and physiologic underpinnings linking sleep, metabolism, and the circadian system. Additional advances will emerge by refining the neuroanatomic map linking circadian and energetic centers with areas regulating sleep and wakefulness; ultimately, such efforts may lead to the development of rational therapeutics for the treatment of obesity and sleep-related metabolic disorders.

Acknowledgments The authors wish to thank A. Wicher for administrative assistance. J. Bass is supported by NIH grants R01 DK090625 and P01 AG011412, the Chicago Biomedical Consortium with support from the Searle Funds at the Chicago Community Trust, and the University of Chicago Diabetes Research and Training Center (DK-20595).

References

1. Balsalobre A, Marcacci L, Schibler U. Multiple signaling pathways elicit circadian gene expression in cultured Rat-1 fibroblasts. Curr Biol. 2000;10: 1291–4.
2. Welsh DK, Yoo SH, Liu AC, Takahashi JS, Kay SA. Bioluminescence imaging of individual fibroblasts reveals persistent, independently phased circadian rhythms of clock gene expression. Curr Biol. 2004;14:2289–95.
3. Yoo SH, Yamazaki S, Lowrey PL, Shimomura K, Ko CH, Buhr ED, Siepka SM, Hong HK, Oh WJ, Yoo OJ, Menaker M, Takahashi JS. PERIOD2::LUCIFERASE real-time reporting of circadian dynamics reveals persistent circadian oscillations in mouse peripheral tissues. Proc Natl Acad Sci USA. 2004;101: 5339–46.
4. Lehman MN, Silver R, Gladstone WR, Kahn RM, Gibson M, Bittman EL. Circadian rhythmicity restored by neural transplant. Immunocytochemical characterization of the graft and its integration with the host brain. J Neurosci. 1987;7:1626–38.
5. Moore RY, Eichler VB. Loss of a circadian adrenal corticosterone rhythm following suprachiasmatic lesions in the rat. Brain Res. 1972;42:201–6.
6. Ralph MR, Foster RG, Davis FC, Menaker M. Transplanted suprachiasmatic nucleus determines circadian period. Science. 1990;247:975–8.
7. Stephan FK, Zucker I. Circadian rhythms in drinking behavior and locomotor activity of rats are eliminated by hypothalamic lesions. Proc Natl Acad Sci USA. 1972;69:1583–6.
8. Sujino M, Masumoto KH, Yamaguchi S, van der Horst GT, Okamura H, Inouye ST. Suprachiasmatic nucleus grafts restore circadian behavioral rhythms of genetically arrhythmic mice. Curr Biol. 2003;13: 664–8.
9. Mohawk JA, Takahashi JS. Cell autonomy and synchrony of suprachiasmatic nucleus circadian oscillators. Trends Neurosci. 2011;34:349–58.
10. Bass J, Takahashi JS. Circadian integration of metabolism and energetics. Science. 2010;330:1349–54.
11. Abrahamson EE, Moore RY. Suprachiasmatic nucleus in the mouse: retinal innervation, intrinsic organization and efferent projections. Brain Res. 2001;916:172–91.
12. Berson DM, Dunn FA, Takao M. Phototransduction by retinal ganglion cells that set the circadian clock. Science. 2002;295:1070–3.
13. Hankins MW, Peirson SN, Foster RG. Melanopsin: an exciting photopigment. Trends Neurosci. 2008;31: 27–36.
14. Saper CB, Scammell TE, Lu J. Hypothalamic regulation of sleep and circadian rhythms. Nature. 2005;437:1257–63.
15. Kramer A, Yang FC, Snodgrass P, Li X, Scammell TE, Davis FC, Weitz CJ. Regulation of daily locomotor activity and sleep by hypothalamic EGF receptor signaling. Science. 2001;294:2511–5.
16. Kraves S, Weitz CJ. A role for cardiotrophin-like cytokine in the circadian control of mammalian locomotor activity. Nat Neurosci. 2006;9:212–9.
17. Cheng MY, Bullock CM, Li C, Lee AG, Bermak JC, Belluzzi J, Weaver DR, Leslie FM, Zhou QY. Prokineticin 2 transmits the behavioural circadian rhythm of the suprachiasmatic nucleus. Nature. 2002;417:405–10.
18. Green CB, Takahashi JS, Bass J. The meter of metabolism. Cell. 2008;134:728–42.
19. Morin LP, Allen CN. The circadian visual system, 2005. Brain Res Rev. 2006;51:1–60.
20. Panda S, Antoch MP, Miller BH, Su AI, Schook AB, Straume M, Schultz PG, Kay SA, Takahashi JS, Hogenesch JB. Coordinated transcription of key pathways in the mouse by the circadian clock. Cell. 2002;109:307–20.
21. Rutter J, Reick M, McKnight SL. Metabolism and the control of circadian rhythms. Annu Rev Biochem. 2002;71:307–31.
22. Ko CH, Takahashi JS. Molecular components of the mammalian circadian clock. Hum Mol Genet. 2006; 15 Spec No 2: R271–7.
23. Lowrey PL, Takahashi JS. Genetics of circadian rhythms in mammalian model organisms. Adv Genet. 2011;74:175–230.
24. Reppert SM, Weaver DR. Coordination of circadian timing in mammals. Nature. 2002;418:935–41.
25. DeBruyne JP, Weaver DR, Reppert SM. CLOCK and NPAS2 have overlapping roles in the suprachiasmatic circadian clock. Nat Neurosci. 2007;10:543–5.

26. DeBruyne JP, Weaver DR, Reppert SM. Peripheral circadian oscillators require CLOCK. Curr Biol. 2007;17:R538–9.
27. Gekakis N, Staknis D, Nguyen HB, Davis FC, Wilsbacher LD, King DP, Takahashi JS, Weitz CJ. Role of the CLOCK protein in the mammalian circadian mechanism. Science. 1998;280:1564–9.
28. Hogenesch JB, Gu YZ, Jain S, Bradfield CA. The basic-helix-loop-helix-PAS orphan MOP3 forms transcriptionally active complexes with circadian and hypoxia factors. Proc Natl Acad Sci USA. 1998;95:5474–9.
29. Jin X, Shearman LP, Weaver DR, Zylka MJ, de Vries GJ, Reppert SM. A molecular mechanism regulating rhythmic output from the suprachiasmatic circadian clock. Cell. 1999;96:57–68.
30. Kume K, Zylka MJ, Sriram S, Shearman LP, Weaver DR, Jin X, Maywood ES, Hastings MH, Reppert SM. mCRY1 and mCRY2 are essential components of the negative limb of the circadian clock feedback loop. Cell. 1999;98:193–205.
31. Lee C, Etchegaray JP, Cagampang FR, Loudon AS, Reppert SM. Posttranslational mechanisms regulate the mammalian circadian clock. Cell. 2001;107:855–67.
32. Preitner N, Damiola F, Lopez-Molina L, Zakany J, Duboule D, Albrecht U, Schibler U. The orphan nuclear receptor REV-ERBalpha controls circadian transcription within the positive limb of the mammalian circadian oscillator. Cell. 2002;110:251–60.
33. Sato TK, Panda S, Miraglia LJ, Reyes TM, Rudic RD, McNamara P, Naik KA, FitzGerald GA, Kay SA, Hogenesch JB. A functional genomics strategy reveals Rora as a component of the mammalian circadian clock. Neuron. 2004;43:527–37.
34. Yin L, Lazar MA. The orphan nuclear receptor Rev-erbalpha recruits the N-CoR/histone deacetylase 3 corepressor to regulate the circadian Bmal1 gene. Mol Endocrinol. 2005;19:1452–9.
35. Akashi M, Takumi T. The orphan nuclear receptor RORalpha regulates circadian transcription of the mammalian core-clock Bmal1. Nat Struct Mol Biol. 2005;12:441–8.
36. Guillaumond F, Dardente H, Giguere V, Cermakian N. Differential control of Bmal1 circadian transcription by REV-ERB and ROR nuclear receptors. J Biol Rhythms. 2005;20:391–403.
37. Liu AC, Tran HG, Zhang EE, Priest AA, Welsh DK, Kay SA. Redundant function of REV-ERBalpha and beta and non-essential role for Bmal1 cycling in transcriptional regulation of intracellular circadian rhythms. PLoS Genet. 2008;4:e1000023.
38. Busino L, Bassermann F, Maiolica A, Lee C, Nolan PM, Godinho SI, Draetta GF, Pagano M. SCFFbxl3 controls the oscillation of the circadian clock by directing the degradation of cryptochrome proteins. Science. 2007;316:900–4.
39. Eide EJ, Woolf MF, Kang H, Woolf P, Hurst W, Camacho F, Vielhaber EL, Giovanni A, Virshup DM. Control of mammalian circadian rhythm by CKIepsilon-regulated proteasome-mediated PER2 degradation. Mol Cell Biol. 2005;25:2795–807.
40. Gallego M, Eide EJ, Woolf MF, Virshup DM, Forger DB. An opposite role for tau in circadian rhythms revealed by mathematical modeling. Proc Natl Acad Sci USA. 2006;103:10618–23.
41. Godinho SI, Maywood ES, Shaw L, Tucci V, Barnard AR, Busino L, Pagano M, Kendall R, Quwailid MM, Romero MR, O'Neill J, Chesham JE, Brooker D, Lalanne Z, Hastings MH, Nolan PM. The after-hours mutant reveals a role for Fbxl3 in determining mammalian circadian period. Science. 2007;316:897–900.
42. Hirota T, Lee JW, Lewis WG, Zhang EE, Breton G, Liu X, Garcia M, Peters EC, Etchegaray JP, Traver D, Schultz PG, Kay SA. High-throughput chemical screen identifies a novel potent modulator of cellular circadian rhythms and reveals CKIalpha as a clock regulatory kinase. PLoS Biol. 2010;8:e1000559.
43. Lee HM, Chen R, Kim H, Etchegaray JP, Weaver DR, Lee C. The period of the circadian oscillator is primarily determined by the balance between casein kinase 1 and protein phosphatase 1. Proc Natl Acad Sci USA. 2011;108:16451–6.
44. Meng QJ, Logunova L, Maywood ES, Gallego M, Lebiecki J, Brown TM, Sladek M, Semikhodskii AS, Glossop NR, Piggins HD, Chesham JE, Bechtold DA, Yoo SH, Takahashi JS, Virshup DM, Boot-Handford RP, Hastings MH, Loudon AS. Setting clock speed in mammals: the CK1 epsilon tau mutation in mice accelerates circadian pacemakers by selectively destabilizing PERIOD proteins. Neuron. 2008;58:78–88.
45. Reischl S, Vanselow K, Westermark PO, Thierfelder N, Maier B, Herzel H, Kramer A. Beta-TrCP1-mediated degradation of PERIOD2 is essential for circadian dynamics. J Biol Rhythms. 2007;22:375–86.
46. Shirogane T, Jin J, Ang XL, Harper JW. SCFbeta-TRCP controls clock-dependent transcription via casein kinase 1-dependent degradation of the mammalian period-1 (Per1) protein. J Biol Chem. 2005;280:26863–72.
47. Siepka SM, Yoo SH, Park J, Song W, Kumar V, Hu Y, Lee C, Takahashi JS. Circadian mutant Overtime reveals F-box protein FBXL3 regulation of cryptochrome and period gene expression. Cell. 2007;129:1011–23.
48. Xu Y, Padiath QS, Shapiro RE, Jones CR, Wu SC, Saigoh N, Saigoh K, Ptacek LJ, Fu YH. Functional consequences of a CKIdelta mutation causing familial advanced sleep phase syndrome. Nature. 2005;434:640–4.
49. Xu Y, Toh KL, Jones CR, Shin JY, Fu YH, Ptacek LJ. Modeling of a human circadian mutation yields insights into clock regulation by PER2. Cell. 2007;128:59–70.
50. Huang W, Ramsey KM, Marcheva B, Bass J. Circadian rhythms, sleep, and metabolism. J Clin Invest. 2011;121:2133–41.

51. Abrahamson EE, Leak RK, Moore RY. The suprachiasmatic nucleus projects to posterior hypothalamic arousal systems. Neuroreport. 2001;12:435–40.
52. Chou TC, Bjorkum AA, Gaus SE, Lu J, Scammell TE, Saper CB. Afferents to the ventrolateral preoptic nucleus. J Neurosci. 2002;22:977–90.
53. Chou TC, Scammell TE, Gooley JJ, Gaus SE, Saper CB, Lu J. Critical role of dorsomedial hypothalamic nucleus in a wide range of behavioral circadian rhythms. J Neurosci. 2003;23:10691–702.
54. Vrang N, Mikkelsen JD, Larsen PJ. Direct link from the suprachiasmatic nucleus to hypothalamic neurons projecting to the spinal cord: a combined tracing study using cholera toxin subunit B and Phaseolus vulgaris-leucoagglutinin. Brain Res Bull. 1997;44:671–80.
55. Lu J, Zhang YH, Chou TC, Gaus SE, Elmquist JK, Shiromani P, Saper CB. Contrasting effects of ibotenate lesions of the paraventricular nucleus and subparaventricular zone on sleep-wake cycle and temperature regulation. J Neurosci. 2001;21:4864–74.
56. Aston-Jones G, Chen S, Zhu Y, Oshinsky ML. A neural circuit for circadian regulation of arousal. Nat Neurosci. 2001;4:732–8.
57. Luo AH, Aston-Jones G. Circuit projection from suprachiasmatic nucleus to ventral tegmental area: a novel circadian output pathway. Eur J Neurosci. 2009;29:748–60.
58. Naylor E, Bergmann BM, Krauski K, Zee PC, Takahashi JS, Vitaterna MH, Turek FW. The circadian clock mutation alters sleep homeostasis in the mouse. J Neurosci. 2000;20:8138–43.
59. Laposky A, Easton A, Dugovic C, Walisser J, Bradfield C, Turek F. Deletion of the mammalian circadian clock gene BMAL1/Mop3 alters baseline sleep architecture and the response to sleep deprivation. Sleep. 2005;28:395–409.
60. Dudley CA, Erbel-Sieler C, Estill SJ, Reick M, Franken P, Pitts S, McKnight SL. Altered patterns of sleep and behavioral adaptability in NPAS2-deficient mice. Science. 2003;301:379–83.
61. Wisor JP, O'Hara BF, Terao A, Selby CP, Kilduff TS, Sancar A, Edgar DM, Franken P. A role for cryptochromes in sleep regulation. BMC Neurosci. 2002; 3:20.
62. Kopp C, Albrecht U, Zheng B, Tobler I. Homeostatic sleep regulation is preserved in mPerl and mPer2 mutant mice. Eur J Neurosci. 2002;16:1099–106.
63. Shiromani PJ, Xu M, Winston EM, Shiromani SN, Gerashchenko D, Weaver DR. Sleep rhythmicity and homeostasis in mice with targeted disruption of mPeriod genes. Am J Physiol Regul Integr Comp Physiol. 2004;287:R47–57.
64. Takahashi JS, Hong HK, Ko CH, McDearmon EL. The genetics of mammalian circadian order and disorder: implications for physiology and disease. Nat Rev Genet. 2008;9:764–75.
65. Jones CR, Campbell SS, Zone SE, Cooper F, DeSano A, Murphy PJ, Jones B, Czajkowski L, Ptacek LJ. Familial advanced sleep-phase syndrome: a short-period circadian rhythm variant in humans. Nat Med. 1999;5:1062–5.
66. Toh KL, Jones CR, He Y, Eide EJ, Hinz WA, Virshup DM, Ptacek LJ, Fu YH. An hPer2 phosphorylation site mutation in familial advanced sleep phase syndrome. Science. 2001;291:1040–3.
67. Ancoli-Israel S, Schnierow B, Kelsoe J, Fink R. A pedigree of one family with delayed sleep phase syndrome. Chronobiol Int. 2001;18:831–40.
68. Iwase T, Kajimura N, Uchiyama M, Ebisawa T, Yoshimura K, Kamei Y, Shibui K, Kim K, Kudo Y, Katoh M, Watanabe T, Nakajima T, Ozeki Y, Sugishita M, Hori T, Ikeda M, Toyoshima R, Inoue Y, Yamada N, Mishima K, Nomura M, Ozaki N, Okawa M, Takahashi K, Yamauchi T. Mutation screening of the human Clock gene in circadian rhythm sleep disorders. Psychiatry Res. 2002;109: 121–8.
69. Archer SN, Robilliard DL, Skene DJ, Smits M, Williams A, Arendt J, von Schantz M. A length polymorphism in the circadian clock gene Per3 is linked to delayed sleep phase syndrome and extreme diurnal preference. Sleep. 2003;26:413–5.
70. Carpen JD, Archer SN, Skene DJ, Smits M, von Schantz M. A single-nucleotide polymorphism in the 5'-untranslated region of the hPER2 gene is associated with diurnal preference. J Sleep Res. 2005;14:293–7.
71. Katzenberg D, Young T, Finn L, Lin L, King DP, Takahashi JS, Mignot E. A CLOCK polymorphism associated with human diurnal preference. Sleep. 1998;21:569–76.
72. Mishima K, Tozawa T, Satoh K, Saitoh H, Mishima Y. The 3111 T/C polymorphism of hClock is associated with evening preference and delayed sleep timing in a Japanese population sample. Am J Med Genet B Neuropsychiatr Genet. 2005;133B:101–4.
73. Pereira DS, Tufik S, Louzada FM, Benedito-Silva AA, Lopez AR, Lemos NA, Korczak AL, D'Almeida V, Pedrazzoli M. Association of the length polymorphism in the human Per3 gene with the delayed sleep-phase syndrome: does latitude have an influence upon it? Sleep. 2005;28:29–32.
74. Robilliard DL, Archer SN, Arendt J, Lockley SW, Hack LM, English J, Leger D, Smits MG, Williams A, Skene DJ, Von Schantz M. The 3111 Clock gene polymorphism is not associated with sleep and circadian rhythmicity in phenotypically characterized human subjects. J Sleep Res. 2002;11:305–12.
75. Takano A, Uchiyama M, Kajimura N, Mishima K, Inoue Y, Kamei Y, Kitajima T, Shibui K, Katoh M, Watanabe T, Hashimotodani Y, Nakajima T, Ozeki Y, Hori T, Yamada N, Toyoshima R, Ozaki N, Okawa M, Nagai K, Takahashi K, Isojima Y, Yamauchi T, Ebisawa T. A missense variation in human casein kinase I epsilon gene that induces functional alteration and shows an inverse association with circadian rhythm sleep disorders. Neuropsychopharmacology. 2004;29:1901–9.

76. Utge SJ, Soronen P, Loukola A, Kronholm E, Ollila HM, Pirkola S, Porkka-Heiskanen T, Partonen T, Paunio T. Systematic analysis of circadian genes in a population-based sample reveals association of TIMELESS with depression and sleep disturbance. PLoS One. 2010;5:e9259.
77. Viola AU, Archer SN, James LM, Groeger JA, Lo JC, Skene DJ, von Schantz M, Dijk DJ. PER3 polymorphism predicts sleep structure and waking performance. Curr Biol. 2007;17:613–8.
78. He Y, Jones CR, Fujiki N, Xu Y, Guo B, Holder Jr JL, Rossner MJ, Nishino S, Fu YH. The transcriptional repressor DEC2 regulates sleep length in mammals. Science. 2009;325:866–70.
79. Akhtar RA, Reddy AB, Maywood ES, Clayton JD, King VM, Smith AG, Gant TW, Hastings MH, Kyriacou CP. Circadian cycling of the mouse liver transcriptome, as revealed by cDNA microarray, is driven by the suprachiasmatic nucleus. Curr Biol. 2002;12:540–50.
80. Duffield GE, Best JD, Meurers BH, Bittner A, Loros JJ, Dunlap JC. Circadian programs of transcriptional activation, signaling, and protein turnover revealed by microarray analysis of mammalian cells. Curr Biol. 2002;12:551–7.
81. Hughes ME, DiTacchio L, Hayes KR, Vollmers C, Pulivarthy S, Baggs JE, Panda S, Hogenesch JB. Harmonics of circadian gene transcription in mammals. PLoS Genet. 2009;5:e1000442.
82. Miller BH, McDearmon EL, Panda S, Hayes KR, Zhang J, Andrews JL, Antoch MP, Walker JR, Esser KA, Hogenesch JB, Takahashi JS. Circadian and CLOCK-controlled regulation of the mouse transcriptome and cell proliferation. Proc Natl Acad Sci USA. 2007;104:3342–7.
83. Oishi K, Miyazaki K, Kadota K, Kikuno R, Nagase T, Atsumi G, Ohkura N, Azama T, Mesaki M, Yukimasa S, Kobayashi H, Iitaka C, Umehara T, Horikoshi M, Kudo T, Shimizu Y, Yano M, Monden M, Machida K, Matsuda J, Horie S, Todo T, Ishida N. Genome-wide expression analysis of mouse liver reveals CLOCK-regulated circadian output genes. J Biol Chem. 2003;278:41519–27.
84. Storch KF, Lipan O, Leykin I, Viswanathan N, Davis FC, Wong WH, Weitz CJ. Extensive and divergent circadian gene expression in liver and heart. Nature. 2002;417:78–83.
85. Doherty CJ, Kay SA. Circadian control of global gene expression patterns. Annu Rev Genet. 2010;44: 419–44.
86. Hughes ME, Hogenesch JB, Kornacker K. JTK_CYCLE: an efficient nonparametric algorithm for detecting rhythmic components in genome-scale data sets. J Biol Rhythms. 2010;25:372–80.
87. Delaunay F, Laudet V. Circadian clock and microarrays: mammalian genome gets rhythm. Trends Genet. 2002;18:595–7.
88. Kohsaka A, Bass J. A sense of time: how molecular clocks organize metabolism. Trends Endocrinol Metab. 2007;18:4–11.
89. Rudic RD, McNamara P, Curtis AM, Boston RC, Panda S, Hogenesch JB, Fitzgerald GA. BMAL1 and CLOCK, two essential components of the circadian clock, are involved in glucose homeostasis. PLoS Biol. 2004;2:e377.
90. Yang X, Downes M, Yu RT, Bookout AL, He W, Straume M, Mangelsdorf DJ, Evans RM. Nuclear receptor expression links the circadian clock to metabolism. Cell. 2006;126:801–10.
91. Kornmann B, Schaad O, Bujard H, Takahashi JS, Schibler U. System-driven and oscillator-dependent circadian transcription in mice with a conditionally active liver clock. PLoS Biol. 2007;5:e34.
92. Asher G, Schibler U. Crosstalk between components of circadian and metabolic cycles in mammals. Cell Metab. 2011;13:125–37.
93. Fontaine C, Dubois G, Duguay Y, Helledie T, Vu-Dac N, Gervois P, Soncin F, Mandrup S, Fruchart JC, Fruchart-Najib J, Staels B. The orphan nuclear receptor Rev-Erbalpha is a peroxisome proliferator-activated receptor (PPAR) gamma target gene and promotes PPARgamma-induced adipocyte differentiation. J Biol Chem. 2003;278:37672–80.
94. Wang J, Lazar MA. Bifunctional role of Rev-erbalpha in adipocyte differentiation. Mol Cell Biol. 2008;28:2213–20.
95. Yin L, Wu N, Curtin JC, Qatanani M, Szwergold NR, Reid RA, Waitt GM, Parks DJ, Pearce KH, Wisely GB, Lazar MA. Rev-erbalpha, a heme sensor that coordinates metabolic and circadian pathways. Science. 2007;318:1786–9.
96. Canaple L, Rambaud J, Dkhissi-Benyahya O, Rayet B, Tan NS, Michalik L, Delaunay F, Wahli W, Laudet V. Reciprocal regulation of brain and muscle Arnt-like protein 1 and peroxisome proliferator-activated receptor alpha defines a novel positive feedback loop in the rodent liver circadian clock. Mol Endocrinol. 2006;20:1715–27.
97. Desvergne B, Wahli W. Peroxisome proliferator-activated receptors: nuclear control of metabolism. Endocr Rev. 1999;20:649–88.
98. Lau P, Nixon SJ, Parton RG, Muscat GE. RORalpha regulates the expression of genes involved in lipid homeostasis in skeletal muscle cells: caveolin-3 and CPT-1 are direct targets of ROR. J Biol Chem. 2004;279:36828–40.
99. Nakahata Y, Sahar S, Astarita G, Kaluzova M, Sassone-Corsi P. Circadian control of the NAD+salvage pathway by CLOCK-SIRT1. Science. 2009; 324:654–7.
100. Ramsey KM, Yoshino J, Brace CS, Abrassart D, Kobayashi Y, Marcheva B, Hong HK, Chong JL, Buhr ED, Lee C, Takahashi JS, Imai SI, Bass J. Circadian clock feedback cycle through NAMPT-mediated NAD+biosynthesis. Science. 2009;324:651–4.
101. Asher G, Gatfield D, Stratmann M, Reinke H, Dibner C, Kreppel F, Mostoslavsky R, Alt FW, Schibler U. SIRT1 regulates circadian clock gene expression through PER2 deacetylation. Cell. 2008;134: 317–28.

102. Nakahata Y, Kaluzova M, Grimaldi B, Sahar S, Hirayama J, Chen D, Guarente LP, Sassone-Corsi P. The NAD+-dependent deacetylase SIRT1 modulates CLOCK-mediated chromatin remodeling and circadian control. Cell. 2008;134:329–40.
103. Canto C, Gerhart-Hines Z, Feige JN, Lagouge M, Noriega L, Milne JC, Elliott PJ, Puigserver P, Auwerx J. AMPK regulates energy expenditure by modulating NAD(+) metabolism and SIRT1 activity. Nature. 2009;458(7241):1056–60.
104. Fulco M, Cen Y, Zhao P, Hoffman EP, McBurney MW, Sauve AA, Sartorelli V. Glucose restriction inhibits skeletal myoblast differentiation by activating SIRT1 through AMPK-mediated regulation of Nampt. Dev Cell. 2008;14:661–73.
105. Cohen HY, Miller C, Bitterman KJ, Wall NR, Hekking B, Kessler B, Howitz KT, Gorospe M, de Cabo R, Sinclair DA. Calorie restriction promotes mammalian cell survival by inducing the SIRT1 deacetylase. Science. 2004;305:390–2.
106. Nemoto S, Fergusson MM, Finkel T. Nutrient availability regulates SIRT1 through a forkhead-dependent pathway. Science. 2004;306:2105–8.
107. Rodgers JT, Lerin C, Haas W, Gygi SP, Spiegelman BM, Puigserver P. Nutrient control of glucose homeostasis through a complex of PGC-1alpha and SIRT1. Nature. 2005;434:113–8.
108. Marcheva B, Ramsey KM, Buhr ED, Kobayashi Y, Su H, Ko CH, Ivanova G, Omura C, Mo S, Vitaterna MH, Lopez JP, Philipson LH, Bradfield CA, Crosby SD, Jebailey L, Wang X, Takahashi JS, Bass J. Disruption of the clock components CLOCK and BMAL1 leads to hypoinsulinaemia and diabetes. Nature. 2010;466:571–2.
109. Turek FW, Joshu C, Kohsaka A, Lin E, Ivanova G, McDearmon E, Laposky A, Losee-Olson S, Easton A, Jensen DR, Eckel RH, Takahashi JS, Bass J. Obesity and metabolic syndrome in circadian Clock mutant mice. Science. 2005;308:1043–5.
110. Bunger MK, Walisser JA, Sullivan R, Manley PA, Moran SM, Kalscheur VL, Colman RJ, Bradfield CA. Progressive arthropathy in mice with a targeted disruption of the Mop3/Bmal-1 locus. Genesis. 2005;41:122–32.
111. Bunger MK, Wilsbacher LD, Moran SM, Clendenin C, Radcliffe LA, Hogenesch JB, Simon MC, Takahashi JS, Bradfield CA. Mop3 is an essential component of the master circadian pacemaker in mammals. Cell. 2000;103:1009–17.
112. Lamia KA, Storch KF, Weitz CJ. Physiological significance of a peripheral tissue circadian clock. Proc Natl Acad Sci USA. 2008;105:15172–7.
113. Zhang EE, Liu Y, Dentin R, Pongsawakul PY, Liu AC, Hirota T, Nusinow DA, Sun X, Landais S, Kodama Y, Brenner DA, Montminy M, Kay SA. Cryptochrome mediates circadian regulation of cAMP signaling and hepatic gluconeogenesis. Nat Med. 2010;16:1152–6.
114. Lucas RJ, Stirland JA, Mohammad YN, Loudon AS. Postnatal growth rate and gonadal development in circadian tau mutant hamsters reared in constant dim red light. J Reprod Fertil. 2000;118:327–30.
115. Oklejewicz M, Hut RA, Daan S, Loudon AS, Stirland AJ. Metabolic rate changes proportionally to circadian frequency in tau mutant Syrian hamsters. J Biol Rhythms. 1997;12:413–22.
116. Douris N, Kojima S, Pan X, Lerch-Gaggl AF, Duong SQ, Hussain MM, Green CB. Nocturnin regulates circadian trafficking of dietary lipid in intestinal enterocytes. Curr Biol. 2011;21:1347–55.
117. Green CB, Douris N, Kojima S, Strayer CA, Fogerty J, Lourim D, Keller SR, Besharse JC. Loss of Nocturnin, a circadian deadenylase, confers resistance to hepatic steatosis and diet-induced obesity. Proc Natl Acad Sci USA. 2007;104:9888–93.
118. Eichner LJ, Giguere V. Estrogen related receptors (ERRs): a new dawn in transcriptional control of mitochondrial gene networks. Mitochondrion. 2011; 11:544–52.
119. Dufour CR, Levasseur MP, Pham NH, Eichner LJ, Wilson BJ, Charest-Marcotte A, Duguay D, Poirier-Heon JF, Cermakian N, Giguere V. Genomic convergence among ERRalpha, PROX1, and BMAL1 in the control of metabolic clock outputs. PLoS Genet. 2011;7:e1002143.
120. Conlon M, Lightfoot N, Kreiger N. Rotating shift work and risk of prostate cancer. Epidemiology. 2007;18:182–3.
121. Drake CL, Roehrs T, Richardson G, Walsh JK, Roth T. Shift work sleep disorder: prevalence and consequences beyond that of symptomatic day workers. Sleep. 2004;27:1453–62.
122. Karlsson B, Knutsson A, Lindahl B. Is there an association between shift work and having a metabolic syndrome? Results from a population based study of 27,485 people. Occup Environ Med. 2001;58: 747–52.
123. Knutsson A. Health disorders of shift workers. Occup Med (Lond). 2003;53:103–8.
124. Kubo T, Ozasa K, Mikami K, Wakai K, Fujino Y, Watanabe Y, Miki T, Nakao M, Hayashi K, Suzuki K, Mori M, Washio M, Sakauchi F, Ito Y, Yoshimura T, Tamakoshi A. Prospective cohort study of the risk of prostate cancer among rotating-shift workers: findings from the Japan collaborative cohort study. Am J Epidemiol. 2006;164:549–55.
125. Lahti TA, Partonen T, Kyyronen P, Kauppinen T, Pukkala E. Night-time work predisposes to non-Hodgkin lymphoma. Int J Cancer. 2008;123: 2148–51.
126. Morikawa Y, Nakagawa H, Miura K, Soyama Y, Ishizaki M, Kido T, Naruse Y, Suwazono Y, Nogawa K. Shift work and the risk of diabetes mellitus among Japanese male factory workers. Scand J Work Environ Health. 2005;31:179–83.
127. Schernhammer ES, Kroenke CH, Laden F, Hankinson SE. Night work and risk of breast cancer. Epidemiology. 2006;17:108–11.
128. Schernhammer ES, Laden F, Speizer FE, Willett WC, Hunter DJ, Kawachi I, Fuchs CS, Colditz GA. Night-

shift work and risk of colorectal cancer in the nurses' health study. J Natl Cancer Inst. 2003;95:825–8.
129. Spiegel K, Tasali E, Leproult R, Van Cauter E. Effects of poor and short sleep on glucose metabolism and obesity risk. Nat Rev Endocrinol. 2009; 5:253–61.
130. Spiegel K, Tasali E, Penev P, Van Cauter E. Brief communication: sleep curtailment in healthy young men is associated with decreased leptin levels, elevated ghrelin levels, and increased hunger and appetite. Ann Intern Med. 2004;141:846–50.
131. Scheer FA, Hilton MF, Mantzoros CS, Shea SA. Adverse metabolic and cardiovascular consequences of circadian misalignment. Proc Natl Acad Sci USA. 2009;106:4453–8.
132. Dupuis J, Langenberg C, Prokopenko I, Saxena R, Soranzo N, Jackson AU, Wheeler E, Glazer NL, Bouatia-Naji N, Gloyn AL, Lindgren CM, Magi R, Morris AP, Randall J, Johnson T, Elliott P, Rybin D, Thorleifsson G, Steinthorsdottir V, Henneman P, Grallert H, Dehghan A, Hottenga JJ, Franklin CS, Navarro P, Song K, Goel A, Perry JR, Egan JM, Lajunen T, Grarup N, Sparso T, Doney A, Voight BF, Stringham HM, Li M, Kanoni S, Shrader P, Cavalcanti-Proenca C, Kumari M, Qi L, Timpson NJ, Gieger C, Zabena C, Rocheleau G, Ingelsson E, An P, O'Connell J, Luan J, Elliott A, McCarroll SA, Payne F, Roccasecca RM, Pattou F, Sethupathy P, Ardlie K, Ariyurek Y, Balkau B, Barter P, Beilby JP, Ben-Shlomo Y, Benediktsson R, Bennett AJ, Bergmann S, Bochud M, Boerwinkle E, Bonnefond A, Bonnycastle LL, Borch-Johnsen K, Bottcher Y, Brunner E, Bumpstead SJ, Charpentier G, Chen YD, Chines P, Clarke R, Coin LJ, Cooper MN, Cornelis M, Crawford G, Crisponi L, Day IN, de Geus EJ, Delplanque J, Dina C, Erdos MR, Fedson AC, Fischer-Rosinsky A, Forouhi NG, Fox CS, Frants R, Franzosi MG, Galan P, Goodarzi MO, Graessler J, Groves CJ, Grundy S, Gwilliam R, Gyllensten U, Hadjadj S, Hallmans G, Hammond N, Han X, Hartikainen AL, Hassanali N, Hayward C, Heath SC, Hercberg S, Herder C, Hicks AA, Hillman DR, Hingorani AD, Hofman A, Hui J, Hung J, Isomaa B, Johnson PR, Jorgensen T, Jula A, Kaakinen M, Kaprio J, Kesaniemi YA, Kivimaki M, Knight B, Koskinen S, Kovacs P, Kyvik KO, Lathrop GM, Lawlor DA, Le Bacquer O, Lecoeur C, Li Y, Lyssenko V, Mahley R, Mangino M, Manning AK, Martinez-Larrad MT, McAteer JB, McCulloch LJ, McPherson R, Meisinger C, Melzer D, Meyre D, Mitchell BD, Morken MA, Mukherjee S, Naitza S, Narisu N, Neville MJ, Oostra BA, Orru M, Pakyz R, Palmer CN, Paolisso G, Pattaro C, Pearson D, Peden JF, Pedersen NL, Perola M, Pfeiffer AF, Pichler I, Polasek O, Posthuma D, Potter SC, Pouta A, Province MA, Psaty BM, Rathmann W, Rayner NW, Rice K, Ripatti S, Rivadeneira F, Roden M, Rolandsson O, Sandbaek A, Sandhu M, Sanna S, Sayer AA, Scheet P, Scott LJ, Seedorf U, Sharp SJ, Shields B, Sigurethsson G, Sijbrands EJ, Silveira A, Simpson L, Singleton A, Smith NL, Sovio U, Swift A, Syddall H, Syvanen AC, Tanaka T, Thorand B, Tichet J, Tonjes A, Tuomi T, Uitterlinden AG, van Dijk KW, van Hoek M, Varma D, Visvikis-Siest S, Vitart V, Vogelzangs N, Waeber G, Wagner PJ, Walley A, Walters GB, Ward KL, Watkins H, Weedon MN, Wild SH, Willemsen G, Witteman JC, Yarnell JW, Zeggini E, Zelenika D, Zethelius B, Zhai G, Zhao JH, Zillikens MC, Borecki IB, Loos RJ, Meneton P, Magnusson PK, Nathan DM, Williams GH, Hattersley AT, Silander K, Salomaa V, Smith GD, Bornstein SR, Schwarz P, Spranger J, Karpe F, Shuldiner AR, Cooper C, Dedoussis GV, Serrano-Rios M, Morris AD, Lind L, Palmer LJ, Hu FB, Franks PW, Ebrahim S, Marmot M, Kao WH, Pankow JS, Sampson MJ, Kuusisto J, Laakso M, Hansen T, Pedersen O, Pramstaller PP, Wichmann HE, Illig T, Rudan I, Wright AF, Stumvoll M, Campbell H, Wilson JF, Bergman RN, Buchanan TA, Collins FS, Mohlke KL, Tuomilehto J, Valle TT, Altshuler D, Rotter JI, Siscovick DS, Penninx BW, Boomsma DI, Deloukas P, Spector TD, Frayling TM, Ferrucci L, Kong A, Thorsteinsdottir U, Stefansson K, van Duijn CM, Aulchenko YS, Cao A, Scuteri A, Schlessinger D, Uda M, Ruokonen A, Jarvelin MR, Waterworth DM, Vollenweider P, Peltonen L, Mooser V, Abecasis GR, Wareham NJ, Sladek R, Froguel P, Watanabe RM, Meigs JB, Groop L, Boehnke M, McCarthy MI, Florez JC, Barroso I. New genetic loci implicated in fasting glucose homeostasis and their impact on type 2 diabetes risk. Nat Genet. 2010;42:105–16.
133. Woon PY, Kaisaki PJ, Braganca J, Bihoreau MT, Levy JC, Farrall M, Gauguier D. Aryl hydrocarbon receptor nuclear translocator-like (BMAL1) is associated with susceptibility to hypertension and type 2 diabetes. Proc Natl Acad Sci USA. 2007;104: 14412–7.
134. Scott EM, Carter AM, Grant PJ. Association between polymorphisms in the Clock gene, obesity and the metabolic syndrome in man. Int J Obes (Lond). 2008;32:658–62.
135. Garaulet M, Corbalan MD, Madrid JA, Morales E, Baraza JC, Lee YC, Ordovas JM. CLOCK gene is implicated in weight reduction in obese patients participating in a dietary programme based on the Mediterranean diet. Int J Obes (Lond). 2010;34:516–23.
136. Garaulet M, Sanchez-Moreno C, Smith CE, Lee YC, Nicolas F, Ordovas JM. Ghrelin, sleep reduction and evening preference: relationships to CLOCK 3111 T/C SNP and weight loss. PLoS One. 2011;6: e17435.
137. Lyssenko V, Nagorny CL, Erdos MR, Wierup N, Jonsson A, Spegel P, Bugliani M, Saxena R, Fex M, Pulizzi N, Isomaa B, Tuomi T, Nilsson P, Kuusisto J, Tuomilehto J, Boehnke M, Altshuler D, Sundler F, Eriksson JG, Jackson AU, Laakso M, Marchetti P, Watanabe RM, Mulder H, Groop L. Common variant in MTNR1B associated with increased risk of

type 2 diabetes and impaired early insulin secretion. Nat Genet. 2009;41:82–8.
138. Mulder H, Nagorny CL, Lyssenko V, Groop L. Melatonin receptors in pancreatic islets: good morning to a novel type 2 diabetes gene. Diabetologia. 2009;52:1240–9.
139. Ralph CL, Mull D, Lynch HJ, Hedlund L. A melatonin rhythm persists in rat pineals in darkness. Endocrinology. 1971;89:1361–6.
140. Ronn T, Wen J, Yang Z, Lu B, Du Y, Groop L, Hu R, Ling C. A common variant in MTNR1B, encoding melatonin receptor 1B, is associated with type 2 diabetes and fasting plasma glucose in Han Chinese individuals. Diabetologia. 2009;52:830–3.
141. Staiger H, Machicao F, Schafer SA, Kirchhoff K, Kantartzis K, Guthoff M, Silbernagel G, Stefan N, Haring HU, Fritsche A. Polymorphisms within the novel type 2 diabetes risk locus MTNR1B determine beta-cell function. PLoS One. 2008;3:e3962.
142. Takeuchi F, Katsuya T, Chakrewarthy S, Yamamoto K, Fujioka A, Serizawa M, Fujisawa T, Nakashima E, Ohnaka K, Ikegami H, Sugiyama T, Nabika T, Kasturiratne A, Yamaguchi S, Kono S, Takayanagi R, Yamori Y, Kobayashi S, Ogihara T, de Silva A, Wickremasinghe R, Kato N. Common variants at the GCK, GCKR, G6PC2-ABCB11 and MTNR1B loci are associated with fasting glucose in two Asian populations. Diabetologia. 2010;53:299–308.
143. Tam CH, Ho JS, Wang Y, Lee HM, Lam VK, Germer S, Martin M, So WY, Ma RC, Chan JC, Ng MC. Common polymorphisms in MTNR1B, G6PC2 and GCK are associated with increased fasting plasma glucose and impaired beta-cell function in Chinese subjects. PLoS One. 2010;5:e11428.
144. Dubocovich ML, Rivera-Bermudez MA, Gerdin MJ, Masana MI. Molecular pharmacology, regulation and function of mammalian melatonin receptors. Front Biosci. 2003;8:d1093–108.
145. Muhlbauer E, Peschke E. Evidence for the expression of both the MT1- and in addition, the MT2-melatonin receptor, in the rat pancreas, islet and beta-cell. J Pineal Res. 2007;42:105–6.
146. Peschke E, Muhlbauer E. New evidence for a role of melatonin in glucose regulation. Best Pract Res Clin Endocrinol Metab. 2010;24:829–41.
147. Pevet P, Challet E. Melatonin: both master clock output and internal time-giver in the circadian clocks network. J Physiol Paris. 2011;105(4–6):170–82.
148. Reiter RJ, Tan DX, Fuentes-Broto L. Melatonin: a multitasking molecule. Prog Brain Res. 2010;181:127–51.
149. Reppert SM, Weaver DR, Godson C. Melatonin receptors step into the light: cloning and classification of subtypes. Trends Pharmacol Sci. 1996;17:100–2.
150. Jenkins JB, Omori T, Guan Z, Vgontzas AN, Bixler EO, Fang J. Sleep is increased in mice with obesity induced by high-fat food. Physiol Behav. 2006;87: 255–62.
151. Laposky AD, Shelton J, Bass J, Dugovic C, Perrino N, Turek FW. Altered sleep regulation in leptin deficient mice. Am J Physiol Regul Integr Comp Physiol. 2006;290:R894–903.
152. Laposky AD, Bradley MA, Williams DL, Bass J, Turek FW. Sleep-wake regulation is altered in leptin-resistant (db/db) genetically obese and diabetic mice. Am J Physiol Regul Integr Comp Physiol. 2008;295: R2059–66.
153. Megirian D, Dmochowski J, Farkas GA. Mechanism controlling sleep organization of the obese Zucker rats. J Appl Physiol. 1998;84:253–6.
154. Mistlberger RE, Lukman H, Nadeau BG. Circadian rhythms in the Zucker obese rat: assessment and intervention. Appetite. 1998;30:255–67.
155. Damiola F, Le Minh N, Preitner N, Kornmann B, Fleury-Olela F, Schibler U. Restricted feeding uncouples circadian oscillators in peripheral tissues from the central pacemaker in the suprachiasmatic nucleus. Genes Dev. 2000;14:2950–61.
156. Hara R, Wan K, Wakamatsu H, Aida R, Moriya T, Akiyama M, Shibata S. Restricted feeding entrains liver clock without participation of the suprachiasmatic nucleus. Genes Cells. 2001;6:269–78.
157. Stokkan KA, Yamazaki S, Tei H, Sakaki Y, Menaker M. Entrainment of the circadian clock in the liver by feeding. Science. 2001;291:490–3.
158. Mistlberger RE. Neurobiology of food anticipatory circadian rhythms. Physiol Behav. 2011;104:535–45.
159. Arble DM, Bass J, Laposky AD, Vitaterna MH, Turek FW. Circadian timing of food intake contributes to weight gain. Obesity (Silver Spring). 2009;17: 2100–2.
160. Deboer T, Vansteensel MJ, Detari L, Meijer JH. Sleep states alter activity of suprachiasmatic nucleus neurons. Nat Neurosci. 2003;6:1086–90.
161. Deboer T, Detari L, Meijer JH. Long term effects of sleep deprivation on the mammalian circadian pacemaker. Sleep. 2007;30:257–62.
162. Van Cauter E, Latta F, Nedeltcheva A, Spiegel K, Leproult R, Vandenbril C, Weiss R, Mockel J, Legros JJ, Copinschi G. Reciprocal interactions between the GH axis and sleep. Growth Horm IGF Res. 2004; 14 Suppl A: S10–7.
163. Kimura F, Tsai CW. Ultradian rhythm of growth hormone secretion and sleep in the adult male rat. J Physiol. 1984;353:305–15.
164. Sassin JF, Parker DC, Mace JW, Gotlin RW, Johnson LC, Rossman LG. Human growth hormone release: relation to slow-wave sleep and sleep-walking cycles. Science. 1969;165:513–5.
165. Takahashi Y, Ebihara S, Nakamura Y, Takahashi K. A model of human sleep-related growth hormone secretion in dogs: effects of 3, 6, and 12 hours of forced wakefulness on plasma growth hormone, cortisol, and sleep stages. Endocrinology. 1981;109:262–72.
166. Van Cauter E, Plat L, Copinschi G. Interrelations between sleep and the somatotropic axis. Sleep. 1998;21:553–66.
167. Van Cauter E, Leproult R, Plat L. Age-related changes in slow wave sleep and REM sleep and relationship with growth hormone and cortisol levels in healthy men. JAMA. 2000;284:861–8.
168. Obal Jr F, Krueger JM. GHRH and sleep. Sleep Med Rev. 2004;8:367–77.

169. Ocampo-Lim B, Guo W, DeMott-Friberg R, Barkan AL, Jaffe CA. Nocturnal growth hormone (GH) secretion is eliminated by infusion of GH-releasing hormone antagonist. J Clin Endocrinol Metab. 1996; 81:4396–9.
170. Bredow S, Taishi P, Obel Jr F, Guha-Thakurta N, Krueger JM. Hypothalamic growth hormone-releasing hormone mRNA varies across the day in rats. Neuroreport. 1996;7:2501–5.
171. Gardi J, Obal Jr F, Fang J, Zhang J, Krueger JM. Diurnal variations and sleep deprivation-induced changes in rat hypothalamic GHRH and somatostatin contents. Am J Physiol. 1999;277:R1339–44.
172. Zhang J, Chen Z, Taishi P, Obal Jr F, Fang J, Krueger JM. Sleep deprivation increases rat hypothalamic growth hormone-releasing hormone mRNA. Am J Physiol. 1998;275:R1755–61.
173. Obal Jr F, Alt J, Taishi P, Gardi J, Krueger JM. Sleep in mice with nonfunctional growth hormone-releasing hormone receptors. Am J Physiol Regul Integr Comp Physiol. 2003;284:R131–9.
174. Obal Jr F, Floyd R, Kapas L, Bodosi B, Krueger JM. Effects of systemic GHRH on sleep in intact and hypophysectomized rats. Am J Physiol. 1996;270: E230–7.
175. Gardi J, Szentirmai E, Hajdu I, Obal Jr F, Krueger JM. The somatostatin analog, octreotide, causes accumulation of growth hormone-releasing hormone and depletion of angiotensin in the rat hypothalamus. Neurosci Lett. 2001;315:37–40.
176. Obal Jr F, Kapas L, Gardi J, Taishi P, Bodosi B, Krueger JM. Insulin-like growth factor-1 (IGF-1)-induced inhibition of growth hormone secretion is associated with sleep suppression. Brain Res. 1999; 818:267–74.
177. Mendelson WB, Slater S, Gold P, Gillin JC. The effect of growth hormone administration on human sleep: a dose-response study. Biol Psychiatry. 1980; 15:613–8.
178. Obal Jr F, Payne L, Kapas L, Opp M, Krueger JM. Inhibition of growth hormone-releasing factor suppresses both sleep and growth hormone secretion in the rat. Brain Res. 1991;557:149–53.
179. Obal Jr F, Payne L, Opp M, Alfoldi P, Kapas L, Krueger JM. Growth hormone-releasing hormone antibodies suppress sleep and prevent enhancement of sleep after sleep deprivation. Am J Physiol. 1992;263:R1078–85.
180. Obal Jr F, Fang J, Taishi P, Kacsoh B, Gardi J, Krueger JM. Deficiency of growth hormone-releasing hormone signaling is associated with sleep alterations in the dwarf rat. J Neurosci. 2001;21: 2912–8.
181. Steiger A, Guldner J, Hemmeter U, Rothe B, Wiedemann K, Holsboer F. Effects of growth hormone-releasing hormone and somatostatin on sleep EEG and nocturnal hormone secretion in male controls. Neuroendocrinology. 1992;56: 566–73.
182. Holsboer F, von Bardeleben U, Steiger A. Effects of intravenous corticotropin-releasing hormone upon sleep-related growth hormone surge and sleep EEG in man. Neuroendocrinology. 1988;48:32–8.
183. Krieger DT, Glick SM. Sleep EEG stages and plasma growth hormone concentration in states of endogenous and exogenous hypercortisolemia or ACTH elevation. J Clin Endocrinol Metab. 1974;39: 986–1000.
184. Shipley JE, Schteingart DE, Tandon R, Starkman MN. Sleep architecture and sleep apnea in patients with Cushing's disease. Sleep. 1992;15:514–8.
185. Linkowski P, Mendlewicz J, Kerkhofs M, Leclercq R, Golstein J, Brasseur M, Copinschi G, Van Cauter E. 24-hour profiles of adrenocorticotropin, cortisol, and growth hormone in major depressive illness: effect of antidepressant treatment. J Clin Endocrinol Metab. 1987;65:141–52.
186. Ehlers CL, Kupfer DJ. Hypothalamic peptide modulation of EEG sleep in depression: a further application of the S-process hypothesis. Biol Psychiatry. 1987;22:513–7.
187. Steiger A, von Bardeleben U, Herth T, Holsboer F. Sleep EEG and nocturnal secretion of cortisol and growth hormone in male patients with endogenous depression before treatment and after recovery. J Affect Disord. 1989;16:189–95.
188. Jedema HP, Grace AA. Corticotropin-releasing hormone directly activates noradrenergic neurons of the locus ceruleus recorded in vitro. J Neurosci. 2004;24: 9703–13.
189. Buckley TM, Schatzberg AF. On the interactions of the hypothalamic-pituitary-adrenal (HPA) axis and sleep: normal HPA axis activity and circadian rhythm, exemplary sleep disorders. J Clin Endocrinol Metab. 2005;90:3106–14.
190. Weitzman ED. Circadian rhythms and episodic hormone secretion in man. Annu Rev Med. 1976;27: 225–43.
191. Steiger A. Sleep and the hypothalamo-pituitary-adrenocortical system. Sleep Med Rev. 2002;6: 125–38.
192. Baumann CR, Bassetti CL. Hypocretins (orexins) and sleep-wake disorders. Lancet Neurol. 2005;4: 673–82.
193. Zeitzer JM, Buckmaster CL, Parker KJ, Hauck CM, Lyons DM, Mignot E. Circadian and homeostatic regulation of hypocretin in a primate model: implications for the consolidation of wakefulness. J Neurosci. 2003;23:3555–60.
194. Zhang S, Zeitzer JM, Yoshida Y, Wisor JP, Nishino S, Edgar DM, Mignot E. Lesions of the suprachiasmatic nucleus eliminate the daily rhythm of hypocretin-1 release. Sleep. 2004;27:619–27.
195. Sakurai T. The neural circuit of orexin (hypocretin): maintaining sleep and wakefulness. Nat Rev Neurosci. 2007;8:171–81.
196. Haynes AC, Jackson B, Chapman H, Tadayyon M, Johns A, Porter RA, Arch JR. A selective orexin-1 receptor antagonist reduces food consumption in male and female rats. Regul Pept. 2000;96:45–51.
197. Funato H, Tsai AL, Willie JT, Kisanuki Y, Williams SC, Sakurai T, Yanagisawa M. Enhanced orexin receptor-2

signaling prevents diet-induced obesity and improves leptin sensitivity. Cell Metab. 2009;9:64–76.
198. Akiyama M, Yuasa T, Hayasaka N, Horikawa K, Sakurai T, Shibata S. Reduced food anticipatory activity in genetically orexin (hypocretin) neuron-ablated mice. Eur J Neurosci. 2004;20:3054–62.
199. Mieda M, Williams SC, Sinton CM, Richardson JA, Sakurai T, Yanagisawa M. Orexin neurons function in an efferent pathway of a food-entrainable circadian oscillator in eliciting food-anticipatory activity and wakefulness. J Neurosci. 2004;24:10493–501.
200. Yamanaka A, Beuckmann CT, Willie JT, Hara J, Tsujino N, Mieda M, Tominaga M, Yagami K, Sugiyama F, Goto K, Yanagisawa M, Sakurai T. Hypothalamic orexin neurons regulate arousal according to energy balance in mice. Neuron. 2003;38:701–13.
201. Kantor S, Mochizuki T, Janisiewicz AM, Clark E, Nishino S, Scammell TE. Orexin neurons are necessary for the circadian control of REM sleep. Sleep. 2009;32:1127–34.
202. Kaur S, Thankachan S, Begum S, Blanco-Centurion C, Sakurai T, Yanagisawa M, Shiromani PJ. Entrainment of temperature and activity rhythms to restricted feeding in orexin knock out mice. Brain Res. 2008;1205:47–54.

Neural Circuitry Responsible for Sleep and Wakefulness

3

Roda Rani Konadhode, Dheeraj Pelluru, and Priyattam J. Shiromani

Abstract

Research over the past 50 years has determined that specific neurons in the brain are responsible for generating waking, non-REM sleep, and REM sleep. Some of the neurons responsible for keeping us awake are also involved in regulating energy metabolism. One such arousal neuronal population contains the neuropeptide hypocretin, also known as orexin. The HCRT neurons are located in the hypothalamus, an area that also contains other neurons regulating energy metabolism. The hypocretin neurons are most active during waking and silent during sleep, and their activity has been shown to regulate brown adipose tissue (BAT) thermogenesis. The hypocretin neurons are also activated by low glucose levels and shut off when the glucose levels increase. Thus, the activity of the hypocretin neurons is linked to energy metabolism. Based on this relationship, it is easy to see how inadequate sleep or even frequent arousals during sleep, as occurs in obstructive sleep apnea, will affect energy metabolism and adiposity.

Introduction

It was initially hypothesized that sleep was a passive process resulting from lack of sensory stimulation to the brain. This hypothesis was rejected when rapid eye movement sleep (REM sleep) was discovered [1, 2], and it was found that during REM sleep the brain was as active as in waking. With that discovery, it became clear that in humans there was a process in the brain that periodically "awakened" the sleeping brain every 90 min at night. We now know that specific neurons are responsible for generating wake, non-REM, and REM sleep (Fig. 3.1). There is a tight link between arousal and feeding, and peripheral signals such as glucose activate arousal neurons, so that the animal can forage for food. Indeed, some of the sleep and arousal neurons reside in areas that regulate feeding. In this chapter, we review the evidence linking arousal, feeding, and sleep.

R.R. Konadhode, PhD • D. Pelluru, PhD
Department of Psychiatry and Behavioral Science, Medical University of South Carolina, 114 Doughty Street, Charleston, SC 29403, USA

P.J. Shiromani, PhD (✉)
Department of Psychiatry and Behavioral Science, Ralph H. Johnson VA and Medical University of South Carolina, 114 Doughty Street, Room 404, Charleston, SC 29403, USA
e-mail: shiroman@musc.edu

P.J. Shiromani et al. (eds.), *Sleep Loss and Obesity: Intersecting Epidemics*,
DOI 10.1007/978-1-4614-3492-4_3, © Springer Science+Business Media, LLC 2012

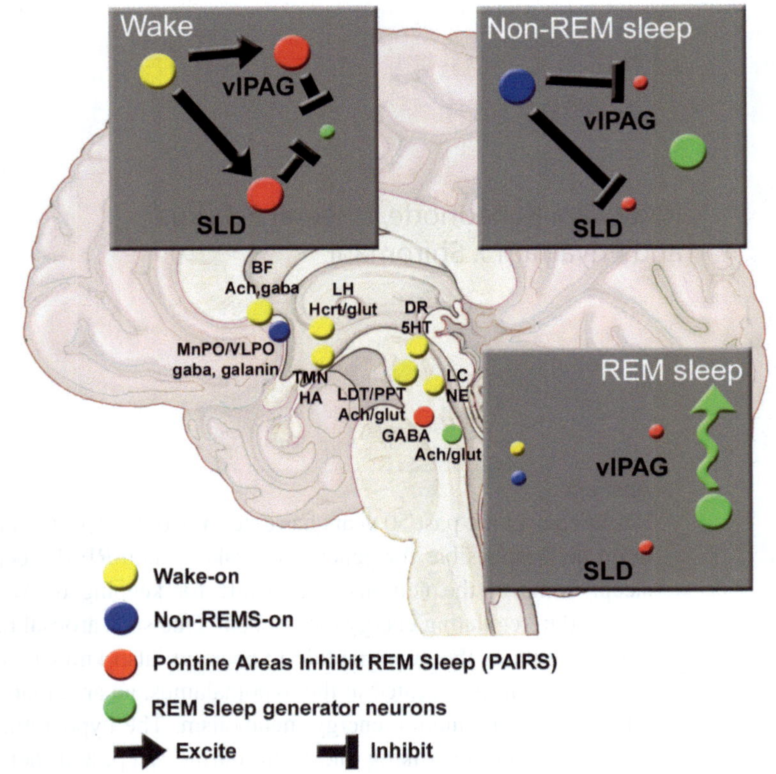

Fig. 3.1 A coordinated interaction between neuronal populations is responsible for wake, non-REM sleep, and REM sleep. Neuronal populations that are considered to generate wakefulness (*yellow*) interact with neurons that generate non-REM sleep (*blue*). Both act on neurons in the pontine brainstem (*red*) and influence the generation of rapid-eye movement sleep (REM sleep). Wake-on neurons inhibit REM sleep by activating pontine GABA neurons in the pons. The strength of the excitatory input to the pontine GABA neurons (*red*) influences REM sleep. A strong input will inhibit REM sleep while a weakened input will facilitate it. This excitatory input to the GABA pontine neurons is from hypocretin and from other sources. The purpose of this excitation is to keep the animal upright and mobile while foraging for food. During non-REM sleep the excitatory input to the pontine GABA neurons is lost and is replaced by a strong inhibitory input. This enables REM sleep generator neurons (*green*) to become active and when sufficient numbers of these are activated then REM sleep ensues. The hypothesis that the pontine areas inhibit REM sleep has been tested in two separate studies from our lab [3–5]. The earlier study tested the hypothesis in rats whereas the second study tested it in mice. In both studies REM sleep increased when the GABA neurons were lesioned with hypocretin-2 saporin thereby supporting the hypothesis of PAIRS. *Ach* acetylcholine, *BF* basal forebrain, *DR* dorsal raphe, *GABA* gamma amino butyric acid, *GLUT* glutamic acid, *HA* histamine, *HCRT* hypocretin, *LC* locus coeruleus, *LH* lateral hypothalamus, *MnPO* median preoptic nucleus, *NE* norepinephrine, *LDT* lateral dorsal pontine tegmentum, *PPT* pedunculopontine tegmentum, *SLD* sub lateral dorsal nucleus, *TMN* tuberomammillary nucleus, *vlPAG* ventral lateral periaquaductal gray, *VLPO* ventral lateral preoptic nucleus. Reprinted with permission from [5]

Neurons Responsible for Sleep

The Ventral Lateral Preoptic Area

This is a small region located in the preoptic area comprising primarily of neurons containing GABA and galanin. These neurons project to and are inhibitory to all of the arousal neurons [6]. The ventral lateral preoptic area (VLPO) neurons are inhibited by the wake-active neurotransmitters, acetylcholine, serotonin, and norepinephrine, but are unaffected by histamine [7]. The VLPO neurons were identified using the immediate-early gene c-FOS [8]. Subsequently, a region dorsal to the VLPO, the median preoptic area (MPO) was identified as sleep active based also on c-FOS immunohistochemistry. Electrophysiology studies have now more definitively determined that VLPO/MPO neurons begin to fire during drowsiness and peak activity is seen during non-REM sleep. The sleep active cells comprise about 25% of the recorded cells in the basal forebrain-preoptic area and are intermixed with wake-active cells which predominate. Electrical stimulation or warming of the VLPO/MPO area induces sleep. Lesions of the VLPO decrease sleep and increase wake [9]. The VLPO and MPO neurons are now linked to sleep pressure [10, 11] and turn on when the wake-active neurons stop firing. Optogenetic methods are currently being used to drive these neurons to initiate sleep. VLPO/MPO neurons project to the pontine regions responsible for gating REM sleep [12].

Recently, sleep-active neurons were identified in the cortex and found to contain GABA, neuronal nitric oxide synthase (nNOS), and some also contained neuropeptide Y [13]. These neurons were diffusely scattered in all cortical layers in the rat, but in the mouse they were predominantly in layers V and VI. Activation of these neurons during sleep would inhibit cortical neurons, while nNOS may dictate sleep pressure since the intensity of sleep was found to be directly related to the number of nNOS expressing GABAergic cortical neurons [13].

Melanin Concentrating Hormone

Neurons containing the neuropeptide melanin-concentrating hormone (MCH) are located primarily in the posterior hypothalamus. These neurons are separate and distinct from the hypocretin neurons, but located in the same region as the hypocretin neurons. They also project to many of the same regions in the brain as the hypocretin neurons. MCH exerts its action via two metabotropic receptors, the MCHR1 and MCHR2, but only the MCHR1 is present in rodents. MCH neurons are active primarily during sleep, especially during REM sleep, and MCH knockout mice have less REM sleep. Intraventricular microinjection of MCH increases both slow wave sleep (SWS) and REM sleep. This evidence suggests an important role of this peptide in sleep.

There is very strong evidence that MCH increases food intake and reduces energy expenditure [14, 15]. For instance, acute infusion of MCH into the lateral ventricles induces feeding in rodents [16]. Mice with deletion of the pre-pro-MCH gene [17] or of the MCH neuron (ataxin-MCH transgenic mice) are lean, hypophagic, and have increased energy expenditure [18]. Mice lacking MCH receptors are hypophagic and lean with increased metabolic rate. MCH-1 receptor antagonists produce significant weight loss in rodent models of obesity and do so by reducing meal size [19].

Neurons Responsible for Arousal

There are many more neuronal populations linked to arousal compared to sleep. The arousal neurons include not only the classical neurotransmitters but also the peptide, hypocretin. Research during the last decade leads us to conclude that one reason for the diverse population of arousal neurons is that some of these populations arouse a sleeping brain in response to specific situation. For instance, the histaminergic and noradrenergic neurons trigger arousal in the event of an alarm (either internal or external) and maintain vigilance in a stressful condition. The basal forebrain cholinergic neurons rapidly bring cognitive functions online when awakened from sleep. Lesions of all three of these neuronal populations do not change overall levels of wakefulness but when the histaminergic or the LC neurons are lesioned then it decreases arousal in unfamiliar conditions [20]. Previously, we reviewed the evidence linking the histamine, noradrenaline, and acetylcholine neurons to arousal [21].

It is also important to have neurons that respond to hunger, which then triggers the overall arousal network to allow foraging for food. A number of peptides and neurotransmitters are located within the hypothalamus, and their role in sleep and wakefulness has been reviewed [22]. In this paper, we will specifically focus on neurons containing the neuropeptide hypocretin, which is located primarily in the posterior hypothalamus. The growing body of evidence, amassed only within the last 10 years, shows that these neurons are linked to sleep and wakefulness and also regulate energy metabolism.

Hypocretin

The peptide hypocretin, also known as orexin, was discovered by two independent groups using different approaches [23–25]. The hypocretins were linked to narcolepsy as a result of the discovery that canines with narcolepsy possess a mutation in the hypocretin-2 receptor [26]. This was then supported by the finding that mice with deletion of the hypocretin/orexin gene exhibit symptoms of narcolepsy [27]. Narcoleptic patients have a loss of the hypocretin-containing neurons [28, 29] and low CSF concentrations of HCRT-1 [30]. Narcoleptic patients, hypocretin knockout mice, and the hypocretin-2 receptor knockout mice weigh significantly more than age-matched controls [27, 31].

The hypocretin neurons project to sites involved in waking and REM sleep. The heaviest projections are to the locus coeruleus (LC) and the tuberomammillary nucleus (TMN). Hypocretin fibers also innervate the dorsal raphe and basal forebrain. Immunohistochemistry indicates that one or both of the receptor subtypes (hypocretin-1 and hypocretin-2 receptors) are expressed in the dorsal raphe, the lateral dorsal tegmental (LDT) nucleus, LC, the locus subcoeruleus, pontis oralis, Barrington's nucleus, the trigeminal complex (mesencephalic trigeminal and motor nucleus of the trigeminal nerve), the dorsal tegmental nucleus of Gudden, the ventral cochlear nucleus, trapezoid nucleus, pontine raphe nucleus, and the pontine reticular nucleus. Complementary analysis of hypocretin receptor mRNA with a cRNA probe shows that hypocretin receptor mRNA is also detected in areas that show protein immunoreactivity, suggesting the mRNA and protein are co-localized.

The receptor is present on neurons involved in mastication, bladder control, gastrointestinal function, and arousal. Given these projection sites and the functions associated with these sites, we hypothesize that hypocretin acts to keep the animal alert and vigilant while engaged in feeding behavior. Hypocretin has a powerful wake-promoting effect and receptor antagonists that block one or both receptors [32] cause sleepiness. In nonhuman primates, inhalation of hypocretin produces arousal even after sleep deprivation [25]. The activity of the HCRT neurons is consistent with promoting arousal. Identified HCRT neurons are active during wake and silent during sleep [33–35]. They begin to fire in anticipation of arousal, and they are easily activated during sleep [35].

Initially it was hypothesized that the hypocretin neurons regulate feeding. However, since then a number of studies, including our own [36] have found that food intake is not changed in hypocretin-KO mice or in mice lacking the hypocretin neurons [37–40]. Hypocretin neurons are sensitive to glucose, ghrelin, and leptin [41]. Hypocretin neurons are also sensitive to and activated by amino acids [42]. The inputs from the ventral medial hypothalamus and the lateral hypothalamus would arouse the animal because glucosensing neurons in these regions [43] respond to a very narrow range of glucose levels (0.1–2.5 mmol/l). A lower glucose level, which is likely at the end of a rodent's sleep period (during the day in nocturnal rodents), would activate arousal neurons so that the animal can forage for food and restore energy balance. Indeed, the HCRT neurons are activated by low glucose levels and shut off when the glucose levels increase [44]. Not surprisingly, hypocretin and MCH neurons respond in a different way to glucose [45]; high glucose inhibits hypocretin neurons but stimulates MCH neurons. In other words, the increased glucose decreases activity of the arousal promoting hypocretin neurons and increases activity of the sleep promoting MCH neurons. This makes perfect sense given that after a meal there is a high tendency to sleep whereas hunger promotes arousal.

SIRT-1, CLOCK, Energy Metabolism, and Hypocretin

As noted earlier, loss of hypocretin or of its receptor increases adiposity. Enhanced orexin receptor 2 signaling prevents obesity [46]. The weight gain may be the result of impaired brown adipose tissue (BAT) thermogenesis [47]. In the brain, orexin input to the raphe pallidus has been shown to regulate BAT thermogenesis [48]. Waking would drive sympathetic outflow resulting in increased energy expenditure and compensatory increase in feeding. Waking at an inappropriate time of day such as what occurs with shift work will also affect energy metabolism.

How might peripheral metabolic signals feedback onto arousal neurons? SIRT-1 may link energy metabolism to hypocretin. SIRT-1 is an enzyme that epigenetically regulates gene expression by using energy stored in nicotinamide adenosine dinucleotide (NAD+) to remove the acetyl group from histones. In so doing, SIRT-1 changes chromatin structure. Recently, it was discovered that SIRT-1 counteracts the activity of CLOCK [49, 50]. As reviewed elsewhere in this book (see Chaps. 1 and 2) CLOCK dimerizes with BMAL1 which then sets into motion an intracellular cascade that regulates circadian rhythms. However, CLOCK acetylates histones, and SIRT-1 associates with the CLOCK:BMAL1 heterodimer to deacetylate histones. SIRT-1 is exquisitely sensitive to glucose [51] and by coupling with CLOCK it may transduce peripheral metabolic signals to the circadian clock. This would then regulate the amplitude of the intracellular cascade underlying circadian rhythms.

SIRT-1 is increased when food is scarce and coordinates the increase in hepatic glucose production [52]. Food restriction also increases SIRT-1 in hypothalamic areas involved in energy metabolism [53]. They also showed that SIRT-1 upregulates the hypocretin-2 receptor. Putting this together, it would appear that reduction in glucose may cause arousal by activating the hypocretin neurons. The intracellular signal involving SIRT-1 upregulates the hypocretin-2 receptor, thereby strengthening the link between the hypocretin ligand and its receptor. This then enables the animal to have the proper locomotor ability to forage for food. Failure of this link, either through loss of the ligand or the hypocretin-2 receptor, causes inadvertent collapse of motor control as in narcolepsy.

Basal Forebrain and Acetylcholine

Acetylcholine was one of the first neurotransmitters linked to arousal. Acetylcholine is released in the cortex during waking and REM sleep [54–57]. The cholinergic neurons in the basal forebrain are the source of the acetylcholine release in the cortex because when these neurons are lesioned there is a decrease in cortical acetylcholine levels [58]. The release of acetylcholine is evident in the wake-active hemisphere even in mammals that display uni-hemispheric sleep [59].

Hypocretin neurons can drive the BF cholinergic neurons. A direct effect of hypocretin-1 on the BF cholinergic neurons has been shown [60]. Hypocretin depolarizes BF cholinergic neurons via the hypocretin type 2 receptor. Moreover, administration of hypocretin-1 into the BF via reverse microdialysis produces a dose-dependent increase in acetylcholine in the prefrontal cortex [61]. In that study, when the hypocretin-1 was applied to the prefrontal cortex no change in acetylcholine was observed indicating that the release was from the BF cholinergic neurons.

Infusion of hypocretin to the BF induces wakefulness [62, 63]. The wakefulness is produced even in rats with lesions of the cholinergic neurons in the BF indicating that hypocretin receptors on the noncholinergic neurons can drive wakefulness [64]. These wake-active noncholinergic neurons might be GABAergic innervating cortical GABA interneurons and may cause arousal through disinhibition [65].

The BF also contains sleep-active neurons, some of which contain neuropeptide Y [66]. These sleep-active neurons would be disinhibited when the wake-active BF neurons become silent [67]. The GABAergic neurons increase activity in conjunction with cortical slow waves [68]. They project to the cortex [69] and to the posterior lateral hypothalamus where the hypocretin neurons are located. Their activity would suppress activity of the hypocretin wake-active neurons and promote sleep.

Hypocretin neurons are active only during wake [34], and we hypothesize that their activation would drive downstream targets such as the BF neurons, which release acetylcholine into the cortex and facilitate cognitive function. In our model, the BF is not regulating daily levels of waking as traditionally hypothesized, but instead its activation during waking is important for memory and cognitive functions. Thus, we hypothesize that one important function of the BF neurons is to rapidly mobilize cognitive function when one is awakened from sleep. In other words, it is important to be cognitively aware of one's surrounding upon awakening from sleep.

The Tuberomammillary Nucleus and Histamine

In the brain, histamine neurons are located exclusively in the TMN [70]. Histamine has a potent arousal effect, and antihistamines produce drowsiness and sedation. Histamine microinjections into projection sites such as the basal forebrain produce a dose-dependent increase in wake [71]. When histamine synthesis in the preoptic area is blocked, sleep increases and wakefulness decreases [71]. Histamine H1 and H2 receptors are postulated to mediate the arousal [71].

Since histamine produces arousal, it is reasonable that histamine neurons should be active during waking. Electrophysiology studies have found that histamine neurons in the TMN region have the highest discharge rate during waking and are virtually silent during sleep [72]. In narcoleptic canines, TMN neurons are also active only during waking and silent during sleep [73].

The hypocretin neurons innervate the TMN and the hypocretin-2 receptor is heavily expressed on these neurons [74]. Hypocretin stimulates identified histamine neurons [75]. However, there does not appear to be a reciprocal histamine/TMN projection to the hypocretin neurons [76], nor is there a direct effect of histamine on identified hypocretin neurons [77]. This suggests that histamine neurons in the TMN are driven by the hypocretin neurons. The histaminergic neurons would then activate the cortex directly via their widespread hypothalamo-cortical projections or, indirectly, by stimulating the basal forebrain cholinergic system. The net effect of the hypocretin-TMN stimulation would be to arouse the cortex.

The Locus Coeruleus and Norepinephrine

The LC contains primarily norepinephrine neurons that innervate virtually the entire brain and spinal cord. Electrophysiology studies have found that the noradrenergic LC neurons are most active during waking, less active during non-REM sleep, and they stop firing during REM sleep [78, 79]. The LC receives an especially heavy innervation of hypocretin fibers [80] but surprisingly, the LC does not project to the hypocretin neurons [76]. LC neurons contain primarily the hypocretin-1 receptor [81]. Hypocretin excites LC neurons and potently increases waking and decreases REM sleep [82, 83].

Serotonin (5-Hydroxytryptamine)

Serotonin neurons are localized in the raphe. These neurons are most active in waking, less active during non-REM sleep, and virtually silent during REM sleep [84]. Thus, these neurons behave very much like the orexin, histamine and LC neurons, in that they are all wake-active. Serotonin exerts its actions via seven distinct receptors (5-HT 1–7), which are G protein-coupled receptors, except 5-HT-3 which is a ligand-gated ion channel. Some of these receptors also have distinct subtypes. Through these receptors, serotonin can act as both an activator and an inhibitor.

Sleep has been examined in specific serotonin receptor knockout mice [85]. 5-HT1A and

5HT1B receptor knockout mice have more REM sleep compared to wild-type mice, and there is no REM sleep rebound after REM sleep deprivation [86, 87]. 5HT1A and 1B receptor agonists block REM sleep whereas the antagonists increase it. These effects are absent in the respective receptor knockouts. These effects can be explained because stimulation of the 5HT1A or 1B receptor hyperpolarizes the neuron and loss or antagonism of these receptors produces disinhibition of the REM sleep generator neurons. This was clearly demonstrated in a study that monitored the activity of individual REM sleep generator neurons in response to local pharmacological stimulation. It was found that administration of 8-OH-DPAT, a selective 5HT1A agonist via reverse microdialysis into the pontine region in rats decreased REM sleep by blocking the activity of REM sleep generator neurons in the pontine tegmentum [88]. This may also explain how selective serotonin reuptake inhibitors (SSRIs) decrease REM sleep.

5-HT2 receptor stimulation depolarizes neurons. 5-HT2A [89] or 2C [90] receptor knockouts are awake more and sleep less. 5-HT2A antagonists such as MDL100907 (0.1, 1.0, and 3.0 mg/kg IP) increase non-REM sleep and delta power without affecting REM sleep [91]. Ritanserine, a broad spectrum antagonist of the 5-HT2 receptor, also increases non-REM sleep [92] . 5HT6 receptor stimulation depolarizes neurons but the effects on non-REM sleep of the antagonist, RO4368554 have been mixed at the highest dose tested (10 mg/kg, IP). One study [91] found increased non-REM sleep at night whereas another study found no effect [93].

With respect to feeding, exogenous 5-HT or drugs that stimulate serotonin signaling decrease food intake while drugs that antagonize serotonin receptor signaling increase food intake. In this regard, D-fenfluramine and phenetermine, which release serotonin produce significant weight loss. These compounds were withdrawn following a number of cases of pulmonary hypertension and cardiac valvulopathy. Another compound that produces significant weight loss is sibutramine. It acts by increasing noradrenaline and serotonin levels in the hypothalamus, nucleus accumbens and the brainstem, all regions associated with energy homeostasis.

Specific serotonin receptors are being targeted for weight loss. For instance, administration of agonists specific to the 5-HT1A receptor increases feeding, whereas agonists specific for the 5-HT1B and 5-HT1C receptors decrease food intake. Selective stimulation of the 5-HT2C receptor decreases food intake and body weight in both lean and obese rodents. These agonists may exert their hypophagic actions via stimulation of receptors located on POMC containing neurons within the arcuate (ARC) nucleus. Mice lacking the 5-HT2C receptor are hyperphagic and weigh more. Selective reintroduction of 5-HT2C receptors on POMC containing neurons rescues the obese phenotype [94–96]. The 5-HT6 receptor is also a potential target for the treatment of obesity [97] since knockout mice do not gain weight when placed on a high-fat diet and the 5-HT6 receptor antagonist, RO 04-6790, significantly reduces weight [98].

Neurons Regulating REM Sleep

Converging data from a variety of studies indicates that REM sleep originates from the pons (summarized in [99]). Neurons that are selectively active during REM sleep and that are likely to generate REM sleep are present in the pons. Using c-FOS we have found that some of the cholinergic neurons in the LDT and pedunculopontine tegmental (PPT) nuclei of the pons are REM sleep-on neurons [100]. These REM sleep-on neurons are inhibited by GABA neurons located in regions of the pons that we refer to as pontine areas inhibiting REM sleep (PAIRS) [3]. One group of these GABAergic neurons resides in the sublateral dorsal tegmental area (just ventral to the LC) and an additional group resides in the ventral lateral periaqueductal gray (vlPAG).

HCRT and other wake-active neurons (such as the LC, serotonin, histamine) activate the pontine GABAergic neurons, which then block the REM sleep-on neurons [101]. If this is the case, then lesion of these pontine GABA neurons should increase REM sleep, which it does. For instance,

HCRT2-saporin-induced lesions of these neurons potently increase REM sleep in both rats [4] and mice [3]. During wake, these pontine GABAergic neurons would be activated and levels of GABA should be high in the pons, which is the case [102]. When the hypocretin neurons are silent, or destroyed as in narcolepsy, then the pontine GABA neurons do not fire effectively which then allows the REM-on neurons to become disinhibited. Mice and rats that lack hypocretin enter into REM sleep often (for review see [76]). Canines with a mutation of the hypocretin-2 receptor also have abnormal onset of REM sleep.

Night-Eating Syndrome

This is a syndrome occurring more often in obese individuals where the individual consumes most of their food from late in the evening into the early hours of the morning [103]. In 1955, Stunkard observed such an eating pattern in 64% of a group of very obese females and established criteria for night eating syndrome (NES) which included nocturnal hyperphagia, insomnia, and morning anorexia [104]. Recent evidence has indicated that patients with NES display a delayed circadian pattern of food intake but retain a normal sleep–wake cycle. A comparative study on the eating and sleep–wake patterns of persons with NES with those of matched control subjects found no difference between the total energy intake, but the pattern of energy intake was different [105]. Food intake after the evening meal, as a proportion of the 24-h intake was more than threefold greater in NES subjects than in controls. NES subjects had sleep onset, offset, and total sleep duration times comparable with those of controls. NES subjects had more nocturnal awakenings than did controls, and their actigraphically monitored arousals occurred earlier during sleep. NES subjects consumed food on 74% of the awakenings vs. 0% for the controls. This suggests a phase delay in energy consumption relative to sleep–wake times in NES.

Another study examined the phase and amplitude of behavioral and neuroendocrine circadian rhythms in patients with NES [106]. Fifteen women with NES (mean age $\pm$ SD, 40.8 ± 8.7 years) and 14 control subjects (38.6 ± 9.5 years) were studied in the laboratory for three nights with food intake measured daily. Blood also was collected for 25 h (every 2 h from 0800 to 2000 h, and then hourly from 2100 to 0900 h) and assayed for glucose, insulin, ghrelin, leptin, melatonin, cortisol, thyroid-stimulating hormone [TSH], and prolactin. Control subjects displayed normal phases and amplitudes for all circadian rhythms. In contrast, patients with NES showed a phase delay in the timing of meals, and delayed circadian rhythms for total caloric, fat, and carbohydrate intake. In addition, phase delays of 1.0–2.8 h were found in two food-regulatory rhythms—leptin and insulin—and in the circadian rhythm of melatonin (with a trend for a delay in the circadian rhythm of cortisol). In contrast, circulating levels of ghrelin, the primary hormone that stimulates food intake, were phase advanced by 5.2 h. The glucose rhythm showed an inverted circadian pattern. Patients with NES also showed reduced amplitudes in the circadian rhythms of food intake, cortisol, ghrelin, and insulin, but increased TSH amplitude. Thus, patients with NES demonstrated significant changes in the timing and amplitude of various behavioral and physiological circadian markers involved in energy metabolism [106, 107].

A neuroimaging study found elevated brain serotonin transporter binding in the midbrain of individuals with NES. Administration of SSRIs restored the circadian rhythm of both food intake and neuroendocrine function [108].

Hypocretin and Obstructive Sleep Apnea

Obstructive sleep apnea (OSA) is a disorder in which during sleep the upper airway collapses and airflow is interrupted, which then forces the individual to awaken and resume breathing. As a result of the cessation of breathing during sleep, the individual is not able to maintain long bouts of sleep and fails to enter into the deeper stages of sleep, i.e., stages 3 and 4. Obesity is a risk factor in that fatty deposits in the upper airway produce a narrowing of the upper airway, which then leads to cessation of airflow during sleep. OSA among obese patients exceeds 30%, reaching as high as 50–98% in the morbidly obese population.

Obese patients have an increase in leptin levels [109, 110]. Although leptin is considered to be a satiety signal to stop eating, it is not clear why the increased leptin levels in obese individuals do not stop food intake. It has been suggested that hyperleptinemia may be an indicator of OSA [111]. Treatment of OSA with nasal CPAP (continuous positive airway pressure) does decrease levels of leptin [112].

Ghrelin stimulates appetite and ghrelin levels are higher in patients with OSA compared to BMI-matched (body matched index) control subjects [113]. Continuous positive airway pressure (CPAP) treatment of 2 days does significantly reduce ghrelin levels in OSA patients [114]. The appetite-stimulating effects of ghrelin may well contribute to increased caloric intake and weight gain in patients with OSA.

The elevated levels of both leptin and ghrelin in OSA suggests that in OSA the feedback regulatory process that controls food intake has been compromised. One possibility is that the increased sleep fragmentation in these patients produces a sleep loss, which increases energy metabolism. As a consequence, the individual eats more, but this may only serve to occlude the upper airway further, leading to more sleep loss. As such a vicious cycle of sleep loss and increased food intake may be occurring in these patients. OSA may be associated with resistance to the weight reducing effects of leptin [115, 116], which may in turn result in increased appetite and weight gain.

Recent data show that hypocretin is involved in the control of upper airway patency [117]. Hypocretin neurons project to respiratory centers in the brainstem, which express hypocretin receptors, and where injection of hypocretin stimulates breathing [118]. In hypocretin knockout mice, there is a 50% reduction in CO_2-induced increases in breathing and these mice have more spontaneous sleep apneas [119, 120]. In wild-type mice, inhalation of CO_2 increases activity of hypocretin neurons as determined by c-FOS labeling [121]. In vitro studies have found increased activity of identified hypocretin neurons in response to acidity [122].

Hypocretin levels have been measured in OSA but with mixed results. One study reported a positive correlation between an apnea–hypopnea index and hypocretin levels [123], whereas another study found that the levels were low in OSA [124]. Levels of histamine have been measured in OSA and not different from controls indicating that CSF histamine is a biomarker reflecting the degree of hypersomnia of central origin [125].

Conclusion

A distributed network of neurons from the preoptic area to the pons is responsible for generating waking, non-REM sleep, and REM sleep. The discovery of hypocretin, its link with waking and the fact that these neurons are located in the part of the brain that also regulates energy metabolism, provides a way to understand disorders where the major symptoms are feeding disturbance, loss of sexual drive, abnormality in endocrine rhythms, hypersomnia, and short REM sleep latency. As we noted in the review, receptors that regulate appetite are found on neurons that generate sleep and wakefulness. As such, pharmacological agents that curb appetite are likely to also impact sleep and wakefulness.

References

1. Aserinsky E, Kleitman N. Regularly occurring periods of eye motility and concomitant phenomenon during sleep. Science. 1953;118:273–4.
2. Morrison AR. The discovery of REM sleep: the death knell of the passive theory of sleep. In: Mallick BN, Pandi-Perumal SR, McCarley RW, Morrison AR, editors. Rapid eye movement sleep: regulation and function. Cambridge: Cambridge University Press; 2011. p. 31–9.
3. Kaur S, Thankachan S, Begum S, Liu M, Blanco-Centurion C, Shiromani PJ. Hypocretin-2 saporin lesions of the ventrolateral periaquaductal gray (vlPAG) increase REM sleep in hypocretin knockout mice. PLoS One. 2009;4(7):e6346.

4. Blanco-Centurion C, Gerashchenko D, Salin-Pascual RJ, Shiromani PJ. Effects of hypocretin2-saporin and antidopamine-beta-hydroxylase-saporin neurotoxic lesions of the dorsolateral pons on sleep and muscle tone. Eur J Neurosci. 2004;19(10):2741–52.
5. Shiromani PJ, Blanco-Centurion C. Pontine areas inhibiting REM sleep. In: Mallick BN, Pandi-Perumal SR, McCarley RW, Morrison AR, editors. Rapid eye movement sleep: regulation and function. Cambridge: Cambridge University Press; 2011. p. 280–4.
6. Chou TC, Bjorkum AA, Gaus SE, Lu J, Scammell TE, Saper CB. Afferents to the ventrolateral preoptic nucleus. J Neurosci. 2002;22(3):977–90.
7. Gallopin T, Fort P, Eggermann E, et al. Identification of sleep-promoting neurons in vitro. Nature. 2000;404:992–5.
8. Sherin JE, Shiromani PJ, McCarley RW, Saper CB. Activation of ventrolateral preoptic neurons during sleep. Science. 1996;271(5246):216–9.
9. Lu J, Greco MA, Shiromani P, Saper CB. Effect of lesions of the ventrolateral preoptic nucleus on NREM and REM sleep. J Neurosci. 2000;20(10): 3830–42.
10. Gvilia I, Turner A, McGinty D, Szymusiak R. Preoptic area neurons and the homeostatic regulation of rapid eye movement sleep. J Neurosci. 2006;26(11):3037–44.
11. Gvilia I, Xu F, McGinty D, Szymusiak R. Homeostatic regulation of sleep: a role for preoptic area neurons. J Neurosci. 2006;26(37):9426–33.
12. Hsieh KC, Gvilia I, Kumar S, et al. c-Fos expression in neurons projecting from the preoptic and lateral hypothalamic areas to the ventrolateral periaqueductal gray in relation to sleep states. Neuroscience. 2011;188:55–67.
13. Gerashchenko D, Wisor JP, Burns D, et al. Identification of a population of sleep-active cerebral cortex neurons. Proc Natl Acad Sci USA. 2008;105(29):10227–32.
14. Nahon JL. The melanocortins and melanin-concentrating hormone in the central regulation of feeding behavior and energy homeostasis. C R Biol. 2006; 329(8):623–38.
15. Saito Y, Nagasaki H. The melanin-concentrating hormone system and its physiological functions. Results Probl Cell Differ. 2008;46:159–79.
16. Qu D, Ludwig DS, Gammeltoft S, et al. A role for melanin-concentrating hormone in the central regulation of feeding behaviour. Nature. 1996;380(6571): 243–7.
17. Shimada M, Tritos NA, Lowell BB, Flier JS, Maratos-Flier E. Mice lacking melanin-concentrating hormone are hypophagic and lean. Nature. 1998;396:670–4.
18. Alon T, Friedman JM. Late-onset leanness in mice with targeted ablation of melanin concentrating hormone neurons. J Neurosci. 2006;26(2):389–97.
19. Shearman LP, Camacho RE, Sloan Stribling D, et al. Chronic MCH-1 receptor modulation alters appetite, body weight and adiposity in rats. Eur J Pharmacol. 2003;475(1–3):37–47.
20. Gompf HS, Mathai C, Fuller PM, et al. Locus ceruleus and anterior cingulate cortex sustain wakefulness in a novel environment. J Neurosci. 2010;30(43):14543–51.
21. Blanco-Centurion C, Gerashchenko D, Shiromani PJ. Effects of saporin-induced lesions of three arousal populations on daily levels of sleep and wake. J Neurosci. 2007;27(51):14041–8.
22. Salin-Pascual R, Gerashchenko D, Greco M, Blanco-Centurion C, Shiromani PJ. Hypothalamic regulation of sleep. Neuropsychopharmacology. 2001;25 Suppl 5:S21–7.
23. De Lecea L, Kilduff TS, Peyron C, et al. The hypocretins: hypothalamus-specific peptides with neuroexcitatory activity. Proc Natl Acad Sci USA. 1998;95(1):322–7.
24. Sakurai T, Amemiya A, Ishii M, et al. Orexins and orexin receptors: a family of hypothalamic neuropeptides and G protein-coupled receptors that regulate feeding behavior [comment]. Cell. 1998;92(5): 573–85.
25. Deadwyler SA, Porrino L, Siegel JM, Hampson RE. Systemic and nasal delivery of orexin-A (hypocretin-1) reduces the effects of sleep deprivation on cognitive performance in nonhuman primates. J Neurosci. 2007;27(52):14239–47.
26. Lin L, Faraco J, Li R, et al. The sleep disorder canine narcolepsy is caused by a mutation in the hypocretin (orexin) receptor 2 gene. Cell. 1999;98(3):365–76.
27. Chemelli RM, Willie JT, Sinton CM, et al. Narcolepsy in orexin knockout mice: molecular genetics of sleep regulation. Cell. 1999;98(4):437–51.
28. Peyron C, Faraco J, Rogers W, et al. A mutation in a case of early onset narcolepsy and a generalized absence of hypocretin peptides in human narcoleptic brains. Nat Med. 2000;6(9):991–7.
29. Thannickal TC, Moore RY, Nienhuis R, et al. Reduced number of hypocretin neurons in human narcolepsy. Neuron. 2000;27(3):469–74.
30. Nishino S, Ripley B, Overeem S, et al. Low cerebrospinal fluid hypocretin (Orexin) and altered energy homeostasis in human narcolepsy. Ann Neurol. 2001;50(3):381–8.
31. Willie JT, Chemelli RM, Sinton CM, et al. Distinct narcolepsy syndromes in Orexin receptor-2 and Orexin null mice: molecular genetic dissection of non-REM and REM sleep regulatory processes. Neuron. 2003;38(5):715–30.
32. Brisbare-Roch C, Dingemanse J, Koberstein R, et al. Promotion of sleep by targeting the orexin system in rats, dogs and humans. Nat Med. 2007;13(2): 150–5.
33. Lee MG, Hassani OK, Jones BE. Discharge of identified orexin/hypocretin neurons across the sleep-waking cycle. J Neurosci. 2005;25:6716–20.
34. Mileykovskiy BY, Kiyashchenko LI, Siegel JM. Behavioral correlates of activity in identified hypocretin/orexin neurons. Neuron. 2005;46(5):787–98.

35. Takahashi K, Lin JS, Sakai K. Neuronal activity of orexin and non-orexin waking-active neurons during wake-sleep states in the mouse. Neuroscience. 2008;153(3):860–70.
36. Kaur S, Thankachan S, Begum S, et al. Entrainment of temperature and activity rhythms to restricted feeding in orexin knock out mice. Brain Res. 2008;1205:47–54.
37. Akiyama M, Yuasa T, Hayasaka N, Horikawa K, Sakurai T, Shibata S. Reduced food anticipatory activity in genetically orexin (hypocretin) neuron-ablated mice. Eur J Neurosci. 2004;20(11):3054–62.
38. Mieda M, Willie JT, Hara J, Sinton CM, Sakurai T, Yanagisawa M. Orexin peptides prevent cataplexy and improve wakefulness in an orexin neuron-ablated model of narcolepsy in mice. Proc Natl Acad Sci USA. 2004;101(13):4649–54.
39. Gooley JJ, Schomer A, Saper CB. The dorsomedial hypothalamic nucleus is critical for the expression of food-entrainable circadian rhythms. Nat Neurosci. 2006;9(3):398–407.
40. Landry GJ, Simon MM, Webb IC, Mistlberger RE. Persistence of a behavioral food-anticipatory circadian rhythm following dorsomedial hypothalamic ablation in rats. Am J Physiol Regul Integr Comp Physiol. 2006;290(6):R1527–34.
41. Kohno D, Gao HZ, Muroya S, Kikuyama S, Yada T. Ghrelin directly interacts with neuropeptide-Y-containing neurons in the rat arcuate nucleus: Ca2+ signaling via protein kinase A and N-type channel-dependent mechanisms and cross-talk with leptin and orexin. Diabetes. 2003;52(4):948–56.
42. Karnani MM, Apergis-Schoute J, Adamantidis A, et al. Activation of central orexin/hypocretin neurons by dietary amino acids. Neuron. 2011;72(4):616–29.
43. Levin BE, Routh VH, Kang L, Sanders NM, Dunn-Meynell AA. Neuronal glucosensing: what do we know after 50 years? Diabetes. 2004;53(10):2521–8.
44. Yamanaka A, Beuckmann CT, Willie JT, et al. Hypothalamic orexin neurons regulate arousal according to energy balance in mice. Neuron. 2003;38(5):701–13.
45. Burdakov D, Luckman SM, Verkhratsky A. Glucose-sensing neurons of the hypothalamus. Philos Trans R Soc Lond B Biol Sci. 2005;360(1464):2227–35.
46. Funato H, Tsai AL, Willie JT, et al. Enhanced orexin receptor-2 signaling prevents diet-induced obesity and improves leptin sensitivity. Cell Metab. 2009;9(1):64–76.
47. Sellayah D, Bharaj P, Sikder D. Orexin is required for brown adipose tissue development, differentiation, and function. Cell Metab. 2011;14(4):478–90.
48. Tupone D, Madden CJ, Cano G, Morrison SF. An orexinergic projection from perifornical hypothalamus to raphe pallidus increases rat brown adipose tissue thermogenesis. J Neurosci. 2011;31(44):15944–55.
49. Yagita K, Tamanini F, van Der Horst GT, Okamura H. Molecular mechanisms of the biological clock in cultured fibroblasts. Science. 2001;292(5515):278–81.
50. Nakahata Y, Kaluzova M, Grimaldi B, et al. The NAD+-dependent deacetylase SIRT1 modulates CLOCK-mediated chromatin remodeling and circadian control. Cell. 2008;134(2):329–40.
51. Murayama A, Ohmori K, Fujimura A, et al. Epigenetic control of rDNA loci in response to intracellular energy status. Cell. 2008;133(4):627–39.
52. Rodgers JT, Lerin C, Haas W, Gygi SP, Spiegelman BM, Puigserver P. Nutrient control of glucose homeostasis through a complex of PGC-1alpha and SIRT1. Nature. 2005;434(7029):113–8.
53. Satoh A, Brace CS, Ben Josef G, et al. SIRT1 promotes the central adaptive response to diet restriction through activation of the dorsomedial and lateral nuclei of the hypothalamus. J Neurosci. 2010;30(30):10220–32.
54. Celesia GG, Jasper HH. Acetylcholine released from cerebral cortex in relation to state of activation. Neurology. 1966;16:1053–64.
55. Szerb JC. Cortical acetylcholine release and electroencephalographic arousal. J Physiol (Lond). 1967;192:329–45.
56. Jasper HH, Tessier J. Acetylcholine liberation from cerbral cortex during paradoxical (REM) sleep. Science. 1971;172:601–2.
57. Rasmusson DD, Clow K, Szerb JC. Frequency-dependent increase in cortical acetylcholine release evoked by stimulation of the nucleus basalis magnocellularis in the rat. Brain Res. 1992;594:150–4.
58. Dekker AJ, Thal LJ. Independent effects of cholinergic and serotonergic lesions on acetylcholine and serotonin release in the neocortex of the rat. Neurochem Res. 1993;18:277–83.
59. Lapierre JL, Kosenko PO, Lyamin OI, Kodama T, Mukhametov LM, Siegel JM. Cortical acetylcholine release is lateralized during asymmetrical slow-wave sleep in northern fur seals. J Neurosci. 2007;27(44):11999–2006.
60. Eggermann E, Serafin M, Bayer L, et al. Orexins/hypocretins excite basal forebrain cholinergic neurones. Neuroscience. 2001;108(2):177–81.
61. Fadel J, Pasumarthi R, Reznikov LR. Stimulation of cortical acetylcholine release by orexin A. Neuroscience. 2005;130(2):541–7.
62. Espana RA, Baldo BA, Kelley AE, Berridge CW. Wake-promoting and sleep-suppressing actions of hypocretin (orexin): basal forebrain sites of action. Neuroscience. 2001;106(4):699–715.
63. Thakkar MM, Ramesh V, Strecker RE, McCarley RW. Microdialysis perfusion of orexin-A in the basal forebrain increases wakefulness in freely behaving rats. Arch Ital Biol. 2001;139(3):313–28.
64. Blanco-Centurion C, Xu M, Murillo-Rodriguez E, et al. Adenosine and sleep homeostasis in the basal forebrain. J Neurosci. 2006;26(31):8092–100.

65. Duque A, Balatoni B, Detari L, Zaborszky L. EEG correlation of the discharge properties of identified neurons in the basal forebrain. J Neurophysiol. 2000;84(3):1627–35.
66. Zaborszky L, Duque A. Sleep-wake mechanisms and basal forebrain circuitry. Front Biosci. 2003;8:d1146–69.
67. Jones BE. Arousal systems. Front Biosci. 2003;8: s438–51.
68. Manns ID, Lee MG, Modirrousta M, Hou YP, Jones BE. Alpha 2 adrenergic receptors on GABAergic, putative sleep-promoting basal forebrain neurons. Eur J Neurosci. 2003;18(3):723–7.
69. Gritti I, Mainville L, Mancia M, Jones BE. GABAergic and other noncholinergic basal forebrain neurons, together with cholinergic neurons, project to the mesocortex and isocortex in the rat. J Comp Neurol. 1997;383(2):163–77.
70. Senba E, Daddona PE, Watanabe T, Wu JY, Nagy JI. Adenosine deaminase is a marker for histamine neurons in the rat. J Neurosci. 1985;5:3393–402.
71. Lin JS, Sakai K, Jouvet M. Hypothalamo-preoptic histaminergic projections in sleep-wake control in the cat. Eur J Neurosci. 1994;6(4):618–25.
72. Takahashi K, Lin JS, Sakai K. Neuronal activity of histaminergic tuberomammillary neurons during wake-sleep states in the mouse. J Neurosci. 2006;26(40):10292–8.
73. John J, Wu MF, Boehmer LN, Siegel JM. Cataplexy-active neurons in the hypothalamus: implications for the role of histamine in sleep and waking behavior. Neuron. 2004;42(4):619–34.
74. Marcus JN, Aschkenasi CJ, Lee CE, et al. Differential expression of orexin receptors 1 and 2 in the rat brain. J Comp Neurol. 2001;435(1):6–25.
75. Eriksson KS, Sergeeva O, Brown RE, Haas HL. Orexin/hypocretin excites the histaminergic neurons of the tuberomammillary nucleus. J Neurosci. 2001;21(23):9273–9.
76. Sakurai T, Nagata R, Yamanaka A, et al. Input of orexin/hypocretin neurons revealed by a genetically encoded tracer in mice. Neuron. 2005;46(2): 297–308.
77. Li Y, Gao XB, Sakurai T, van den Pol AN. Hypocretin/Orexin excites hypocretin neurons via a local glutamate neuron-A potential mechanism for orchestrating the hypothalamic arousal system. Neuron. 2002;36(6):1169–81.
78. Hobson JA, McCarley RW, Wyzinski PW. Sleep cycle oscillation: reciprocal discharge by two brainstem neuronal groups. Science. 1975;189:55–8.
79. McGinty DJ, Harper RM. Dorsal raphe neurons: depression of firing during sleep in cats. Brain Res. 1976;101(3):569–75.
80. Peyron C, Tighe DK, Van den Pol AN, et al. Neurons containing hypocretin (orexin) project to multiple neuronal systems. J Neurosci. 1998;18(23): 9996–10015.
81. Greco MA, Shiromani PJ. Hypocretin receptor protein and mRNA expression in the dorsolateral pons of rats. Brain Res Mol Brain Res. 2001;88(1–2): 176–82.
82. Hagan JJ, Leslie RA, Patel S, et al. Orexin A activates locus coeruleus cell firing and increases arousal in the rat. Proc Natl Acad Sci USA. 1999;96(19): 10911–6.
83. Horvath TL, Peyron C, Diano S, et al. Hypocretin (orexin) activation and synaptic innervation of the locus coeruleus noradrenergic system. J Comp Neurol. 1999;415(2):145–59.
84. Harper RM. A technique for recording single neurons from unrestrained animals. In: Mi P, editor. Brain unit activity during behavior. Springfield: Thomas; 1973. p. 80–104.
85. Adrien J. Sleep and waking in mutant mice that do not express various proteins involved in serotonergic neurotransmission such as the serotonergic transporter, monoamine oxidase A, and 5-HT1A, 5-HT1B, 5-HT2A, 5-HT2C and 5-HT7 receptors. In: Monti JM, Pandi-Perumal SR, Jacobs BL, Nutt DJ, editors. Serotonin and sleep: molecular, functional and clinical aspects. Basel: Birkhauser; 2008. p. 457–76.
86. Boutrel B, Franc B, Hen R, Hamon M, Adrien J. Key role of 5-HT1B receptors in the regulation of paradoxical sleep as evidenced in 5-HT1B knock-out mice. J Neurosci. 1999;19(8):3204–12.
87. Boutrel B, Monaca C, Hen R, Hamon M, Adrien J. Involvement of 5-HT1A receptors in homeostatic and stress-induced adaptive regulations of paradoxical sleep: studies in 5-HT1A knock-out mice. J Neurosci. 2002;22(11):4686–92.
88. Thakkar MM, Strecker RE, McCarley RW. Behavioral state control through differential serotonergic inhibition in the mesopontine cholinergic nuclei: a simultaneous unit recording and microdialysis study. J Neurosci. 1998;18(14):5490–7.
89. Popa D, Lena C, Fabre V, et al. Contribution of 5-HT2 receptor subtypes to sleep-wakefulness and respiratory control, and functional adaptations in knock-out mice lacking 5-HT2A receptors. J Neurosci. 2005;25(49):11231–8.
90. Frank MG, Stryker MP, Tecott LH. Sleep and sleep homeostasis in mice lacking the 5-HT2c receptor. Neuropsychopharmacology. 2002;27(5):869–73.
91. Morairty SR, Hedley L, Flores J, Martin R, Kilduff TS. Selective 5HT2A and 5HT6 receptor antagonists promote sleep in rats. Sleep. 2008;31(1):34–44.
92. Dugovic C, Wauquier A, Leysen JE, Marrannes R, Janssen PA. Functional role of 5-HT2 receptors in the regulation of sleep and wakefulness in the rat. Psychopharmacology (Berl). 1989;97(4):436–42.
93. Monti JM, Jantos H. Effects of the 5-HT receptor antagonists SB-399885 and RO-4368554 and of the 5-HT(2A) receptor antagonist EMD 281014 on sleep and wakefulness in the rat during both phases of the light-dark cycle. Behav Brain Res. 2011;216(1): 381–8.
94. Qiu J, Xue C, Bosch MA, et al. Serotonin 5-hydroxytryptamine2C receptor signaling in hypothalamic proopiomelanocortin neurons: role in

energy homeostasis in females. Mol Pharmacol. 2007;72(4):885–96.
95. Heisler LK, Jobst EE, Sutton GM, et al. Serotonin reciprocally regulates melanocortin neurons to modulate food intake. Neuron. 2006;51(2):239–49.
96. Zhou L, Sutton GM, Rochford JJ, et al. Serotonin 2 C receptor agonists improve type 2 diabetes via melanocortin-4 receptor signaling pathways. Cell Metab. 2007;6(5):398–405.
97. Heal DJ, Smith SL, Fisas A, Codony X, Buschmann H. Selective 5-HT6 receptor ligands: progress in the development of a novel pharmacological approach to the treatment of obesity and related metabolic disorders. Pharmacol Ther. 2008;117(2):207–31.
98. Woolley ML, Bentley JC, Sleight AJ, Marsden CA, Fone KC. A role for 5-ht6 receptors in retention of spatial learning in the Morris water maze. Neuropharmacology. 2001;41(2):210–9.
99. Jones BE, Paradoxical REM. Sleep promoting and permitting neuronal networks. Arch Ital Biol. 2004;142(4):379–96.
100. Shiromani PJ, Winston S, McCarley RW. Pontine cholinergic neurons show Fos-like immunoreactivity associated with cholinergically induced REM sleep. Brain Res Mol Brain Res. 1996;38(1):77–84.
101. Thankachan S, Kaur S, Shiromani PJ. Activity of pontine neurons during sleep and cataplexy in hypocretin knock-out mice. J Neurosci. 2009;29(5):1580–5.
102. Vanini G, Wathen BL, Lydic R, Baghdoyan HA. Endogenous GABA levels in the pontine reticular formation are greater during wakefulness than during rapid eye movement sleep. J Neurosci. 2011;31(7):2649–56.
103. Kuldau JM, Rand CSW. The night eating syndrome and bulimia nervosa in the morbidly obese. Int J Eat Disord. 1986;5:143–8.
104. Stunkard AJ, Grace WJ, Wolff HG. The night-eating syndrome; a pattern of food intake among certain obese patients. Am J Med. 1955;19(1):78–86.
105. O'Reardon JP, Ringel BL, Dinges DF, et al. Circadian eating and sleeping patterns in the night eating syndrome. Obes Res. 2004;12(11):1789–96.
106. Goel N, Stunkard AJ, Rogers NL, et al. Circadian rhythm profiles in women with night eating syndrome. J Biol Rhythms. 2009;24(1):85–94.
107. Allison KC, Ahima RS, O'Reardon JP, et al. Neuroendocrine profiles associated with energy intake, sleep, and stress in the night eating syndrome. J Clin Endocrinol Metab. 2005;90(11):6214–7.
108. Stunkard AJ, Allison KC, Lundgren JD, O'Reardon JP. A biobehavioural model of the night eating syndrome. Obes Rev. 2009;10 Suppl 2:69–77.
109. Halle M, Berg A, Garwers U, Grathwohl D, Knisel W, Keul J. Concurrent reductions of serum leptin and lipids during weight loss in obese men with type II diabetes. Am J Physiol. 1999;277(2 Pt 1):E277–82.
110. Scheen AJ, Luyckx FH. Medical aspects of obesity. Acta Chir Belg. 1999;99(3):135–9.
111. Ozturk L, Unal M, Tamer L, Celikoglu F. The association of the severity of obstructive sleep apnea with plasma leptin levels. Arch Otolaryngol Head Neck Surg. 2003;129(5):538–40.
112. Sanner BM, Kollhosser P, Buechner N, Zidek W, Tepel M. Influence of treatment on leptin levels in patients with obstructive sleep apnoea. Eur Respir J. 2004;23(4):601–4.
113. Ulukavak Ciftci T, Kokturk O, Bukan N, Bilgihan A. Leptin and ghrelin levels in patients with obstructive sleep apnea syndrome. Respiration. 2005;72(4):395–401.
114. Harsch IA, Konturek PC, Koebnick C, et al. Leptin and ghrelin levels in patients with obstructive sleep apnoea: effect of CPAP treatment. Eur Respir J. 2003;22(2):251–7.
115. Herlein JA, Fink BD, Morgan DA, Phillips BG, Haynes WG, Sivitz WI. Leptin administration to normal rats does not alter catecholamine responsiveness to insulin-induced hypoglycemia. Metabolism. 2003;52(11):1484–90.
116. Phillips BG, Somers VK. Hypertension and obstructive sleep apnea. Curr Hypertens Rep. 2003;5(5):380–5.
117. Kuwaki T. Hypothalamic modulation of breathing. Adv Exp Med Biol. 2010;669:243–7.
118. Williams RH, Burdakov D. Hypothalamic orexins/hypocretins as regulators of breathing. Expert Rev Mol Med. 2008;10:e28.
119. Nakamura A, Zhang W, Yanagisawa M, Fukuda Y, Kuwaki T. Vigilance state-dependent attenuation of hypercapnic chemoreflex and exaggerated sleep apnea in orexin knockout mice. J Appl Physiol. 2007;102(1):241–8.
120. Terada J, Nakamura A, Zhang W, et al. Ventilatory long-term facilitation in mice can be observed during both sleep and wake periods and depends on orexin. J Appl Physiol. 2008;104(2):499–507.
121. Sunanaga J, Deng BS, Zhang W, Kanmura Y, Kuwaki T. CO2 activates orexin-containing neurons in mice. Respir Physiol Neurobiol. 2009;166(3):184–6.
122. Williams RH, Jensen LT, Verkhratsky A, Fugger L, Burdakov D. Control of hypothalamic orexin neurons by acid and CO2. Proc Natl Acad Sci USA. 2007;104(25):10685–90.
123. Igarashi N, Tatsumi K, Nakamura A, et al. Plasma orexin-A levels in obstructive sleep apnea-hypopnea syndrome. Chest. 2003;124(4):1381–5.
124. Busquets X, Barbe F, Barcelo A, et al. Decreased plasma levels of orexin-A in sleep apnea. Respiration. 2004;71(6):575–9.
125. Kanbayashi T, Kodama T, Kondo H, et al. CSF histamine contents in narcolepsy, idiopathic hypersomnia and obstructive sleep apnea syndrome. Sleep. 2009;32(2):181–7.

Feeding as a Reward Mechanism

4

Ralph J. DiLeone, Nandakumar S. Narayanan, and Douglas J. Guarnieri

Abstract

Rates of obesity are increasing worldwide and pose a significant threat to individual health and to health-care systems. The natural drive to eat, combined with a surplus of readily available food, is together partly responsible for this modern epidemic. Recent research has better defined the molecular and neural mechanisms by which the brain regulates food intake. While much of this research focused on the hypothalamus, it has long been recognized that reward pathways have an important role in food intake. Here, an overview of the role of dopamine reward systems in regulating food intake is presented, with emphasis on regulation by peripheral metabolic signals. Moreover, there are emerging results that better connect regulation of sleep and reward circuits. The orexin (hypocretin) neuropeptide is an example of this and work on its role in addiction is also highlighted in this chapter.

Introduction

The acquisition of food requires a complex array of behaviors, including motivation, goal-directed behavior, motor control, and the recognition of rewarding stimuli. Dopamine-containing brain systems have been implicated in these behavioral processes, as best studied in drug addiction. Reward is intimately involved with projections of dopamine-containing neurons in the midbrain to the ventral striatum (or nucleus accumbens, NAc), cortex, and subcortical nuclei [1]. Similarly, it has been proposed that dopamine mediates the motivational and rewarding aspects of food seeking via specific dopaminergic projections from the ventral tegmental area (VTA) to the NAc (the

R.J. DiLeone, PhD (✉)
Department of Psychiatry and Neurobiology, Yale School of Medicine, 34 Park Street, New Haven, CT 06508, USA
e-mail: ralph.dileone@yale.edu

N.S. Narayanan, MD, PhD
Department of Neurobiology, Yale Medical School, 20 York Street, New Haven, CT 06508, USA

D.J. Guarnieri, PhD
Department of Psychiatry, Yale University, Park Street, New Haven, CT 06519, USA

P.J. Shiromani et al. (eds.), *Sleep Loss and Obesity: Intersecting Epidemics*, DOI 10.1007/978-1-4614-3492-4_4,

"mesolimbic" projection), and that dysfunction of reward processing may contribute to the pathogenesis of obesity [2].

One approach to determine the role of brain regions in feeding is to assess food intake following lesions or genetic mutations. For example, lesions of the lateral hypothalamus cause dramatic losses in body weight [3], while stimulating the lateral hypothalamus causes an increase in food intake [4]. Based on the results from these early studies, most mechanistic and molecular work has focused on the hypothalamus, where peripheral signals can influence the brain. More recently, genetic studies have revealed that *obese* (*ob/ob*) mice lacking a functional version of leptin, an adipocyte-derived hormone, have increased feeding and dramatic weight gain [5, 6]. Work on leptin has traditionally focused on the hypothalamus but is now being recognized as acting in multiple brain regions. Moreover, while hypothalamic lesions produce the most dramatic effects on feeding and weight, manipulations of other regions has been shown to influence food intake. For example, neurochemical disruptions in the NAc can influence food intake [7], and it is likely that food intake involves a diverse neuronal ensemble that engages a number of extrahypothalamic brain areas.

Disrupting Dopamine and Feeding

Mesolimbic circuits have been implicated in many behaviors, such as reward prediction, hedonia, reinforcement and motivation. It seems that such processes would be intimately involved in regulating food intake [2]. If dopaminergic mesolimbic circuitry is necessary for feeding, then disrupting dopamine in the mesolimbic system should affect intake. This work and challenges in interpreting it are discussed below.

The hypothesis that midbrain dopamine provides a specific motivation to eat is based on dopamine neuron depletion studies using the neurotoxin 6-hydroxydopamine, which irreversibly depletes catecholamines (including dopamine) and eventually leads to neuronal degeneration [8]. Intraventricular administration of 6-hydroxydopamine impairs free feeding and decreases body weight [9, 10]. Interestingly, dopamine neuron depleted rats gradually recover the ability to eat and drink despite a permanent depletion of dopamine from central dopaminergic nuclei [11]. In addition, systemic dopamine blockade can block feeding elicited by electrical stimulation of the lateral hypothalamus [12]. Some investigators have described intact movement in these animals [13], but no movement controls were reported in original studies [9]. Peripheral dopamine blockade also can disrupt feeding elicited via hunger [14–17], and this work was influential in bringing attention to the role of dopamine neural systems in feeding behavior. However, as reviewed below, effects on feeding due to specific dopamine receptor antagonists in the NAc have not produced as dramatic results.

Dopamine's role in feeding was further advanced by a unique genetic form of animals in which tyrosine hydroxylase is removed [18, 19]. These animals cannot synthesize dopamine, have dramatically decreased eating and drinking, and only survive a few weeks after birth. However, dopamine replenishment (via injections of L-DOPA,3,4-dihydroxy-L-phenylalanine; a precursor of dopamine synthesis converted from L-tyrosine) enable temporary survival and analysis of animals lacking detectable dopamine. While these animals are hypophagic, they also exhibit clear deficits in movement. Subsequent experiments suggested that dopamine-deficient animals can exhibit preferences for sucrose [20], raising the possibility that preference and reward-related decisions are possible without dopamine. Taken together, studies in which dopaminergic transmission is disrupted point to a clear role for dopamine in feeding.

Experiments exploring the role of VTA dopamine in feeding suggest that mesolimbic dopamine influences, but is not necessary for, feeding behavior. Initial studies that depleted dopamine neurons in the VTA [21] reported no effect on body weight despite changes locomotor activity and deficits in passive-avoidance behavior. While VTA dopamine neuron depletions can influence a wide range of behaviors, such as drug self-administration, perseveration, and memory, no data suggest that mesolimbic dopamine neuron depletions selectively impact feeding. However, negative data in dopamine neuron depletion experiments are difficult to interpret, as small amounts of residual dopamine function may be sufficient for critical feeding-related signals. As reviewed below, some studies suggest that general lesions in the region do affect food intake. Moreover, more specific pharmacological or viral manipulation of neurons in the VTA can result in changes in food intake. This approach, which modulates specific populations of VTA neurons, has demonstrated clear effects on feeding behavior [22–24], in contrast to the negative data seen with traditional lesion strategies.

Supporting a role for the VTA in feeding is extensive literature establishing dopamine release in the NAc during feeding. These data derive chiefly from microdialysis studies, in which changes in extracellular dopamine and breakdown products are seen in the NAc in response to both feeding and hypothalamic stimulation [25]. Food anticipation corresponds with dopamine release in the shell, whereas food stimuli promote dopamine release in the core [26]. Voltammetry studies, which detect changes in dopamine efflux with high temporal resolution, reported rapid dopamine release (<100 ms) in response to food cues and food seeking behavior [27].

In-vivo recordings also suggest that mesolimbic dopaminergic neurons are modulated in pursuit of food rewards. Such experiments are difficult to interpret outside of specific behavioral conditions making it challenging to study free-feeding in detail. However, the work of Wolfram Schultz and colleagues has shed light on the activity of VTA dopaminergic neurons in behavior [28–30]. These studies evaluated nonhuman primate dopaminergic VTA neurons and reported phasic modulations in activity following liquid or food rewards [28]. Food rewards also change activity in midbrain dopaminergic neurons in rodents [31]. Increases in activity are seen in response to stimuli that predict rewards, while decreases in activity are observed when rewards are omitted [32]. Dopaminergic neurons increase in firing rate when an animal contacts food and when liquid reward is administered to the animal outside of a behavioral context. Such neurons respond primarily to novel rewards, and when outcomes are unexpected. These data have led to the hypothesis that dopamine neurons in the midbrain encode reward prediction errors [28–30]. The experiments suggest that mesolimbic dopamine neuron response is more consistent with a limited role free feeding and a potentially greater role in signaling unexpected reward, cue associations, or reward valence. This is also consistent with the finding that novel or unexpected food causes more dopamine release in the NAc shell [26].

The Nucleus Accumbens and Food Intake

The NAc is the major target of VTA dopamine neurons, and the region plays an important role in drug addiction. However, blockade of dopamine in the NAc does not reduce the amount of food consumed in a free-feeding paradigm [20, 33]. Previous data with the dopamine-deficient genetic models suggests a role for striatal dopamine in food intake and NAc in locomotor function [34]. Another study elegantly rescued midbrain dopamine in dopamine-deficient genetic models [35], resulting in normalized locomotor behavior but only partially normalized food intake. Taken together, these studies implicate both the dorsal and ventral striatum (NAc) in controlling food intake. However, while pharmacological studies have shown little effect of D1 receptor antagonism on ad libitum food intake [33], D1 receptor antagonists do result in attenuation of operant performance under high workloads (i.e., many lever presses required to obtain food) [36, 37]. Indeed, dopamine in the NAc is necessary for seeking and exerting effort to obtain food, but may not be required for intake under free feeding, or other low effort, conditions. Other studies suggest additional roles for NAc dopamine, some of which could be relevant to excessive food intake. For example, intra-NAc amphetamine increases the seeking of sucrose in response to a conditioned cue [38], and dopamine in the NAc drives behavioral responses to cues that predict reward

[39]. These data have provoked discussion beyond the scope of this chapter about what dopamine does and how important it is for food reward [1, 40, 41].

While there are many ways to evaluate the importance of molecular pathways in biology, another approach is to look for regulation under different physiological states. For example, voltammetry experiments have revealed peaks in dopamine release during lever pressing and food seeking with a return to baseline after food ingestion [27], consistent with a potential function at different points of the behavior. The recent development of optogenetic techniques for stimulation of selected neuronal populations will likely enable higher resolution tests of behavior in these timescales.

Leptin: A Link Between the Body and the Brain

The greatest progress in the field has come from the identification of metabolic signals that influence brain function. While peripheral signals that communicate metabolic state, such as insulin, had already been isolated and well characterized, the cloning of the mouse *obese* (*ob*) mutation initiated a new stage of feeding research. The gene product mutated in *ob/ob* mice was shown to be a small protein expressed in adipocytes and was given the name leptin [42]. Genetic and molecular studies, including the identification of the leptin receptor (lepr) [43, 44], confirmed the importance of the hypothalamus in regulating feeding behavior and metabolism, and suggest that it is a primary site of action for circulating factors. Additional work has led to a neuronal hypothesis whereby lepr activity inhibits the orexigenic while exciting the anorexigenic neurons within the arcuate nucleus [45].

While the isolation of critical peripheral metabolic signals has invigorated the field of feeding research, the mechanism by which leptin acts on brain circuitry to alter animal behavior remains only partly understood. Current work in the field is mainly focused on identifying which second order feeding centers, such as the paraventricular nucleus (PVN) or lateral hypothalamus (LH), are important for mediating leptin responses downstream of the arcuate nucleus. However, it is not clear which neural circuits are modified to result in a change in the motivation to seek food. Strikingly, cortico-striatal circuits integral for motivated behavior and inhibitory control have not been well integrated with metabolic hormones that are known to influence food intake. Food intake is a complex behavior that is clearly influenced by many nonhomeostatic mechanisms. It seems likely that leptin-initiated signaling circuits eventually converge with circuits that regulate reward and motivation, including the dopamine circuits and cortico-striatal pathways.

Administration of leptin has been shown to modulate behaviors that are dependent upon the mesolimbic dopamine circuit. For example, leptin administration alters intracranial self-stimulation (ICSS) [46], suggesting an interaction with dopamine circuits that are thought to underlie the behavior. Other data have demonstrated that leptin attenuates the increased propensity to heroin relapse caused by food restriction [47], and modulates conditioned place preference (CPP) for sucrose or high fat food [48, 49]. These data suggest that leptin can modify reward-based behavior and that this may occur via alteration in the function of dopamine pathways [50]. While this interaction with dopamine pathways could occur downstream from leptin's effects on the hypothalamus, data also suggest direct action of leptin on dopamine centers of the brain.

The VTA as a Novel Target for Leptin

While many studies have investigated leptin signaling in the hypothalamus, recent evidence suggests that leptin communicates directly to many other brain regions. There is extensive lepr expression in extrahypothalamic brain regions including the hippocampus, cortex, and the midbrain [51]. Grill and colleagues have shown that leptin targets in the brainstem are important for controlling food intake [52], indicating that leptin's effects could be mediated via direct signaling to regions beyond the hypothalamus. The lepr is expressed in most dopamine neurons of the VTA [53], and plays a role in regulation of

VTA neuronal function and feeding behavior [23]. Interestingly, other work showed that leptin administration can modify responses to psychostimulants [54], suggesting implications for drug addiction. The work also demonstrated that the VTA neurons are responsive to leptin and that *ob/ob* mice without leptin have attenuated amphetamine sensitization, which is reversed with leptin administration [54]. While both papers establish effects of leptin in the VTA, the Fulton et al. study demonstrates that genetic loss of leptin results in attenuated mesolimbic function, while Hommel et al. suggest that leptin directly attenuates activity of dopamine neurons. A possible explanation for these contrasting data is that acute and chronic leptin may have distinct effects on VTA dopamine neurons. The *ob/ob* mutant mice could have altered dopamine neuron synaptic structure or function that was effectively reversed by the long-term (10 days) leptin treatment [54]. Leptin has been shown to be critical for normal development of hypothalamic neuronal projections [55], and leptin administration can result in dramatic rearrangement of synapses [56]. Whereas the acute suppression of dopamine neuron function is consistent with previous data demonstrating reduced dopamine release [57] and attenuated dopamine-dependent behaviors [46–49] following leptin administration. A recent fMRI study also shows that leptin deficient people demonstrate hyperactivated ventral striatal regions in response to food images. Leptin administration suppressed this hyperactivity while also leading to reduced preference ratings of the images [58]. Finally, the hormone ghrelin activates VTA dopamine neurons [59], suggesting opposite regulation by orexigenic factors and the idea that the midbrain integrates a larger set of metabolic signals.

It should also be emphasized that there likely exists a number of potential mechanisms by which leptin modulates dopamine neurons leading to changes in animal behavior. In fact, earlier studies demonstrated that leptin can have opposing effects on reinforcement depending upon the specific brain circuits being stimulated [46]. More generally, dopamine modulation of the midbrain may influence a number of brain areas, with potentially contrasting effects on behavior. While VTA dopamine neurons prominently project to the NAc, the VTA also projects to the prefrontal cortex. This circuit might in turn influence drug addiction and related behavior via action on the NAc. Other projections, such as those to the amygdala, could also influence reward-related behavior. The lepr is also expressed in the substantia nigra, which would likely influence the dorsal striatum. Finally, a more recent study suggests that dopamine neurons with unconventional firing properties may influence many of these regions and deserve further study [60].

Direct action of leptin on dopamine circuits has profound implications for both feeding and addiction research, as it provides a molecular and neuronal mechanism for how the state of the body may modulate the drive to eat. Moreover, this direct connection to peripheral signals places the mesolimbic system within a larger physiological framework of eating regulation. As noted above, other results have shown that additional metabolic factors, such as insulin [53, 61] and ghrelin [24, 59], appear to act on the VTA to modulate neuron function or food intake and reward. Thus, as with the hypothalamus and brain stem, midbrain regions could be considered a region for integration of peripheral metabolic signals.

The full behavioral consequences of metabolic hormones acting on dopamine neurons will require more work to appreciate. Imaging data suggesting that obese patients resemble drug addicts [62] has been influential and motivated attempts to better connect the animal models and human studies. At this point, it is even more critical that we fully describe the molecular and neural players so that the findings from animal studies can be better connected to the human conditions.

Nondopamine Signaling in the Nucleus Accumbens and Hedonic Responses

The VTA dopamine neurons show prominent projections to NAc, which has been well studied for its role in drug responses and drug addiction [63]. Because of this, it is often assumed that this pathway will influence food intake via the "pleasure" of ingesting palatable food. However,

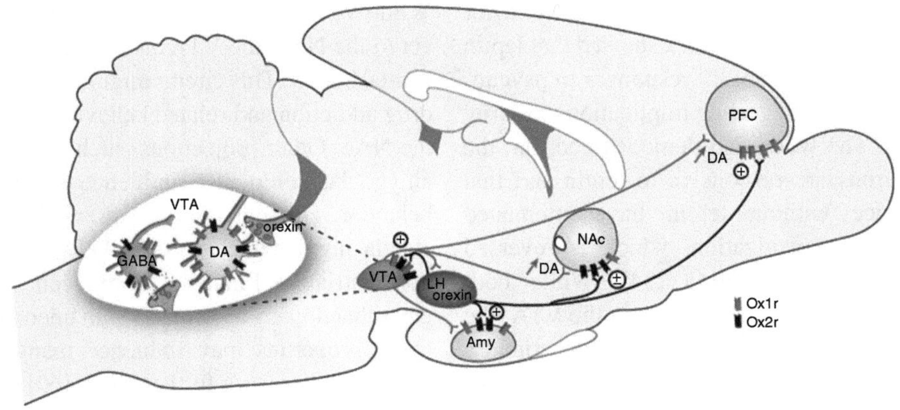

Fig. 4.1 Schematic representation of the orexin system and its interactions with the dopamineic reward pathways. Orexin projections activate the ventral tegmental area (VTA), amygdala (Amy), and prefrontal cortex (PFC). Conflicting data suggest both activation and inhibition of the nucleus accumbens (NAc) by orexin. In the VTA, orexin activates both dopamine and GABA neurons. Orexin activation of the VTA, via Ox1r, results in increased dopamine (DA) in via projections to the NAc and PFC. Orexin's action in the PFC is likely due to stimulation of Ox1r whereas Ox2r may mediate effects in the NAc.

animal model data suggest that the mesolimbic reward centers may influence both hedonic (related to pleasure) and nonhedonic behaviors. A large set of studies in the NAc has shown a role for glutamate, GABA, and opioids in regulation of food intake. While μ-opioid receptor agonists lead to a selective increase in intake of highly palatable food [64], glutamate antagonists and GABA agonists lead to an increase in overall food intake [65, 66]. As with GABA and glutamate, it does not appear that NAc dopamine, or dopamine in general, influences hedonic responses or "liking" of food [38, 67]. In fact, this is a core component of the incentive-sensitization theory of addiction, where liking and wanting are dissociable with dopamine playing a role in the latter [68]. It is also notable that infusion of μ-opioid agonists into the NAc shell dramatically increases the effort, assessed by lever pressing, that animals will make to obtain food [69], suggesting that opioids in the NAc can modulate multiple components of behavior relevant to food intake. Opioid microinfusion studies have suggested that a small region of the NAc shell mediates this enhancement of hedonic "liking," whereas a larger area of the shell mediates general feeding responses [70].

In all, these data suggest that a central role of the NAc shell is of a behavioral switch whereby silencing of the local neurons would allow feeding to occur without clear effects on hedonic components [71]. While the studies have relied on relatively blunt pharmacological approaches, in vivo recordings support this model since most NAc neurons are inhibited before food intake [72, 73]. Interestingly, the lateral hypothalamic peptide MCH also promotes general food intake [74], and appears to inhibit neurons of the NAc [75].

Orexin, Sleep, and Drug Addiction

The hypothalamic neuropeptide orexin, also called hypocretin, provides an opportune connection between sleep and reward pathways. Orexin influences a range of behaviors ranging from feeding to sleep and arousal (for details, see Chap. 3). While genetic studies in rodents first demonstrated a role in narcolepsy (for details, see Chap. 3), other work has brought attention to the role of orexin in drug addiction. Orexin influences many aspects of addiction, including drug dependence, reward, and reinstatement. Orexin acts through its receptors to influence the mesocortical and mesolimbic dopamine pathways that are known to underlie responses to drugs and the development of addiction (Fig. 4.1).

Importantly, orexin may also influence positive and negative (stress) states that can influence addiction. The orexin neurons project broadly and the receptors are widely expressed making it challenging to determine the neural circuits that are central to these effects. It may be that overlapping, or distinct circuits are mediating orexin's influence on reward/addiction versus arousal/narcolepsy.

Orexin, Receptors, and Connections with Dopamine Circuits

The orexin peptides, orexin A, and orexin B, are derived from proteolytic processing of the precursor prepro-orexin in hypothalamic neurons of the perifornical area (PFA), dorsomedial hypothalamus (DMH), and lateral hypothalamus (LH) [76, 77]. Orexin neurons represent a distinct population from MCH-expressing neurons that are also localized in the PFA and LH [78]. The G-protein coupled orexin receptors have different selectivity to orexin ligands whereby orexin 1 receptor (Ox1r) is selective for orexin A whereas the orexin 2 receptor (Ox2r) is less selective. The Ox1r couples to G_q while Ox2r is primarily $G_{i/o}$-coupled [76], and both Ox1r and Ox2r have been found on soma as well as pre- and post-synaptic processes [79].

Most abused drugs increase extracellular DA levels at the level of the NAc and adaptation in this system is thought to underlie many elements of addiction. The orexin system has a number of potential connections with this well-studied dopamine circuitry. Orexin receptors are broadly expressed and the neurons innervate many regions, including prominent axonal projections to the VTA which expresses orexin receptors [77, 80, 81]. As with many neuropeptides, it is not clear if orexin is released synaptically, extra-synaptically, or via both mechanisms. While low power analysis has shown that orexin terminals contact tyrosine-hydroxylase (TH) positive cells [81, 82], electron microscopic studies indicated that orexin-positive axons infrequently synapse onto dopamine and GABA neurons within the VTA [83].

Drugs of Abuse Regulate Orexin Neuron Excitation and Gene Expression

c-Fos is noted in orexin neurons following acute administration of both methamphetamine [84] and amphetamine [81]. Amphetamine increases c-Fos (a marker of neural activation) expression in neurons specifically in the DMH, but not in neurons of the PFA or LH [81]. Acute administration of cocaine does not affect either of the orexin receptors [85, 86] or peptide levels [85, 86] and acute morphine fails to alter orexin mRNA levels [87].

In addition to effects of acute drugs, chronic drugs of abuse increase orexin neural activity and mRNA levels of orexin and/or its receptors. Chronic amphetamine exposure results in increased c-Fos expression in orexin neurons [88]. In contrast, chronic morphine treatment [87, 89, 90] and steady-state chronic cocaine [86] fail to change orexin mRNA levels or c-Fos expression in orexin neurons in the LH. Nonetheless, Ox2r levels are increased in the NAc following chronic cocaine; a change that persists even months after the cessation of cocaine exposure [85].

Orexin's Effects on Neuronal Excitation and Plasticity Within the VTA

Orexin excites both DA and non-DA neurons [82, 91] in the VTA. Intra-VTA application of orexin results in increased Fos expression in DA neurons located specifically in the caudomedial portion of the VTA [92], and increases DA in the nucleus accumbens shell and medial prefrontal cortex [93, 94]. Intra-VTA orexin also results in potentiation of NMDA receptor-mediated excitatory post-synaptic currents [95] indicating a role for orexin in neural plasticity.

In addition to effects on the midbrain dopamine neurons of the VTA, orexin may also directly influence the target regions of dopamine neurons, such as the NAc and prefrontal cortex. Orexin is excitatory in the cortex [96], while some reports suggest inhibition [97], and others excitation [98] in the NAc.

The Role of Orexin in Withdrawal from Drugs on Abuse

The first study establishing a functional role of orexin in addiction demonstrated an involvement for orexin in morphine withdrawal [89]. Precipitated morphine withdrawal leads to the induction of cAMP response element (CRE) activity and Fos expression in orexin cells [89, 90]. The increases in Fos expression were restricted to the DMH and PFA, and were not seen in the LH orexin neurons [90]. Likewise, spontaneous morphine withdrawal increases orexin mRNA in the LH [87]. Activation of orexin neurons in response to withdrawal appears to be opiate specific and is not seen following spontaneous withdrawal after chronic cocaine [86].

The orexin system appears to be necessary for the expression of withdrawal. Treatment with the selective Ox1r antagonist SB-334867 prior to naloxone-precipitated withdrawal attenuates somatic morphine withdrawal symptoms [90]. The somatic withdrawal intensity correlates with cFos expression in the NAc shell, and blockade of Ox1r indirectly attenuates NAc cellular activation [90]. Interestingly, the VTA does not show changes in cFos expression in the same animals, suggesting a possible VTA-independent mechanism by which orexin alters NAc activity.

Effects of Orexin on Locomotor Response and Sensitization

Drug-induced hyperlocomotion and behavioral sensitization is argued to represent a form of drug induced neural plasticity [99] that may contribute to dependence and addictive processes. Thus far, investigation of orexin's role in drug hyperlocomotion and sensitization has been limited and conflicted. Orexin knock-out mice show reduced locomotor response in response to acute morphine [94]. Systemic administration of SB-334867 does not affect acute cocaine-induced hyperlocomotion and activity in previously sensitized animals [95]. However, systemic and intra-VTA blockade of Ox1r blocks the development of cocaine sensitization [95]. However, recent data suggests that SB-334867 treated mice and OKO mice respond normally following an acute morphine treatment and no changes in behavioral sensitization are seen following chronic morphine administration [100].

Orexin in CPP for Drugs of Abuse

CPP is a widely used paradigm for assessing reward-related behavior. An animal is placed in an enclosed environment, which is divided into two or more distinct compartments, each containing distinctive environmental cues that differentiate each side. After the animal is allowed to move freely among the compartments, it is confined to one of the compartments while being presented with a rewarding stimulus. The animal forms an association between the rewarding stimulus and the environmental cues during the conditioning session. After the session, when allowed to roam freely among the compartments, the animal will spend more time in the reward-paired chamber if the drug or experience was reinforcing. In the CPP test, animals are evaluated in a drug-free state and preference is indicative of the rewarding effects of the context rather than of the drug itself. It is also notable that discrepancy has been reported between CPP data and drug self-administration data [101, 102]. Nonetheless, CPP remains a useful experimental protocol in the investigation of neural mechanisms underlying drug addiction.

Animals that exhibit a preference for an environment previously paired with food, morphine, or cocaine show increased Fos expression in LH orexin neurons [103]. Interestingly, administration of SB-334867 reduces the expression of morphine place preference [100, 103], but not cocaine place preference [100]. Previous reports also suggest a role for orexin in the formation of place preference to a morphine-paired environment. Unilateral lesions of the LH and intra-VTA administration of SB-334867 on the contralateral side also block the development of morphine CPP [104], suggesting that orexin acting in the VTA is essential for the development of morphine

CPP. A role for the VTA in the expression of morphine CPP was also suggested from intra-VTA suppression of morphine place preference by orexin receptor antagonists [94].

Orexin in Self-Administration, Extinction, and Reinstatement

Self-administration studies are powerful measures that best mimic conditions of addiction since the subject is controlling the drug delivery and drug seeking behavior. SB-334867 administration results in decreased self-administration of nicotine [105] and alcohol [106]. In contrast, intracerebroventricular (ICV) administration of orexin A [107] or treatment with SB-334867 [108] fails to affect responding during cocaine self-administration. Orexin does play a role in drug relapse as modeled in extinction and reinstatement paradigms. Following self-administration training, animals continue to lever-press or nose-poke, but no drug is delivered following previously correct responses until eventually animals decrease responding ("extinction"). Animals are then tested for reinstatement, or relapse, by assessing lever pressing after extinction. Reinstatement of the extinguished response can be induced by presentation of various stimuli, including drugs (drug-primed reinstatement) or cues paired with the previously active response (cue-primed reinstatement). While ICV administration of orexin A reinstates extinguished responses [107], blockade of orexin suggests a more specific role for drug reinstatement. With both alcohol [106] and cocaine [108], treatment with orexin antagonists fails to affect drug-primed reinstatement, but attenuates cue-primed reinstatement. It is possible that orexin plays a more general role in mediating cue- and/or context-drug associations.

Orexin's involvement in reinstatement of extinguished drug seeking may reflect a role in mediating stress responses. Stress induced reinstatement of cocaine self-administration is abolished by previous orexin antagonist SB-334867 administration [107]. These data suggest that drug seeking induced by activation of the stress pathway depends on orexin. Interestingly, intra-VTA antagonists of corticotropin-releasing factor (CRF, increased in stress) do not block orexin-A induced reinstatement of cocaine-seeking and foot-shock stress induced cocaine reinstatement is not blocked by VTA orexin antagonism [109], suggesting that orexin and CRF have independent actions and that blockade of stress induced reinstatement by orexin receptor antagonism is not mediated by the VTA. It has been proposed that distinct populations of hypothalamic neurons mediate orexin's role in stress and reward [103, 110].

Orexin and Reward

The self-administration data thus far suggests a role for orexin in the rewarding properties of some drugs of abuse. Additional evidence for the role of orexin in reward processing comes from CPP and cue-induced reinstatement data showing that orexin function contributes to drug seeking when the rewarding stimulus is no longer present. These data are suggestive of a reward component but the CPP data more directly suggest that processing of environmental cues and drug-associated contexts, and drug-seeking induced by the presence of such cues, is mediated by the orexin system. However, this fails to explain data demonstrating that activation of the orexin system results in spontaneous reinstatement of an extinguished behavioral response.

In brain stimulation reward (BSR), animals lever-press for the delivery of intracranial electrical stimulation and animals will robustly respond for stimulation in the LH [111]. Interestingly, orexin A administration has been shown to elevate ICSS thresholds, suggesting a paradoxical decrease in reward sensitivity [107]. In contrast, orexin antagonism (SB-334867) abolishes nicotine-mediated reductions in BSR thresholds, suggesting that blockade of Ox1r results in reduced sensitivity to reward [105]. Based on these findings, it remains inconclusive how the orexin mediated alterations in drug self-administration are related to reward processing.

Orexin and Addiction: Implications for Narcolepsy

It is clear that the orexin system is engaged by drugs of abuse and that orexin plays a role in drug responses and behaviors relevant to addiction. Orexin can alter drug self-administration, drug-associated cue processing and reward, and stress responses in drug-abstinent and drug-dependent animals (somatic withdrawal and reinstatement of extinguished drug seeking). Although orexin can modulate all of these effects, it is possible that they are mediated by specific anatomical substrates, and distinct orexin neuronal subpopulations may serve disparate functions. Reward- and cue-processing appears to be mediated by LH orexin neurons [103, 110] acting in the VTA [104], whereas drug-related stress or aversive responses may be mediated by DMH and PFA orexin neurons acting directly or indirectly through other brain regions. Although much evidence supports a role for orexin functioning in VTA in reward processing [104, 110], other brain regions are likely to be involved, such as the basolateral amygdala and the prefrontal cortex.

Orexin has also been shown to mediate sleep and arousal processes (for additional details see Chap. 3). Canine narcolepsy has been attributed to a genetic disruption of the Ox2r gene [112] and orexin-deficient mice demonstrate behavioral arrests similar to those seen in human narcoleptic episodes [113]. Are the same neural circuits relevant for drugs of abuse and for narcolepsy? One possibility is that the neural circuits mediating orexin's effects with drugs of abuse are independent of those mediating sleep function and narcolepsy. Since the orexin axonal projections, and receptor expression, is so broad, this seems not only possible, but likely. On the other hand, some of the pathways implicated in addiction (e.g., VTA) would be expected to play a role in orexin's ability to activate behaviors and thus contribute to a state of arousal. In fact, orexin itself can modulate locomotor activity [82], suggesting a general role in this process independent of drugs of abuse.

However, it is important to note that comparisons between an evolved role for the molecule and that molecule's role under pharmacological challenge (drugs of abuse) are difficult to make. That is, the orexin neuropeptide system likely evolved to balance a complex set of animal behaviors. As with all neurochemicals and pathways that mediate addiction, it may be that drugs of abuse act via a specific subcomponent of this neuropeptide system that otherwise serves to integrate arousal, motivation and the formation of cue associations. In addition, orexin has been implicated in other biological functions such as regulation of body temperature and heart rate [114], supporting a broader role for the molecule in animal physiology. More research is needed to elucidate specific neuroanatomical substrates and receptor mechanisms of orexin's role in drug dependence and drug addiction. Moreover, more analysis of orexin's role in nondrug rewards (e.g., food and sex) will help to more completely describe orexin's influence on a diverse set of motivated and reward-related behaviors. For example, while orexin's role in food intake is likely complex, it is clear that it can modulate the motivation for food [115].

References

1. Wise RA. Dopamine, learning and motivation. Nat Rev Neurosci. 2004;5:483–94.
2. Volkow ND, Wise RA. How can drug addiction help us understand obesity? Nat Neurosci. 2005;8:555–60.
3. Margules DL, Olds J. Identical "feeding" and "rewarding" systems in the lateral hypothalamus of rats. Science. 1962;135:374–5.
4. Delgado JM, Anand BK. Increase of food intake induced by electrical stimulation of the lateral hypothalamus. Am J Physiol. 1953;172:162–8.
5. Ingalls AM, Dickie MM, Snell GD. Obese, a new mutation in the house mouse. J Hered. 1950;41:317–8.
6. Friedman JM, Halaas JL. Leptin and the regulation of body weight in mammals. Nature. 1998;395:763–70.
7. Baldo BA, Gual-Bonila L, Sijapati K, et al. Activation of a subpopulation of orexin/hypocretin-containing hypothalamic neurons by GABAA receptor-mediated inhibition of the nucleus accumbens shell, but not by exposure to a novel environment. Eur J Neurosci. 2004;19:376–86.

8. Ungerstedt U. 6-Hydroxy-dopamine induced degeneration of central monoamine neurons. Eur J Pharmacol. 1968;5:107–10.
9. Zigmond MJ, Stricker EM. Deficits in feeding behavior after intraventricular injection of 6-hydroxydopamine in rats. Science. 1972;177:1211–4.
10. Ungerstedt U. Stereotaxic mapping of the monoamine pathways in the rat brain. Acta Physiol. 1971;367:1–48.
11. Zigmond MJ, Stricker EM. Recovery of feeding and drinking by rats after intraventricular 6-hydroxydopamine or lateral hypothalamic lesions. Science. 1973;182:717–20.
12. Phillips AG, Nikaido RS. Disruption of brain stimulation-induced feeding by dopamine receptor blockade. Nature. 1975;258:750–1.
13. Price MT, Fibiger HC. Discriminated escape learning and response to electric shock after 6-hydroxydopamine lesions of the nigro-neostriatal dopaminergic projection. Pharmacol Biochem Behav. 1975;3:285–90.
14. Wise RA, Colle LM. Pimozide attenuates free feeding: best scores analysis reveals a motivational deficit. Psychopharmacology (Berl). 1984;84:446–51.
15. Wise RA. Catecholamine theories of reward: a critical review. Brain Res. 1978;152:215–47.
16. Wise RA, Spindler J, deWit H, et al. Neuroleptic-induced "anhedonia" in rats: pimozide blocks reward quality of food. Science. 1978;201:262–4.
17. Wise RA, Spindler J, Legault L. Major attenuation of food reward with performance-sparing doses of pimozide in the rat. Can J Psychol. 1978;32:77–85.
18. Zhou QY, Palmiter RD. Dopamine-deficient mice are severely hypoactive, adipsic, and aphagic. Cell. 1995;83:1197–209.
19. Palmiter RD. Is dopamine a physiologically relevant mediator of feeding behavior? Trends Neurosci. 2007;30:375–81.
20. Cannon CM, Palmiter RD. Reward without dopamine. J Neurosci. 2003;23:10827–31.
21. Galey D, Simon H, Le Moal M. Behavioral effects of lesions in the A10 dopaminergic area of the rat. Brain Res. 1977;124:83–97.
22. MacDonald AF, Billington CJ, Levine AS. Alterations in food intake by opioid and dopamine signaling pathways between the ventral tegmental area and the shell of the nucleus accumbens. Brain Res. 2004;1018:78–85.
23. Hommel JD, Trinko, R, Sears RM, et al. Leptin receptor signaling in midbrain dopamine neurons regulates feeding. Neuron. 2006;51:801–10.
24. Naleid AM, et al. Ghrelin induces feeding in the mesolimbic reward pathway between the ventral tegmental area and the nucleus accumbens. Peptides. 2005;26:2274–9.
25. Hernandez L, Hoebel BG. Feeding and hypothalamic stimulation increase dopamine turnover in the accumbens. Physiol Behav. 1988;44:599–606.
26. Bassareo V, Di Chiara G. Differential responsiveness of dopamine transmission to food-stimuli in nucleus accumbens shell/core compartments. Neuroscience. 1999;89:637–41.
27. Roitman MF, Stuber GD, Phillips PE, et al. Dopamine operates as a subsecond modulator of food seeking. J Neurosci. 2004;24:1265–71.
28. Schultz W. Dopamine neurons and their role in reward mechanisms. Curr Opin Neurobiol. 1997;7:191–7.
29. Schultz W. Reward signaling by dopamine neurons. Neuroscientist. 2001;7:293–302.
30. Schultz W. Behavioral dopamine signals. Trends Neurosci. 2007;30:203–10.
31. Hyland BI, Reynolds JNJ, Hay J, et al. Firing modes of midbrain dopamine cells in the freely moving rat. Neuroscience. 2002;114:475–92.
32. Schultz W, Apicella P, Ljungberg T. Responses of monkey dopamine neurons to reward and conditioned stimuli during successive steps of learning a delayed response task. J Neurosci. 1993;13:900–13.
33. Baldo BA, Sadeghian K, Basso AM, et al. Effects of selective dopamine D1 or D2 receptor blockade within nucleus accumbens subregions on ingestive behavior and associated motor activity. Behav Brain Res. 2002;137:165–77.
34. Szczypka MS, Kwok K, Brot MD, et al. Dopamine production in the caudate putamen restores feeding in dopamine-deficient mice. Neuron. 2001;30:819–28.
35. Hnasko TS, Perez FA, Scouras AD, et al. Cre recombinase-mediated restoration of nigrostriatal dopamine in dopamine-deficient mice reverses hypophagia and bradykinesia. Proc Natl Acad Sci USA. 2006;103:8858–63.
36. Nowend KL, Arizzi M, Carlson BB, et al. D1 or D2 antagonism in nucleus accumbens core or dorsomedial shell suppresses lever pressing for food but leads to compensatory increases in chow consumption. Pharmacol Biochem Behav. 2001;69:373–82.
37. Salamone JD, Correa M, Mingote S, et al. Nucleus accumbens dopamine and the regulation of effort in food-seeking behavior: implications for studies of natural motivation, psychiatry, and drug abuse. J Pharmacol Exp Ther. 2003;305:1–8.
38. Wyvell CL, Berridge KC. Intra-accumbens amphetamine increases the conditioned incentive salience of sucrose reward: enhancement of reward "wanting" without enhanced "liking" or response reinforcement. J Neurosci. 2000;20:8122–30.
39. Nicola SM, Taha SA, Kim SW, et al. Nucleus accumbens dopamine release is necessary and sufficient to promote the behavioral response to reward-predictive cues. Neuroscience. 2005;135:1025–33.
40. Salamone JD, Correa M, MIngote SM, et al. Beyond the reward hypothesis: alternative functions of nucleus accumbens dopamine. Curr Opin Pharmacol. 2005;5:34–41.
41. Wise RA. Role of brain dopamine in food reward and reinforcement. Philos Trans R Soc Lond B Biol Sci. 2006;361:1149–58.

42. Zhang Y, Proenca R, Maffei M, et al. Positional cloning of the mouse obese gene and its human homologue. Nature. 1994;372:425–32.
43. Tartaglia LA, Dembski M, Weng X, et al. Identification and expression cloning of a leptin receptor OB-R. Cell. 1995;83:1263–71.
44. Chen H, Charlat O, Tartaglia LA, et al. Evidence that the diabetes gene encodes the leptin receptor: identification of a mutation in the leptin receptor gene in db/db mice. Cell. 1996;84:491–5.
45. Cowley MA, Smart JL, Rubinstein M, et al. Leptin activates anorexigenic POMC neurons through a neural network in the arcuate nucleus. Nature. 2001;411:480–4.
46. Fulton S, Woodside B, Shizgal P. Modulation of brain reward circuitry by leptin. Science. 2000;287: 125–8.
47. Shalev U, Yap J, Shaham Y. Leptin attenuates acute food deprivation-induced relapse to heroin seeking. J Neurosci. 2001;21:RC129.
48. Figlewicz DP, Bennet J, Evans SB, et al. Intraventricular insulin and leptin reverse place preference conditioned with high-fat diet in rats. Behav Neurosci. 2004;118:479–87.
49. Figlewicz DP, HIggins MS, Ng-Evans SB, et al. Leptin reverses sucrose-conditioned place preference in food-restricted rats. Physiol Behav. 2001;73:229–34.
50. Figlewicz DP, Woods SC. Adiposity signals and brain reward mechanisms. Trends Pharmacol Sci. 2000;21:235–6.
51. Elmquist JK, Bjorbaek C, Ahima RS, et al. Distribution of leptin receptor mRNA isoforms in the rat brain. J Comp Neurol. 1998;535–47.
52. Grill HJ, Schwartz MW, Kaplan JM, et al. Evidence that the caudal brainstem is a target for the inhibitory effect of leptin on food intake. Endocrinology. 2002; 143:239–46.
53. Figlewicz DP, Evans SB, Murphy J, et al. Expression of receptors for insulin and leptin in the ventral tegmental area/substantia nigra (VTA/SN) of the rat. Brain Res. 2003;964:107–15.
54. Fulton S, Pissios P, Manchon RP, et al. Leptin regulation of the mesoaccumbens dopamine pathway. Neuron. 2006;51:811–22.
55. Bouret SG, Draper SJ, Simerly RB. Trophic action of leptin on hypothalamic neurons that regulate feeding. Science. 2004;304:108–10.
56. Pinto S, Roseberry AG, Liu H, et al. Rapid rewiring of arcuate nucleus feeding circuits by leptin. Science. 2004;304:110–5.
57. Krugel U, Schraft T, Kittner H, et al. Basal and feeding-evoked dopamine release in the rat nucleus accumbens is depressed by leptin. Eur J Pharmacol. 2003;482:185–7.
58. Farooqi IS, Bullmore E, Keogh J, et al. Leptin regulates striatal regions and human eating behavior. Science. 2007;317:1355.
59. Abizaid A, Liu ZW, Andrews ZB, et al. Ghrelin modulates the activity and synaptic input organization of midbrain dopamine neurons while promoting appetite. J Clin Invest. 2006;116:3229–39.
60. Lammel S, Hetzel A, Hackel O, et al. Unique properties of mesoprefrontal neurons within a dual mesocorticolimbic dopamine system. Neuron. 2008;57: 760–73.
61. Figlewicz DP, Szot P, Chavez M, et al. Intraventricular insulin increases dopamine transporter mRNA in rat VTA/substantia nigra. Brain Res. 1994;644:331–4.
62. Wang GJ, Volkow ND, Logan J, et al. Brain dopamine and obesity. Lancet. 2001;357:354–7.
63. Nestler EJ. Is there a common molecular pathway for addiction? Nat Neurosci. 2005;8:1445–9.
64. Zhang M, Gosnell BA, Kelley AE. Intake of high-fat food is selectively enhanced by mu opioid receptor stimulation within the nucleus accumbens. J Pharmacol Exp Ther. 1998;285:908–14.
65. Stratford TR, Kelley AE, Simansky KJ. Blockade of GABAA receptors in the medial ventral pallidum elicits feeding in satiated rats. Brain Res. 1999;825: 199–203.
66. Kelley AE, Swanson CJ. Feeding induced by blockade of AMPA and kainate receptors within the ventral striatum: a microinfusion mapping study. Behav Brain Res. 1997;89:107–13.
67. Pecina S, Cagniard B, Berridge KC, et al. Hyperdopaminergic mutant mice have higher "wanting" but not "liking" for sweet rewards. J Neurosci. 2003;23:9395–402.
68. Robinson TE, Berridge KC. Incentive-sensitization and addiction. Addiction. 2001;96:103–14.
69. Zhang M, Balmadrid C, Kelley AE. Nucleus accumbens opioid, GABaergic, and dopaminergic modulation of palatable food motivation: contrasting effects revealed by a progressive ratio study in the rat. Behav Neurosci. 2003;117:202–11.
70. Pecina S, Berridge KC. Hedonic hot spot in nucleus accumbens shell: where do mu-opioids cause increased hedonic impact of sweetness? J Neurosci. 2005;25:11777–86.
71. Kelley AE, Baldo BA, Pratt WE, et al. Corticostriatal-hypothalamic circuitry and food motivation: integration of energy, action and reward. Physiol Behav. 2005;86:773–95.
72. Taha SA, Fields HL. Encoding of palatability and appetitive behaviors by distinct neuronal populations in the nucleus accumbens. J Neurosci. 2005;25:1193–202.
73. Taha SA, Fields HL. Inhibitions of nucleus accumbens neurons encode a gating signal for reward-directed behavior. J Neurosci. 2006;26:217–22.
74. Clegg DJ, Air AL, Woods SC, et al. Eating elicited by orexin-A, but not melanin-concentrating hormone, is opioid mediated. Endocrinology. 2002;143: 2995–3000.
75. Sears RM, Liu RJ, Narayanan NS, et al. Regulation of nucleus accumbens activity by the hypothalamic neuropeptide melanin-concentrating hormone. J Neurosci. 2010;30:8263–73.
76. de Lecea L, Kilduff, TS, Peyron C, et al. The hypocretins: hypothalamus-specific peptides with neuroexcitatory activity. Proc Natl Acad Sci USA. 1998;95:322–7.

77. Date Y, Ueta Y, Yamashita Y, et al. Orexins, orexigenic hypothalamic peptides, interact with autonomic, neuroendocrine, and neuroregulatory systems. Proc Natl Acad Sci USA. 1999;96:748–53.
78. Broberger C, de Lecea L, Sutcliffe JG, et al. Hypocretin/orexin- and melanin-concentrating hormone-expressing cells form distinct populations in the rodent lateral hypothalamus: relationship to the neuropeptide Y and agouti gene-related protein systems. J Comp Neurol. 1998;402:460–74.
79. van den Pol AN, Gao XB, Obrietan K, et al. Presynaptic and postsynaptic actions and modulation of neuroendocrine neurons by a new hypothalamic peptide, hypocretin/orexin. J Neurosci. 1998;18:7962–71.
80. Baldo BA, Daniel RA, Berridge CW, et al. Overlapping distributions of orexin/hypocretin- and dopamine-beta-hydroxylase immunoreactive fibers in rat brain regions mediating arousal, motivation, and stress. J Comp Neurol. 2003;464:220–37.
81. Fadel J, Deutch AY. Anatomical substrates of orexin-dopamine interactions: lateral hypothalamic projections to the ventral tegmental area. Neuroscience. 2002;111:379–87.
82. Nakamura T, Uramur K, Nambu T, et al. Orexin-induced hyperlocomotion and stereotypy are mediated by the dopaminergic system. Brain Res. 2000;873:181–7.
83. Balcita-Pedicino JJ, Sesack SR. Orexin axons in the rat ventral tegmental area synapse infrequently onto dopamine and gamma-aminobutyric acid neurons. J Comp Neurol. 2007;503:668–84.
84. Estabrooke IV, McCarthy MT, Ko E, et al. Fos expression in orexin neurons varies with behavioral state. J Neurosci. 2001;21:1656–62.
85. Zhang GC, Mao LM, Liu XY, et al. Long-lasting up-regulation of orexin receptor type 2 protein levels in the rat nucleus accumbens after chronic cocaine administration. J Neurochem. 2007;103:400–7.
86. Zhou Y, Cui CL, Schlussman SD, et al. Effects of cocaine place conditioning, chronic escalating-dose "binge" pattern cocaine administration and acute withdrawal on orexin/hypocretin and preprodynorphin gene expressions in lateral hypothalamus of Fischer and Sprague-Dawley rats. Neuroscience. 2008;153:1225–34.
87. Zhou Y, Bendor J, Hofmann L, et al. Mu opioid receptor and orexin/hypocretin mRNA levels in the lateral hypothalamus and striatum are enhanced by morphine withdrawal. J Endocrinol. 2006;191:137–45.
88. Morshedi MM, Meredith GE. Repeated amphetamine administration induces Fos in prefrontal cortical neurons that project to the lateral hypothalamus but not the nucleus accumbens or basolateral amygdala. Psychopharmacology. 2008;197:179–89.
89. Georgescu D, Zachariou V, Barrot M, et al. Involvement of the lateral hypothalamic peptide orexin in morphine dependence and withdrawal. J Neurosci. 2003;23:3106–11.
90. Sharf R, Sarhan M, DiLeone RJ. Orexin mediates the expression of precipitated morphine withdrawal and concurrent activation of the nucleus accumbens shell. Biol Psychiatry. 2008;64:175–83.
91. Korotkova TM, Sergeeva OA, Eriksson KS, et al. Excitation of ventral tegmental area dopaminergic and nondopaminergic neurons by orexins/hypocretins. J Neurosci. 2003;23:7–11.
92. Vittoz N, Schmeichel B, Berridge C. Hypocretin/orexin preferentially activates caudomedial ventral tegmental area dopamine neurons. Eur J Neurosci. 2008;28:1629–40.
93. Vittoz NM, Berridge CW. Hypocretin/orexin selectively increases dopamine efflux within the prefrontal cortex: involvement of the ventral tegmental area. Neuropsychopharmacology. 2006;31:384–95.
94. Narita M, Nagumo Y, Hashimoto S, et al. Direct involvement of orexinergic systems in the activation of the mesolimbic dopamine pathway and related behaviors induced by morphine. J Neurosci. 2006;26:398–495.
95. Borgland SL, Taha SA, Sarti F, et al. Orexin A in the VTA is critical for the induction of synaptic plasticity and behavioral sensitization to cocaine. Neuron. 2006;49:589–601.
96. Xia J, Chen X, Song C, et al. Postsynaptic excitation of prefrontal cortical pyramidal neurons by hypocretin-1/orexin A through the inhibition of potassium currents. J Neurosci Res. 2005;82:729–36.
97. Martin G, Febre V, Siggins GR, et al. Interaction of the hypocretins with neurotransmitters in the nucleus accumbens. Regul Pept. 2002;104:111–7.
98. Mukai K, Kim J, Nakajima K, et al. Electrophysiological effects of orexin/hypocretin on nucleus accumbens shell neurons in rats: an in vitro study. Peptides. 2009;30:1487–96.
99. Robinson TE, Becker JB, Priesty SK. Long-term facilitation of amphetamine-induced rotational behavior and striatal dopamine release produced by a single exposure to amphetamine: sex differences. Brain Res. 1982;253:231–41.
100. Sharf R, Guarnieri DJ, Taylor JR, et al. Orexin mediates morphine place preference, but not morphine-induced hyperactivity or sensitization. Brain Res. 2010;1317:24–32.
101. Bardo MT, Valone JM, Bevins RA. Locomotion and conditioned place preference produced by acute intravenous amphetamine: role of dopamine receptors and individual differences in amphetamine self-administration. Psychopharmacology (Berl). 1999;143:39–46.
102. Deroche V, Le Moal M, Piazza PV. Cocaine self-administration increases the incentive motivational properties of the drug in rats. Eur J Neurosci. 1999;11:2731–6.
103. Harris GC, Wimmer M, Aston-Jones G. A role for lateral hypothalamic orexin neurons in reward seeking. Nature. 2005;437:556–9.
104. Harris GC, Wimmer M, Randall-Thompson JF, et al. Lateral hypothalamic orexin neurons are critically involved in learning to associate an environment with morphine reward. Behav Brain Res. 2007;183:43–51.
105. Hollander JA, Lu Q, Cameron MD, et al. Insular hypocretin transmission regulates nicotine reward. Proc Natl Acad Sci USA. 2008;105:19480–5.

106. Lawrence AJ, Cowen MS, Yang HJ, et al. The orexin system regulates alcohol-seeking in rats. Br J Pharmacol. 2006;148:752–9.
107. Boutrel B, Kenny PJ, Specio SE, et al. Role for hypocretin in mediating stress-induced reinstatement of cocaine-seeking behavior. Proc Natl Acad Sci USA. 2005;102:19168–73.
108. Aston-Jones G, Smith RJ, Moorman DE, et al. Role of lateral hypothalamic orexin neurons in reward processing and addiction. Neuropharmacology. 2009;56:112–21.
109. Wang B, You ZB, Wise RA. Reinstatement of cocaine seeking by hypocretin (orexin) in the ventral tegmental area: independence from the local corticotropin-releasing factor network. Biol Psychiatry. 2009;65:857–62.
110. Harris GC, Aston-Jones G. Arousal and reward: a dichotomy in orexin function. Trends Neurosci. 2006;29:571–7.
111. Olds J, Milner P. Positive reinforcement produced by electrical stimulation of septal area and other regions of rat brain. J Comp Physiol Psychol. 1954;47:419–27.
112. Lin L, Faraco J, Li R, et al. The sleep disorder canine narcolepsy is caused by a mutation in the hypocretin (orexin) receptor 2 gene. Cell. 1999;98:365–76.
113. Chemelli RM, Willie JT, Sinton CM, et al. Narcolepsy in orexin knockout mice: molecular genetics of sleep regulation. Cell. 1999;98:437–51.
114. Zheng H, Patterson LM, Berthoud HR. Orexin-A projections to the caudal medulla and orexin-induced c-Fos expression, food intake, and autonomic function. J Comp Neurol. 2005;485:127–42.
115. Sharf R, Sarhan M, Brayton CE, et al. Orexin signaling via the orexin 1 receptor mediates operant responding for food reinforcement. Biol Psychiatry. 2010;67:753–60.

Plasticity of Brain Feeding Circuits in Response to Food

5

Tamas Horvath

Abstract

Feeding and energy expenditure are regulated by neuronal circuits of the brain. Feeding, by definition, occurs in most mammals during waking mandating a tight relationship between systems regulating sleep/wake cycles and energy metabolism. This is accomplished not only by neuronal interactions between brain regions responsible for these functions, but also by sharing intracellular and intercellular signaling modalities. Peripheral hormones associated with energy metabolism not only affect the brain structures that have been classically associated with endocrine and autonomic functions, but also alter the function of higher brain regions, including the hippocampus and cerebral cortex. A better understanding of the mechanism of feeding behavior and energy expenditure associated with brain structures will also enhance our ability to combat disorders such as diabetes and obesity, which are among the most prevalent medical problems of both developed and developing societies.

Introduction

Feeding, and energy metabolism in general, is the corner stone of individual and, consequently, species survival. In terms of its significance, feeding equals, and by default is upstream, to reproductive behavior. It could be argued that the purpose behind the emergence and evolution of the central nervous system was to promote the most effective management of metabolism in support of survival. Both common sense and the experimental data are in agreement with this basic assertion [1]. Perhaps then, even higher brain regions, such as the archio- and neocortex, emerged under environmental pressures to support behaviors that more effectively deal with available energy resources. Despite this, most of the attention regarding the understanding of feeding behavior and brain regulation of energy homeostasis has focused on more primitive regions of the brain such as the hypothalamus. Through the description of the various hypothalamic regions, synaptic plasticity will be discussed as a potentially fundamental component of brain regulation of energy metabolism.

T. Horvath, DVM, PhD (✉)
Section of Comparative Medicine, Yale University School of Medicine, 300 George Street, Suite 203, New Haven, CT 06511, USA
e-mail: tamas.horvath@yale.edu

P.J. Shiromani et al. (eds.), *Sleep Loss and Obesity: Intersecting Epidemics*,
DOI 10.1007/978-1-4614-3492-4_5, © Springer Science+Business Media, LLC 2012

Over the last century, a vast number of experimental observations have been collected, particularly within the last decade due to the cloning of the product of the *ob* gene (its mutation causes obesity in rodents), leptin. This resulted in a wave of discovery of other metabolic hormones, neuropeptides and signaling pathways, and established their relationship to key neuronal structures in the hypothalamus [2–5]. Recently, the focus has shifted to studies of more complex signal integrations and neuronal adaptations that actually initiate the changes in ingestive behavior and metabolism. This review will follow the history behind this line of research and highlight some important new advances.

The Link Between the Periphery and the Brain

Brain regions implicated in the regulation of long-term energy balance, by definition, need to be able to monitor the amount of available fuel in the body and, then, use this knowledge to modulate energy intake and expenditure. What might be the signal(s) that conveys such information to the brain? Various hypotheses speculated about the nature of these signal(s). One such proposal, the "glucostatic hypothesis," [6] postulated that small changes in plasma glucose levels trigger meal initiation or termination. This model correlates well with the initiation or termination of feeding per se (the "depletion–repletion" model of energy intake); however, it does not provide information about or a mechanism by which to determine the amount of food to be consumed, and it correlates poorly with energy expenditure [7]. Another paradigm that linked food intake to the amount of stored energy (fat mass) in the body is the "lipostatic model" that hypothesized that signal(s) proportional to the amount of fat in the body modulates the amount of food eaten at each meal to maintain overall energy balance [8]. More recent developments have supported this hypothesis that food intake is controlled by a lipostatic system to maintain energy homeostasis.

Two sets of experiments appear to have been critical in identifying the specific brain structures that are directly involved in energy homeostasis and in linking the "lipostatic" signal to these brain regions. First, various lesion studies in which the hypothalamic ventromedial (VMH), paraventricular (PVN), or dorsomedial (DMH) nuclei were destroyed induced hyperphagia and obesity [9–13], whereas lesions in the lateral hypothalamus (LH) led to hypophagia [14]. This led to the proposal of a dual center model that identified the VMH as the "satiety center", while the LH was deemed the "hunger center" [15]. In retrospect to our current knowledge of specific hypothalamic neuronal populations, these lesion studies are strikingly precise in distinguishing between subregions that contain circuits that either promote or suppress feeding, and has since been used as a roadmap for the studies of neurobiology on feeding and metabolism.

Second, the initial parabiosis studies on hypothalamic lesioned rats concluded the existence of humoral signal(s) in relation to an animal's lipostatics [16]. In these studies, two live animals, one with a lesion in the VMH, were joined by surgical means, which allowed humoral factors to pass from one animal to the other. As obesity developed in the lesioned rat, its partner became hypophagic and lost weight, suggesting that a signal, in proportion to the amount of fat mass, is highly potent in inhibiting food intake. Additional parabiosis studies on genetically obese mutant mice, *ob/ob* and *db/db* [17, 18], concluded that *ob/ob* mice lack this lipostatic signal, whereas *db/db* mice are insensitive to it. These hypotheses were later confirmed by the discoveries of the *ob* gene that encodes leptin and the *db* gene that encodes leptin receptor.

The discovery of leptin [19] and its receptor soon led to the recognition that all the hypothalamic nuclei associated with energy regulation, i.e., the VMH, DMH, PVN, and LH, are the regions where leptin receptors are highly expressed [20–22]. This also promoted a new wave of gene discoveries, which identified novel neuropeptides, their receptors and transcription factors that are crucial in mediating leptin function. This then led to the detection of the hypothalamic melanocortin system as the key neuronal system involved in the control of energy balance by leptin signaling. Other metabolic signals including insulin, ghrelin, estrogen, prolactin, glucocorticoids, resistin, and interleukins, to name a few, as well as nutrients such as glucose and free fatty acids [23–25] also target the hypothalamic melanocortin system to affect food intake and energy balance presumably in concert with leptin function [3, 7, 26–28]. All of these peripheral signals have been found to trigger rapid rewiring of neuronal circuits in the hypothalamus and beyond [29–38].

The Arcuate Nucleus Melanocortin System

The hypothalamic arcuate nucleus (ARC) melanocortin system has been considered as the most successful model in explaining the neuronal control of long-term energy balance. The neurons within the ARC are located in an anatomically strategic place, in that they are in close proximity to fenestrated capillaries at the base of the hypothalamus, giving them access to various humoral signals that are restricted to other portions of the brain [39, 40]. These neurons are innervated by axons containing almost all major neurotransmitters, and synthesize receptors for most metabolic hormones [3, 39, 41, 42]. ARC cells appear to respond rapidly to nutritional signals such as glucose and fatty acids as well [4, 43].

Two subsets of neurons have been identified within the ARC that have opposite effects on feeding; both contain the inhibitory neurotransmitter GABA [44, 45]. One of these groups of neurons expresses proopiomelanocortin (POMC) and cocaine-amphetamine-regulated transcript (CART); when stimulated, they produce anorectic effects [46–48]. The POMC precursor is cleaved into melanocyte-stimulating hormones (α-, β-, and γ-MSH), β-endorphin and adrenocorticotrophic hormone (ACTH) [48]. Of these, α-and β-MSH reduce food intake and body weight, and increase energy expenditure in animals and humans [49–51]. Both α and β-MSH act on melanocortin receptor subtype 3 and 4 (MC3/4R) [52] that are particularly abundant in the ARC, PVN, LH, DMH, and anterior hypothalamic regions [53].

The second group of cells in which increased activity leads to orexigenic responses contains neuropeptide Y (NPY) and the agouti-gene-related transcript (AgRP) [54, 55]. NPY potently stimulates food intake and reduces energy expenditure [56, 57]. It is found throughout the central nervous system [58], with its highest concentration in the ARC [59]. In genetically obese animals, e.g., *ob/ob* and *db/db* mice, as well as animals in a negative energy state, e.g., fasted or lactating animals, the ARC NPY mRNA and protein content are elevated [60–63]. AgRP is also orexigenic. It acts as a natural antagonist of MC3/4R, thereby, reducing the anorectic effect of α-MSH [54, 64–66]. An ectopic expression of Agouti, an AgRP-like peptide, results in an obese phenotype of the *yA* agouti mouse [54].

Substantial neuroanatomical and electrophysiological data have been generated, in particular through the use of fluorescent reporter genes that are selectively expressed in specific neurons, thus, allowing the visualization of hypothalamic neuronal subpopulations so that they can be studied in the context of feeding circuits while the cells are alive. By employing this method, results demonstrate that NPY neurons contact nearby POMC cells and inhibit them through the release of GABA [67]. The unidirectional NPY to POMC input [68] is an interesting aspect of the anatomical basis of energy regulation because it may represent a wiring blueprint that favors the tonic inhibition of satiety signals, a feature that may have been derived through natural selection to not only promote feeding, but also overfeeding in situations when food is available in excess [69].

The POMC and NPY neurons of the ARC function through broad projections to various brain regions that include the PVN, LH, and perifornical hypothalamic region, all of which contain substantial numbers of MC3/4R [53, 70]. The projection from the ARC to PVN is important for the regulation of neurons that produce corticotropin- and thyrotropin-releasing hormones (CRH and TRH, respectively) and for the modulation of sympathetic activity, both of which are important physiological mechanisms in energy metabolism [71–73]. A study using a *cre/loxP* genetic approach showed that mice with mutations to the MC4R gene (loxTB *mc4*) weighed and ate less if the MC4R gene was restored selectively in the PVN via viral or genetic manipulations [74]. However, the restoration of the MC4R in the PVN was not sufficient to bring energy expenditure levels to those of controls, suggesting that energy expenditure is regulated by melanocortins elsewhere in the hypothalamus or through receptors other than MC4R [74].

The intact melanocortin neuronal circuitry is mandatory for acute regulation of feeding [75–78]. The studies originated from the fact that neither NPY nor AgRP single gene knockout mice [79] nor NPY/AgRP double knockout mice [80] exhibited the expected hypophagia.

The Lateral Hypothalamic Arousal/Feeding System

Two subsets of neuron populations have been identified within the LH: one group contains hypocretin (orexin), a peptide implicated in arousal and feeding, the other, melanin-concentrating hormone (MCH), another potent stimulator of food intake. Both hypocretin and MCH neurons have a wide projection field and modulate a variety of behavioral responses related to learning, memory, emotion, motivation, and motor responses in association with changes in the energy state [81–84]. Although the projections of hypocretin and MCH neurons exhibit significant overlap, their overall effects and actual targets are quite different [81, 85]. The activity of hypocretin and MCH neurons is regulated by numerous hormones that include leptin and ghrelin, as well as by practically every neurotransmitter system [86–90]. Within the LH, hypocretin and MCH neurons have reciprocal connections with each other, and with nearby neurons [91]. Electrophysiological studies in brain slices or isolated neurons indicated that hypocretin, in general, has a stimulatory effect on LH neurons including MCH neurons [92, 93], whereas MCH depresses the synaptic activity of glutamate and GABA neurons from the rat LH [94]. It is not clear whether such electrical interactions between hypocretin and MCH neurons, at the intensities and dynamics observed, are relevant to feeding and long-term homeostasis, or if they are more related to arousal behavior.

Hypocretin and MCH neurons are capable of directly integrating metabolic signals to modulate energy balance, and they do so, more or less, independent of each other, despite their proximate location and physiological interaction in the LH. For example, hypocretin neurons are rapidly activated by fasting in rodents and nonhuman primates [95] and exhibit leptin-dependent synaptic plasticity during fasting [96]. Hypocretin mRNA levels in the LH are upregulated upon fasting as well [97]. Notably, all of the hypothalamic neurons that are activated by fasting receive strong hypocretin input. These observations, together with the earlier demonstration of a massive hypocretin input to the ARC, particularly to the NPY neurons [98], as well as the dependence of hypocretin function on NPY signaling (in particular on the NPY Y1 and Y5 receptors) [99, 100], and a synergistic action between NPY and hypocretin (at low concentrations) to induce feeding [101], argue that hypocretin neurons may be upstream to the NPY system with regard to feeding and metabolism. Mice with deletions of the hypocretin gene are hypophagic, but maintain normal growth curves, suggesting that they possess a reduced metabolic rate [102]. This, in association with the critical role of hypocretin neurons to promote arousal [103] through direct projections to the brain stem including the locus coeruleus [99], makes it likely that the hypocretin neurons provide a link between obesity and insomnia. Interestingly, NPY appears to inhibit hypocretin neurons to exert tonic depression of the hypothalamic arousal system [104].

MCH, on the other hand, shows no or little interaction with NPY or hypocretin in inducing food intake when injected together into the third ventricle of the rat [102]. These observations provide a physiological component concomitant to the preexisting morphological data that shows that the strength of MCH projections to the ARC is limited compared to that of the hypocretin neurons. Thus, the action of MCH on feeding is likely independent of NPY and hypocretin action. However, like NPY, MCH does exhibit characteristics typical of orexigenic genes: its mRNA levels are increased in obese mutant animals and fasting further increases its expression in both normal and obese animals [105]; and it has a potent orexigenic effect. Targeted deletion of the MCH gene resulted in a phenotype of hypophagia and leanness with an inappropriately high metabolic rate [106], suggesting that MCH is a typical "thrift gene" that increases energy intake and reduces energy expenditure simultaneously. Consistent with this idea, MCH suppresses thyroid-stimulating hormone (TSH) release [107].

In contrast to hypocretin that acts through NPY, MCH appears to compete with the action of α-MSH to produce its effect (a mechanism conserved from skin color regulation in fish to hypothalamic control of energy balance in mammals), such that MCH administration increases feeding, while α-MSH acts to decrease it. When the peptides are administered together, depending on the relative dose, one antagonizes the action of the other [82]. Recently, a mouse model of MCH neuron ablation was generated by expressing a toxin gene, ataxin-3, targeted at MCH neurons [103]. Mice that express this gene have a chronic loss of MCH neurons. Interestingly, the phenotype of these mice highly resembles that of mice lacking only the MCH gene, that is they exhibit reduced food intake and increased energy expenditure. Moreover, the ablation of MCH neurons in mice with an *ob/ob* mutant background resulted in improved obesity and glucose tolerance. Thus, these results suggest that the function of MCH cells in energy regulation may be limited to the MCH system itself, but not to other aspects of the cells such as their classic neurotransmitter function and/or their synaptic plasticity, which are distinct from NPY cells (see below).

Circadian Input to the Endocrine Hypothalamus

The suprachiasmatic nucleus (SCN), also considered as the master circadian clock of the body, and the DMH also express receptors for leptin and ghrelin. They express high levels of UCP2 suggesting that these hormones directly target the SCN and DMH to generate additional behavioral and physiological responses that complement those mediated by cells in the ARC [87, 109]. The SCN and the extended visual system send direct and indirect projections to hypothalamic neuroendocrine cells [110, 111] (for details regarding the SCN, see Chaps. 1 and 2). The DMH appears to modulate glucocorticoid secretion, body temperature, arousal, and circadian rhythms of locomotor activity [112]. It receives inputs from cells in the ARC and regions in the brainstem associated with feeding [113]. Lesions restricted to the DMH typically result in hypophagia, although animals can still maintain their body composition [113]. In a recent study, the DMH was found to be critical for the entrainment of circadian rhythms to feeding schedules [114]. The DMH of animals with restricted access to food (4 h/day) exhibited an increased expression of c-Fos, indicating an increased cellular activation at a time when the food was presented compared to animals that had free access to food throughout the day. Ibotenic lesions of the DMH resulted in reduced levels of locomotor activity and decreased food intake. When lesioned rats were placed in the restricted feeding schedule, they showed less preprandial increases in food anticipatory locomotor activity than those of sham-operated animals. DMH lesions also blocked the rise in body temperature that is entrained to the timing of food presentation [114]. The nature of the cells within the DMH remains obscure. A number of these cells are glutamatergic and project to the PVN and preoptic area; both are involved in the circadian regulation of corticosteroid secretion and body temperature [113, 115]. Projections from the DMH to the LH and ventrolateral preoptic area may be involved in the control of sleep and arousal and may be related to the enhanced activity of animals in restricted feeding schedules [116].

Neuronal Responses to Peripheral Hormones Nutritional States

A large body of data has emerged regarding the intracellular signaling of various peripheral hormones within the hypothalamus. In addition, transcriptional events have been described to support overall alterations in cellular capabilities. However, Cajal's neuronal doctrine predicts that functional outcomes of circuit alteration should stem from altered firing and/or connectivity (synaptology) of neuronal networks. Mounting evidence suggests that peripheral hormones accomplish just this as well.

Acute Neuronal Responses to Peripheral Hormones

With the emergence of transgenic technology that allows selective labeling of particular neuropeptide-producing cell groups, it has become easier to study the acute responses of subpopulations of ARC neurons to peripheral hormones, such as leptin and ghrelin. These studies revealed that the two critical components of the melanocortin system, the orexigenic NPY/AgRP-producing cells and the anorexigenic POMC/α-MSH-expressing neurons respond to peripheral hormones in an acute fashion in slice preparations. It was shown that firing of POMC cells is enhanced by leptin via both pre- and postsynaptic modes of action [67], while the firing frequency of orexigenic NPY/AgRP neurons is diminished by leptin [117]. On the other hand, the gut-derived, appetite-stimulating hormone, ghrelin, was observed to enhance the firing rate of NPY/AgRP neurons via a direct mechanism while it diminished the frequency of action potential of POMC cells predominantly by a presynatpic mode of action [118]. Leptin's acute action on neuronal function was tied to a nonspecific cation channel [119], whereas the mechanism of action of ghrelin is yet to be determined, although it may require GHSR signaling [33]. The pancreas-derived hormone, insulin, was also shown to affect neuronal firing in the ARC, an event that appears to be mediated by ATP-sensitive potassium channels [120]. Of course, considering the limitations of slice electrophysiology, it is a reasonable yet unresolved question whether any of these peripheral signals are present physiologically at high enough quantities to alter action potential generation in the above described circuitry in vivo.

Synaptic Plasticity Driven by Peripheral Metabolic Hormones

Cajal's neuronal doctrine also predicts that the output of a neuronal population is greatly influenced by its input organization. Thus, the inquiry into whether various metabolic states might correlate with different organizations of synaptic input (synaptic plasticity) on the melanocortin cells was a logical line of pursuit. This concept raised the possibility that metabolic signals, leptin and ghrelin in particular, may have acute effects on synaptic plasticity within the appetite center. Indeed, the hypothalamus has been known to retain various forms of plasticity throughout life, and, immature synapses can frequently be found in the adult hypothalamus. For example, the magnocellular system shows plasticity during changes in water homeostasis [121–123]; in the ARC, the input organization of unidentified neuronal systems exhibited changes in response to a varying gonadal steroid milieu [124, 125]; and the input organization of the perikarya of luteinizing hormone-releasing hormone neurons varies during changes in the gonadal steroid milieu [126] or in photoperiod lengths [127]. Nevertheless, such synaptic plasticity was not considered as a potentially important aspect in the regulation of daily energy control. Recent observations [29–31, 35, 36, 96], however, now clearly indciate that it may be a regulatory component in the hypothalamic control of energy homeostasis. First, results on leptin replacement in *ob/ob* mice indicated [29] that synaptic rearrangements of feeding circuits is part of an ongoing general phenomena. Further support came from results that showed robust effects of peripheral ghrelin injections on the input organization of POMC neurons of mice leading to a wiring different from that induced by leptin [29]. Additionally, leptin was found to robustly regulate the synaptic organization of lateral hypothalamic hypocretin/orexin neurons [96], and, ghrelin was shown to have a rapid and robust effect on synapse formation in the hippocampus [34] and ventral tegmental area [33]. In these latter studies, ghrelin was found to promote the propagation of long-term potentiation in the hippocampal formation enhancing learning and memory consolidation [34], and it caused the release of dopamine in the nucleus accumbens associated with synaptic changes in the ventral tegmentum [33]. From a methodological perspective, it needs to be emphasized that conclusions made on the presence of synaptic plasticity relies on the combination of light and electron microscopical assays and slice electrophysiology.

An important issue remains unresolved, however, and that is whether these synaptic alterations have an impact on neuronal activity and, consequently, on the metabolic phenotype. While this remains to be proven, a task that is still daunting even for long-term potentiation and long-term depression [128], it is important to recognize that these changes occur both preceding and concomitant to the varying behavioral and endocrine outputs [29, 32–34]. Even if it does not directly affect action potentials, the changing input organization of NPY and POMC perikarya may have an impact on the "set point" of these cell populations under various circumstances [31, 32]. The intracellular signaling modality that brings about these rapid changes as well as the parent cells of origin of the altered inputs in wiring will need further investigation. Another fundamental question regarding these synaptic changes is whether ghrelin/leptin signaling is required on the pre- and/or postsynaptic sites? Based on the association between ghrelin-induced synaptic plasticity and the dependence of ghrelin's electrophysiological action in the VTA [33] and hypothalamus [29] on presynaptic mechanisms, it is tempting to speculate that these morphological changes represent activity-dependent synaptic plasticities. One important conclusion, however, of these studies is that synaptic plasticity is inherent to neuronal circuits that govern appetite. Because this network of neurons is becoming well characterized showing a direct relationship between a small, identifiable subset of neurons and a primitive behavior independent of conditioning (for example, learning and memory consolidation paradigms), and feeding [75, 76], these sets of neurons may become more attractive for an experimental paradigm that could directly prove the causal relationship between synaptic plasticity and behavior [128].

Fuel Sensing, Energy Balance and Synaptic Plasticity

Hypothalamic neurons, particularly those in the ARC and VMH, modulate their activity directly in response to fluctuations in the levels of body metabolic fuels such as glucose and free fatty acids [24, 129–138]. Brain cells, like other cells, have developed mechanisms that monitor energy availability in the extracellular space. One of these mechanisms includes an increase in adenosine monophosphate kinase (AMPK) activity in response to a decrease in the ratio of AMP/adenosine triphosphate (AMP/ATP) [24]. The activation of AMPK favors cellular responses generated to increase ATP levels including increases in the synthesis and uptake of glucose and fatty acids by most cells [24]. Recent studies suggest that hypothalamic neurons have a similar nutrient sensing mechanism [24, 36, 131–133]. Fasting or treatment with 2-deoxy-D-glucose (2-DG) increases the activity of AMPK in the hypothalamus [139].

In addition to sensing glucose availability, hypothalamic cells are also sensitive to circulating levels of free fatty acids and their response may be mediated by AMPK [140, 141]. Free fatty acids diffuse into hypothalamic neurons where they are esterified and transferred into the mitochondria for oxidation, a process that ultimately results in increased feeding [36, 43, 142]. The transfer of esterified fatty acids into the mitochondria is mediated by the activity of carnitine-dependent acyltransferases 1 and 2 (CPT1 and 2) [23, 43]. Interestingly, malonyl-CoA, the activity of which is increased by glucose utilization, decreases the activity of CPT1 and thus the oxidation of fatty acids [23]. Furthermore, increases in AMPK activity lead to decreased formation of malonyl-CoA and ultimately increased fatty acid oxidation [23].

The activation of AMPK may also gate hypothalamic responses to peripheral hormones such as leptin and ghrelin, given that leptin decreases and ghrelin increases the activity of AMPK in the hypothalamus [36, 143, 144]. Like ghrelin, AgRP, or mutations in the MC4 receptor also result in increased hypothalamic AMPK activity, whereas

MTII, an α-MSH agonist decreases it, supporting the idea that the melanocortin system is critical for this enzyme to have its downstream effects in the control of energy balance [132].

Cells within the LH also change their activity in response to the fluctuations of extracellular ATP. For example, the local application of ATP depolarized hypocretin neurons, increasing their firing frequency [145]. More studies are needed to determine if this effect is mediated by mechanisms of fuel sensing similar to those seen in the ARC.

The relationship between intracellular energy sensing in hypothalamic neurons and the cascade of events that follows energy substrate oxidation is still poorly understood, but it was shown that mitochondrial uncoupling proteins such as UCP2 may play a role in the generation of cellular responses within the ARC that ultimately lead to increased activity of the melanocortin system [31, 98]. It has been suggested that UCP2 may modulate the efficiency of metabolic processes in hypothalamic cells, increase neurotransmission and modulate synaptic remodeling [146, 147]. Two lines of results show that fuel sensing and consequent neuronal responses may hinge on the different fuel preferences for neuronal firing of the ARC NPY/AgRP and POMC neurons and they determine synaptic plasticity. Several studies have shown that POMC neurons increase their firing when glucose levels rise [148–150]. While most of these studies emphasized the mechanism of glucose sensing by POMC neurons because their consequent effect of liver glucose homeostasis and insulin action, by default, all of these results implicate glucose as the main driver of POMC neuronal firing. This is entirely consistent with the putative role of POMC cells as satiety signaling neurons, as at a time of satiety, circulating glucose levels are elevated and POMC neurons have increased firing [151]. In contrast, during negative energy balance, glucose levels are lower and NPY/AgRP neurons have increased firing [151].

Our recent results [36] revealed the following chain of intracellular events that contributes to the appropriate neuronal responses to ghrelin on the part of ARC NPY/AgRP neurons during negative energy balance: ghrelin induces a rapid increase of NPY/AgRP neuronal firing via activation of GHSR. GHSR activation, likely via ATP utilization during action potential generation, results in AMPK activation. AMPK activation suppresses ACC activity eliminating the inhibitory effect of malonyl-CoA on CPT1 activity. CPT1 activation enhances long chain fatty acid entry into the mitochondrial matrix that, in turn, undergoes beta oxidation. Fatty acid beta oxidation promotes generation of ROS, which together with fatty acids promote UCP2 transcription and activity. UCP2 activity neutralizes ROS thereby allowing continuous CPT1-promoted fatty acid beta oxidation that enables continuous support of the bioenergetic needs of sustained firing of NPY/AgRP cells. Sustained firing of NPY/AgRP neurons, the efferents of which are GABAergic onto POMC neurons, results in activity-dependent synaptic plasticity promoting an organization of increased inhibitory input onto POMC neurons. Ghrelin indirectly suppresses POMC neuronal firing through elevated NPY/AgRP neuronal activity and thereby promotes feeding. Our results also indicate that fatty acids, which are elevated in the circulation concurrently with ghrelin during fasting, are acutely increased in the hypothalamus in response to ghrelin administration so they can serve as continuous fuel supply for NPY/AgRP neurons during negative energy balance. Engagement of neuronal UCP2 was found to be critical for synaptic plasticity of diverse circuits in the brain [36, 152], and the expression levels of UCP2 was positively correlated with longevity of mice and rats [153]. Thus, it is reasonable to argue that differential fuel utilization by various subsets of neurons contributes not only to the regulation of energy homeostasis, but to healthy lifespan as well.

Conclusion

Feeding and energy expenditure is regulated by neuronal circuits of the brain. Feeding, by definition, occurs in most mammals during waking mandating a tight relationship between systems regulating sleep/wake cycles and energy metabolism. This is accomplished not only by neuronal interactions between brain regions responsible for these functions, but also by sharing intracellular and intercellular signaling modalities. Peripheral hormones associated with energy metabolism not only affect brain structures that have been classically associated with endocrine and autonomic functions, but also alter the function of higher brain regions, including the hippocampus and cerebral cortex. A better understanding of the mechanism of feeding behavior and energy expenditure associated with brain structures will also enhance our ability to combat disorders such as diabetes and obesity, which are among the most prevalent medical problems of both developed and developing societies.

References

1. Clifton PG, Vickers SP, Somerville EM. Little and often: ingestive behavior patterns following hippocampal lesions in rats. Behav Neurosci. 1998;112:502–11.
2. Friedman JM, Halaas JL. Leptin and the regulation of body weight in mammals. Nature. 1998;395:763–70.
3. Kalra SP, Dube MG, Pu S, Xu B, Horvath TL, Kalra PS. Interacting appetite-regulating pathways in the hypothalamic regulation of body weight. Endocr Rev. 1999;20:68–100.
4. Schwartz MW, Woods SC, Porte Jr D, Seeley RJ, Baskin DG. Central nervous system control of food intake. Nature. 2000;404:661–71.
5. Seeley RJ, Woods SC. Monitoring of stored and available fuel by the CNS: implications for obesity. Nat Rev Neurosci. 2003;4:901–9.
6. Mayer J, Thomas DW. Regulation of food intake and obesity. Science. 1967;156:328–37.
7. Woods SC, Seeley RJ, Porte Jr D, Schwartz MW. Signals that regulate food intake and energy homeostasis. Science. 1998;280:1378–83.
8. Kennedy GC. The role of depot fat in the hypothalamic control of food intake in the rat. Proc R Soc Lond B. 1953;140:579–92.
9. Hetherington AW. The relation of various hypothalamic lesions to adiposity and other phenomena in the rat. Am J Physiol. 1941;133:326.
10. Hetherington AW, Ranson SW. Hypothalamic lesions and adipocity in the rat. Anat Rec. 1940;78:149.
11. Hetherington AW, Ranson SW. The relation of various hypothalamic lesions to adiposity in the rat. J Comp Neurol. 1942;76:475–99.
12. Brobeck JR, Tepperman J, Long CNH. Experimental hypothalamic hyperphagia in the albino rat. Yale J Biol Med. 1943;15:831–53.
13. Brobeck JR. Mechanisms of the development of obesity in animals with hypothalamic lesions. Physiol Rev. 1946;26:541–59.
14. Anand BK, Brobeck JR. Localization of a feeding center in the hypothalamus of the rat. Proc Soc Exp Biol Med. 1951;77:323–4.
15. Stellar E. The physiology of motivation. Psychol Rev. 1954;61:5–22.
16. Hervey GR. The effects of lesions in the hypothalamus in parabiotic rats. J Physiol. 1959;145:336–52.
17. Coleman DL, Hummel KP. Effects of parabiosis of normal with genetically diabetic mice. Am J Physiol. 1969;217:1298–304.
18. Coleman DL. Obese and diabetes: two mutant genes causing diabetes-obesity syndromes in mice. Diabetologia. 1978;14:141–8.
19. Zhang Y, Proenca R, Maffei M, Barone M, Leopold L, Friedman JM. Positional cloning of the mouse obese gene and its human homologue. Nature. 1994;372:425–32.
20. Mercer JG, Hoggard N, Williams LM, Lawrence CB, Hannah LT, Trayhurn P. Localization of leptin receptor mRNA and the long form splice variant (Ob-Rb) in mouse hypothalamus and adjacent brain regions by in situ hybridization. FEBS Lett. 1996;387:113–6.
21. Huang XF, Koutcherov I, Lin S, Wang HQ, Storlien L. Localization of leptin receptor mRNA expression in mouse brain. Neuroreport. 1996;7:2635–8.
22. Couce ME, Burguera B, Parisi JE, Jensen MD, Lloyd RV. Localization of leptin receptor in the human brain. Neuroendocrinology. 1997;66:145–50.
23. Lam TK, Schwartz GJ, Rossetti L. Hypothalamic sensing of fatty acids. Nat Neurosci. 2005;8:579–84.
24. Lee K, Li B, Xi X, Suh Y, Martin RJ. Role of neuronal energy status in the regulation of adenosine 5'-monophosphate-activated protein kinase, orexigenic neuropeptides expression, and feeding behavior. Endocrinology. 2005;146:3–10.
25. Seeley RJ, York DA. Fuel sensing and the central nervous system (CNS): implications for the regulation of energy balance and the treatment for obesity. Obes Rev. 2005;6:259–65.
26. Kojima M, Hosoda H, Date Y, Nakazato M, Matsuo H, Kangawa K. Ghrelin is a growth-hormone-releasing acylated peptide from stomach. Nature. 1999;402:656–60.
27. Bray GA. Afferent signals regulating food intake. Proc Nutr Soc. 2000;59:373–84.
28. Woods SC, Schwartz MW, Baskin DG, Seeley RJ. Food intake and the regulation of body weight. Annu Rev Psychol. 2000;51:255–77.
29. Pinto S, Roseberry AG, Liu H, Diano S, Shanabrough M, Cai X, Friedman JM, Horvath TL. Rapid rewiring of arcuate nucleus feeding circuits by leptin. Science. 2004;304:110–5.

30. Sternson SM, Shepherd GM, Friedman JM. Topographic mapping of VMH -> arcuate nucleus microcircuits and their reorganization by fasting. Nat Neurosci. 2005;8:1356–63.
31. Horvath TL, Diano S. The floating blueprint of hypothalamic feeding circuits. Nat Rev Neurosci. 2004;5:662–7.
32. Horvath TL. The hardship of obesity: a soft-wired brain to feed and retain. Nat Neurosci. 2005;8:561–5.
33. Abizaid A, Liu Z-W, Andrews ZB, Shanabrough M, Borok E, Elsworth J, Roth RH, Sleeman M, Picciotto MR, Tschop M, Gao X-B, Horvath TL. Ghrelin modulates the activity and synaptic input organization of midbrain dopamine neurons while promoting appetite. J Clin Invest. 2006;116:3229–39.
34. Diano S, Farr SA, Benoit SC, McNay EC, da Silva I, Horvath B, Gaskin FS, Nonaka N, Jaeger LB, Banks WA, Morley JE, Pinto S, Sherwin RS, Xu L, Yamada KA, Sleeman MW, Tschop MH, Horvath TL. Ghrelin controls hippocampal spine synapse density and memory performance. Nat Neurosci. 2006;9:381–8.
35. Gao Q, Mezei G, Rao Y, Nie Y, Choi CS, Bechmann I, Naftolin F, Leranth C, Torran-Allerand D, Priest CA, Roberts JL, Gao XB, Mobbs C, Shulman G, Diano S, Horvath TL. Anorexigenic estradiol mimics leptin's effect on re-wiring of melanocortin cells in obese animals. Nat Med. 2007;13(1):89–94.
36. Andrews ZB, Liu Z-W, Wallingford N, Erion DM, Borok E, Friedman JM, Tschop MH, Shanabrough M, Cline G, Shulman GI, Coppola A, Gao X-B, Horvath TL, Diano S. UCP2 mediates ghrelin's action on NPY/AgRP neurons by lowering free radicals. Nature. 2008;454(7206):846–51.
37. Gyengesi E, Liu ZW, D'Agostino G, Gan G, Horvath TL, Gao XB, Diano S. Corticosterone regulates synaptic input organization of POMC and NPY/AgRP neurons in adult mice. Endocrinology. 2010;151:5395–401.
38. Horvath TL, Sarman B, García-Cáceres C, Enriori PJ, Shanabrough M, Borok E, Argente J, Chowen JA, Perez-Tilve D, Pfluger PT, Brönneke HS, Levin BE, Diano S, Cowley MA, Tschöp MH. Synaptic input organization of the melanocortin system predicts diet-induced hypothalamic reactive gliosis and obesity. PNAS. 2010;107(33):14875–80.
39. Benoit S, Schwartz M, Baskin D, Woods SC, Seeley RJ. CNS melanocortin system involvement in the regulation of food intake. Horm Behav. 2000;37:299–305.
40. Cone RD, Cowley MA, Butler AA, Fan W, Marks DL, Low MJ. The arcuate nucleus as a conduit for diverse signals relevant to energy homeostasis. Int J Obes Relat Metab Disord. 2001;25 Suppl 5:S63–7.
41. Seeley RJ, Drazen DL, Clegg DJ. The critical role of the melanocortin system in the control of energy balance. Annu Rev Nutr. 2004;24:133–49.
42. van den Pol AN. Weighing the role of hypothalamic feeding neurotransmitters. Neuron. 2003;40:1059–61.
43. Obici S, Rossetti L. Minireview: nutrient sensing and the regulation of insulin action and energy balance. Endocrinology. 2003;144:5172–8.
44. Horvath TL, Bechmann I, Kalra SP, Naftolin F, Leranth C. Heterogeneity in the neuropeptide Y-containing neurons of the rat arcuate nucleus: GABAergic and non-GABAergic subpopulations. Brain Res. 1997;756:283–6.
45. Hentges ST, Nishiyama M, Overstreet LS, Stenzel-Poore M, Williams JT, Low MJ. GABA release from proopiomelanocortin neurons. J Neurosci. 2004;24:1578–83.
46. Boston BA, Blaydon KM, Varnerin J, Cone RD. Independent and additive effects of central POMC and leptin pathways on murine obesity. Science. 1997;278:1641–4.
47. Ellacott KL, Cone RD. The central melanocortin system and the integration of short- and long-term regulators of energy homeostasis. Recent Prog Horm Res. 2004;59:395–408.
48. Cone RD. Anatomy and regulation of the central melanocortin system. Nat Neurosci. 2005;8:571–8.
49. Fan W, Boston BA, Kesterson RA, Hruby VJ, Cone RD. Role of melanocortinergic neurons in feeding and the agouti obesity syndrome. Nature. 1997;385:165–8.
50. Biebermann H, Castaneda TR, van Landeghem F, von Deimling A, Escher F, Brabant G, Hebebrand J, Hinney A, Tschop MH, Gruters A, Krude H. A role for beta-melanocyte-stimulating hormone in human body-weight regulation. Cell Metab. 2006;3:141–6.
51. Lee YS, Challis BG, Thompson DA, Yeo GS, Keogh JM, Madonna ME, Wraight V, Sims M, Vatin V, Meyre D, Shield J, Burren C, Ibrahim Z, Cheetham T, Swift P, Blackwood A, Hung CC, Wareham NJ, Froguel P, Millhauser GL, O'Rahilly S, Farooqi IS. A POMC variant implicates beta-melanocyte-stimulating hormone in the control of human energy balance. Cell Metab. 2006;3:135–40.
52. Adan RA, Cone RD, Burbach JP, Gispen WH. Differential effects of melanocortin peptides on neural melanocortin receptors. Mol Pharmacol. 1994;46:1182–90.
53. Mountjoy KG, Mortrud MT, Low MJ, Simerly RB, Cone RD. Localization of the melanocortin-4 receptor (MC4-R) in neuroendocrine and autonomic control circuits in the brain. Mol Endocrinol. 1994;8:1298–308.
54. Ollmann MM, Wilson BD, Yang YK, Kerns JA, Chen Y, Gantz I, Barsh GS. Antagonism of central melanocortin receptors in vitro and in vivo by agouti-related protein. Science. 1997;278:135–8.
55. Baskin DG, Breininger JF, Schwartz MW. Leptin receptor mRNA identifies a subpopulation of neuropeptide Y neurons activated by fasting in rat hypothalamus. Diabetes. 1999;48:828–33.
56. Clark JT, Kalra PS, Crowley WR, Kalra SP. Neuropeptide Y and human pancreatic polypeptide stimulate feeding behavior in rats. Endocrinology. 1984;115:427–9.

57. Stanley BG, Leibowitz SF. Neuropeptide Y: stimulation of feeding and drinking by injection into the paraventricular nucleus. Life Sci. 1984;35:2635–42.
58. Sahu A, Kalra SP, Crowley WR, Kalra PS. Evidence that NPY-containing neurons in the brainstem project into selected hypothalamic nuclei: implication in feeding behavior. Brain Res. 1988;457:376–8.
59. Chronwall BM, Chase TN, O'Donohue TL. Coexistence of neuropeptide Y and somatostatin in rat and human cortical and rat hypothalamic neurons. Neurosci Lett. 1984;52:213–7.
60. Morley JE, Hernandez EN, Flood JF. Neuropeptide Y increases food intake in mice. Am J Physiol. 1987;253:R516–22.
61. Sahu A, Kalra PS, Kalra SP. Food deprivation and ingestion induce reciprocal changes in neuropeptide Y concentrations in the paraventricular nucleus. Peptides. 1988;9:83–6.
62. Sanacora G, Kershaw M, Finkelstein JA, White JD. Increased hypothalamic content of preproneuropeptide Y messenger ribonucleic acid in genetically obese Zucker rats and its regulation by food deprivation. Endocrinology. 1990;127:730–7.
63. Wilding JP, Gilbey SG, Bailey CJ, Batt RA, Williams G, Ghatei MA, Bloom SR. Increased neuropeptide-Y messenger ribonucleic acid (mRNA) and decreased neurotensin mRNA in the hypothalamus of the obese (ob/ob) mouse. Endocrinology. 1993;132:1939–44.
64. Quillan JM, Sadee W, Wei ET, Jimenez C, Ji L, Chang JK. A synthetic human Agouti-related protein-(83-132)-NH2 fragment is a potent inhibitor of melanocortin receptor function. FEBS Lett. 1998;428:59–62.
65. Rossi M, Kim MS, Morgan DG, Small CJ, Edwards CM, Sunter D, Abusnana S, Goldstone AP, Russell SH, Stanley SA, Smith DM, Yagaloff K, Ghatei MA, Bloom SR. A C-terminal fragment of Agouti-related protein increases feeding and antagonizes the effect of alpha-melanocyte stimulating hormone in vivo. Endocrinology. 1998;139:4428–31.
66. Tritos NA, Elmquist JK, Mastaitis JW, Flier JS, Maratos-Flier E. Characterization of expression of hypothalamic appetite-regulating peptides in obese hyperleptinemic brown adipose tissue-deficient (uncoupling protein-promoter-driven diphtheria toxin A) mice. Endocrinology. 1998;139:4634–41.
67. Cowley MA, Smart JL, Rubinstein M, Cerdan MG, Diano S, Horvath TL, Cone RD, Low MJ. Leptin activates anorexigenic POMC neurons through a neural network in the arcuate nucleus. Nature. 2001;411:480–4.
68. Horvath TL, Naftolin F, Kalra SP, Leranth C. Neuropeptide-Y innervation of beta-endorphin-containing cells in the rat mediobasal hypothalamus: a light and electron microscopic double immunostaining analysis. Endocrinology. 1992;131: 2461–7.
69. Bates SH, Myers Jr MG. The role of leptin receptor signaling in feeding and neuroendocrine function. Trends Endocrinol Metab. 2003;14:447–52.
70. Sahm UG, Qarawi MA, Olivier GW, Ahmed AR, Branch SK, Moss SH, Pouton CW. The melanocortin (MC3) receptor from rat hypothalamus: photoaffinity labelling and binding of alanine-substituted alpha-MSH analogues. FEBS Lett. 1994;350:29–32.
71. Diano S, Naftolin F, Goglia F, Horvath TL. Segregation of the intra- and extrahypothalamic neuropeptide Y and catecholaminergic inputs on paraventricular neurons, including those producing thyrotropin-releasing hormone. Regul Pept. 1998;75–76:117–26.
72. Diano S, Naftolin F, Goglia F, Csernus V, Horvath TL. Monosynaptic pathway between the arcuate nucleus expressing glial type II iodothyronine 5'-deiodinase mRNA and the median eminence-projective TRH cells of the rat paraventricular nucleus. J Neuroendocrinol. 1998;10:731–42.
73. Cowley MA, Pronchuk N, Fan W, Dinulescu DM, Colmers WF, Cone RD. Integration of NPY, AGRP, and melanocortin signals in the hypothalamic paraventricular nucleus: evidence of a cellular basis for the adipostat. Neuron. 1999;24:155–63.
74. Balthasar N, Dalgaard LT, Lee CE, Yu J, Funahashi H, Williams T, Ferreira M, Tang V, McGovern RA, Kenny CD, Christiansen LM, Edelstein E, Choi B, Boss O, Aschkenasi C, Zhang CY, Mountjoy K, Kishi T, Elmquist JK, Lowell BB. Divergence of melanocortin pathways in the control of food intake and energy expenditure. Cell. 2005;123:493–505.
75. Gropp E, Shanabrough M, Borok E, Xu AW, Janoshek R, Buch T, Plum L, Balthasar N, Hampel B, Waisman A, Barsh GS, Horvath TL, Bruning J. Agouti-related peptide-expressing neurons are mandatory for feeding. Nat Neurosci. 2005;8:1289–91.
76. Luquet S, Perez FA, Hnasko TS, Palmiter RD. NPY/AgRP neurons are essential for feeding in adult mice but can be ablated in neonates. Science. 2005;310:683–5.
77. Wu Q, Boyle MP, Palmiter RD. Loss of GABAergic signaling by AgRP neurons to the parabrachial nucleus leads to starvation. Cell. 2009;137(7): 1225–34.
78. Dietrich MO, Horvath TL. GABA keeps up an appetite for life. Cell. 2009;137:1177–9.
79. Erickson JC, Clegg KE, Palmiter RD. Sensitivity to leptin and susceptibility to seizures of mice lacking neuropeptide Y. Nature. 1996;381:415–21.
80. Qian S, Chen H, Weingarth D, Trumbauer ME, Novi DE, Guan X, Yu H, Shen Z, Feng Y, Frazier E, Chen A, Camacho RE, Shearman LP, Gopal-Truter S, MacNeil DJ, Van der Ploeg LH, Marsh DJ. Neither agouti-related protein nor neuropeptide Y is critically required for the regulation of energy homeostasis in mice. Mol Cell Biol. 2002;22:5027–35.
81. Broberger C, De Lecea L, Sutcliffe JG, Hokfelt T. Hypocretin/orexin- and melanin-concentrating hormone-expressing cells form distinct populations in the rodent lateral hypothalamus: relationship to the neuropeptide Y and agouti gene-related protein systems. J Comp Neurol. 1998;402:460–74.

82. Ludwig DS, Mountjoy KG, Tatro JB, Gillette JA, Frederich RC, Flier JS, Maratos-Flier E. Melanin-concentrating hormone: a functional melanocortin antagonist in the hypothalamus. Am J Physiol. 1998;274:627–33.
83. Peyron C, Tighe DK, van den Pol AN, de Lecea L, Heller HC, Sutcliffe JG, Kilduff TS. Neurons containing hypocretin (orexin) project to multiple neuronal systems. J Neurosci. 1998;18:9996–10015.
84. Saito Y, Cheng M, Leslie FM, Civelli O. Expression of the melanin-concentrating hormone (MCH) receptor mRNA in the rat brain. J Comp Neurol. 2001;435:26–40.
85. Horvath TL, Diano S, van den Pol AN. Synaptic interaction between hypocretin (orexin) and neuropeptide Y cells in the rodent and primate hypothalamus: a novel circuit implicated in metabolic and endocrine regulations. J Neurosci. 1999;19:1072–87.
86. Elias CF, Saper CB, Maratos-Flier E, Tritos NA, Lee C, Kelly J, Tatro JB, Hoffman GE, Ollmann MM, Barsh GS, Sakurai T, Yanagisawa M, Elmquist JK. Chemically defined projections linking the mediobasal hypothalamus and the lateral hypothalamic area. J Comp Neurol. 1998;402:442–59.
87. Hakansson ML, Brown H, Ghilardi N, Skoda RC, Meister B. Leptin receptor immunoreactivity in chemically defined target neurons of the hypothalamus. J Neurosci. 1998;18:559–7278.
88. Torrealba F, Yanagisawa M, Saper CB. Colocalization of orexin and glutamate immunoreactivity in axon terminals in the tuberomammillary nucleus in rats. Neuroscience. 2003;119:1033–44.
89. Toshinai K, Date Y, Murakami N, Shimada M, Mondal MS, Shimbara T, Guan JL, Wang QP, Funahashi H, Sakurai T, Shioda S, Matsukura S, Kangawa K, Nakazato M. Ghrelin induced food intake is mediated via the orexin pathway. Endocrinology. 2003;144:1506–12.
90. Sakurai T, Nagata R, Yamanaka A, Kawamura H, Tsujino N, Muraki Y, Kageyama H, Kunita S, Takahashi S, Goto K, Koyama Y, Shioda S, Yanagisawa M. Input of orexin/hypocretin neurons revealed by a genetically encoded tracer in mice. Neuron. 2005;46:297–308.
91. Guan JL, Uehara K, Lu S, Wang QP, Funahashi H, Sakurai T, Yanagizawa M, Shioda S. Reciprocal synaptic relationships between orexin- and melanin-concentrating hormone-containing neurons in the rat lateral hypothalamus: a novel circuit implicated in feeding regulation. Int J Obes Relat Metab Disord. 2002;26:1523–32.
92. van den Pol AN, Acuna-Goycolea C, Clark KR, Ghosh PK. Physiological properties of hypothalamic MCH neurons identified with selective expression of reporter gene after recombinant virus infection. Neuron. 2004;42:635–52.
93. Li Y, Gao XB, Sakurai T, van den Pol AN. Hypocretin/Orexin excites hypocretin neurons via a local glutamate neuron-A potential mechanism for orchestrating the hypothalamic arousal system. Neuron. 2002;36:1169–81.
94. Gao XB, van den Pol AN. Melanin concentrating hormone depresses synaptic activity of glutamate and GABA neurons from rat lateral hypothalamus. J Physiol. 2001;533:237–52.
95. Diano S, Horvath B, Urbanski HF, Sotonyi P, Horvath TL. Fasting activates the non-human primate hypocretin (orexin) system and its postsynaptic targets. Endocrinology. 2003;144:3774–8; first published July 3 2003 as doi:10.1210/en.2003-0274.
96. Horvath TL, Gao XB. Inpur organization and plasticityof hypocretin neurons: possible clues for obesity's association with insomnia. Cell Metab. 2005;1: 279–86.
97. Sakurai T, Amemiya A, Ishii M, Matsuzaki I, Chemelli RM, Tanaka H, Williams SC, Richardson JA, Kozlowski GP, Wilson S, Arch JR, Buckingham RE, Haynes AC, Carr SA, Annan RS, McNulty DE, Liu WS, Terrett JA, Elshourbagy NA, Bergsma DJ, Yanagisawa M. Orexins and orexin receptors: a family of hypothalamic neuropeptides and G protein-coupled receptors that regulate feeding behavior. Cell. 1998;92:573–85.
98. Horvath TL, Peyron C, Diano S, Ivanov A, Aston-Jones G, Kilduff TS, van den Pol AN. Hypocretin (orexin) activation and synaptic innervation of the locus coeruleus noradrenergic system. J Comp Neurol. 1999;415:145–59.
99. Jain MR, Horvath TL, Kalra PS, Kalra SP. Evidence that NPY Y1 receptors are involved in stimulation of feeding by orexins (hypocretins) in sated rats. Regul Pept. 1999;87:19–24.
100. Yamanaka A, Kunii K, Nambu T, Tsujino N, Sakai A, Matsuzaki I, Miwa Y, Goto K, Sakurai T. Orexin-induced food intake involves neuropeptide Y pathway. Brain Res. 2000;859:404–9.
101. Sahu A. Interactions of neuropeptide Y, hypocretin-I (orexin A) and melanin-concentrating hormone on feeding in rats. Brain Res. 2002;944:232–8.
102. Willie JT, Chemelli RM, Sinton CM, Yanagisawa M. To eat or to sleep? Orexin in the regulation of feeding and wakefulness. Annu Rev Neurosci. 2001;24: 429–58.
103. Chemelli RM, Willie JT, Sinton CM, Elmquist JK, Scammell T, Lee C, Richardson JA, Williams SC, Xiong Y, Kisanuki Y, Fitch TE, Nakazato M, Hammer RE, Saper CB, Yanagisawa M. Narcolepsy in orexin knockout mice: molecular genetics of sleep regulation. Cell. 1999;98:437–51.
104. Fu LY, Acuna-Goycolea C, van den Pol AN. Neuropeptide Y inhibits hypocretin/orexin neurons by multiple presynaptic and postsynaptic mechanisms: tonic depression of the hypothalamic arousal system. J Neurosci. 2004;24:8741–51.
105. Qu D, Ludwig DS, Gammeltoft S, Piper M, Pelleymounter MA, Cullen MJ, Mathes WF, Przypek R, Kanarek R, Maratos-Flier E. A role for melanin-concentrating hormone in the central regulation of feeding behaviour. Nature. 1996;380:243–7.

106. Shimada M, Tritos NA, Lowell BB, Flier JS, Maratos-Flier E. Mice lacking melanin-concentrating hormone are hypophagic and lean. Nature. 1998;396:670–4.
107. Kennedy AR, Todd JF, Stanley SA, Abbott CR, Small CJ, Ghatei MA, Bloom SR. Melanin-concentrating hormone (MCH) suppresses thyroid stimulating hormone (TSH) release, in vivo and in vitro, via the hypothalamus and the pituitary. Endocrinology. 2001;142:3265–8.
108. Alon T, Friedman JM. Late-onset leanness in mice with targeted ablation of melanin concentrating hormone neurons. J Neurosci. 2006;26:389–97.
109. Zigman JM, Jones JE, Lee CE, Saper CB, Elmquist JK. Expression of ghrelin receptor mRNA in the rat and the mouse brain. J Comp Neurol. 2006;494: 528–48.
110. Horvath TL. Suprachiasmatic efferents avoid phenestrated capillaries but innervate neuroendocrine cells including those producing dopamine. Endocrinology. 1997;138:1312–20.
111. Horvath TL. An alternate pathway for visual signal integration into the hypothalamo-pituitary axis: retinorecipient intergeniculate neurons project to various regions of the hypothalamus and innervate neuroendocrine cells including those producing dopamine. J Neurosci. 1998;18:1546–58.
112 Chou TC, Scammell TE, Gooley JJ, Gaus SE, Saper CB, Lu J. Critical role of dorsomedial hypothalamic nucleus in a wide range of behavioral circadian rhythms. J Neurosci. 2003;23:10691–702.
113. Bellinger LL, Bernardis LL. The dorsomedial hypothalamic nucleus and its role in ingestive behavior and body weight regulation: lessons learned from lesioning studies. Physiol Behav. 2002;76:431–42.
114. Gooley JJ, Schomer A, Saper CB. The dorsomedial hypothalamic nucleus is critical for the expression of food-entrainable circadian rhythms. Nat Neurosci. 2006;9:398–407.
115. ter Horst GJ, Luiten PG. The projections of the dorsomedial hypothalamic nucleus in the rat. Brain Res Bull. 1986;16:231–48.
116. Chou TC, Bjorkum AA, Gaus SE, Lu J, Scammell TE, Saper CB. Afferents to the ventrolateral preoptic nucleus. J Neurosci. 2002;22:977–90.
117. van den Top M, Lee K, Whyment AD, Blanks AM, Spanswick D. Orexigen-sensitive NPY/AgRP pacemaker neurons in the hypothalamic arcuate nucleus. Nat Neurosci. 2004;7:493–4.
118. Cowley MA, Smith RG, Diano S, Tschöp M, Pronchuk N, Grove KL, Strasburger CJ, Bidlingmaier M, Esterman M, Heiman ML, Garcia-Segura LM, Nillni EA, Mendez P, Low MJ, Sotonyi P, Friedman JM, Liu H, Pinto S, Colmers WF, Cone RD, Horvath TL. The distribution and mechanism of action of ghrelin in the CNS demonstrates a novel hypothalamic circuit regulating energy homeostasis. Neuron. 2003;37:649–61.
119. Spanswick D, Smith MA, Groppi VE, Logan SD, Ashford ML. Leptin inhibits hypothalamic neurons by activation of ATP-sensitive potassium channels. Nature. 1997;390:521–5.
120. Spanswick D, Smith MA, Mirshamsi S, Routh VH, Ashford ML. Insulin activates ATP-sensitive K+channels in hypothalamic neurons of lean, but not obese rats. Nat Neurosci. 2000;3:757–8.
121. Miyata S, Nakashima T, Kiyohara T. Structural dynamics of neural plasticity in the supraoptic nucleus of the rat hypothalamus during dehydration and rehydration. Brain Res Bull. 1994;34: 169–75.
122. Stern JE, Armstrong WE. Reorganization of the dendritic trees of oxytocin and vasopressin neurons of the rat supraoptic nucleus during lactation. J Neurosci. 1998;18:841–53.
123. Theodosis DT, Rougon G, Poulain DA. Retention of embryonic features by an adult neuronal system capable of plasticity: polysialylated neural cell adhesion molecule in the hypothalamo-neurohypophysial system. Proc Natl Acad Sci USA. 1991;88:5494–8.
124. Garcia-Segura LM, Chowen JA, Parducz A, Naftolin F. Gonadal hormones as promoters of structural synaptic plasticity: cellular mechanisms. Prog Neurobiol. 1994;44:279–307.
125. Parducz A, Zsarnovszky A, Naftolin F, Horvath TL. Estradiol affects axo-somatic contacts of neuroendocrine cells in the arcuate nucleus of adult rats. Neuroscience. 2003;117:791–4.
126. Zsarnovszky A, Horvath TL, Garcia-Segura LM, Horvath B, Naftolin F. Oestrogen-induced changes in the synaptology of the monkey (Cercopithecus aethiops) arcuate nucleus during gonadotropin feedback. J Neuroendocrinol. 2001;13:22–8.
127. Xiong JJ, Karsch FJ, Lehman MN. Evidence for seasonal plasticity in the gonadotropin-releasing hormone (GnRH) system of the ewe: changes in synaptic inputs onto GnRH neurons. Endocrinology. 1997;138:1240–50.
128. Malenka RC, Bear MF. LTP and LTD; an embarrassment of riches. Neuron. 2004;44:5–21.
129. Oomura Y, Yoshimatsu H. Neural network of glucose monitoring system. J Auton Nerv Syst. 1984;10:359–72.
130. Storlien LH. The role of the ventromedial hypothalamic area in periprandial glucoregulation. Life Sci. 1985;36:505–14.
131. McCrimmon RJ, Fan X, Ding Y, Zhu W, Jacob RJ, Sherwin RS. Potential role for AMP-activated protein kinase in hypoglycemia sensing in the ventromedial hypothalamus. Diabetes. 2004;53:1953–8.
132. Minokoshi Y, Alquier T, Furukawa N, Kim YB, Lee A, Xue B, Mu J, Foufelle F, Ferre P, Birnbaum MJ, Stuck BJ, Kahn BB. AMP-kinase regulates food intake by responding to hormonal and nutrient signals in the hypothalamus. Nature. 2004;428: 569–74.
133. Kahn BB, Alquier T, Carling D, Hardie DG. AMP-activated protein kinase: ancient energy gauge provides clues to modern understanding of metabolism. Cell Metab. 2005;1:15–25.

134. Lam TK, Gutierrez-Juarez R, Pocai A, Rossetti L. Regulation of blood glucose by hypothalamic pyruvate metabolism. Science. 2005;309:943–7.
135. Pocai A, Lam TK, Gutierrez-Juarez R, Obici S, Schwartz GJ, Bryan J, Aguilar-Bryan L, Rossetti L. Hypothalamic K(ATP) channels control hepatic glucose production. Nature. 2005;434:1026–31.
136. Pocai A, Obici S, Schwartz GJ, Rossetti L. A brain-liver circuit regulates glucose homeostasis. Cell Metab. 2005;1:53–61.
137. King BM. The rise, fall, and resurrection of the ventromedial hypothalamus in the regulation of feeding behavior and body weight. Physiol Behav. 2006;87:221–44.
138. Pocai A, Lam TK, Obici S, Gutierrez-Juarez R, Muse ED, Arduini A, Rossetti L. Restoration of hypothalamic lipid sensing normalizes energy and glucose homeostasis in overfed rats. J Clin Invest. 2006;116:1081–91.
139. Kim EK, Miller I, Aja S, Landree LE, Pinn M, McFadden J, Kuhajda FP, Moran TH, Ronnett GV. C75, a fatty acid synthase inhibitor, reduces food intake via hypothalamic AMP-activated protein kinase. J Biol Chem. 2004;279:19970–6.
140. Kim MS, Park JY, Namkoong C, Jang PG, Ryu JW, Song HS, Yun JY, Namgoong IS, Ha J, Park IS, Lee IK, Viollet B, Youn JH, Lee HK, Lee KU. Anti-obesity effects of alpha-lipoic acid mediated by suppression of hypothalamic AMP-activated protein kinase. Nat Med. 2004;10:727–33.
141. Landree LE, Hanlon AL, Strong DW, Rumbaugh G, Miller IM, Thupari JN, Connolly EC, Huganir RL, Richardson C, Witters LA, Kuhajda FP, Ronnett GV. C75, a fatty acid synthase inhibitor, modulates AMP-activated protein kinase to alter neuronal energy metabolism. J Biol Chem. 2004;279:3817–27.
142. Horvath TL, Andrews ZB, Diano S. Fuel utilization by hypothalamic neurons: roles for ROS. Trends Endocrinol Metab. 2009;20(2):78–87.
143. Minokoshi Y, Kim YB, Peroni OD, Fryer LG, Muller C, Carling D, Kahn BB. Leptin stimulates fatty-acid oxidation by activating AMP-activated protein kinase. Nature. 2002;415:339–43.
144. Kola B, Hubina E, Tucci SA, Kirkham TC, Garcia EA, Mitchell SE, Williams LM, Hawley SA, Hardie DG, Grossman AB, Korbonits M. Cannabinoids and ghrelin have both central and peripheral metabolic and cardiac effects via AMP-activated protein kinase. J Biol Chem. 2005;280:25196–201.
145. Wollmann G, Acuna-Goycolea C, van den Pol AN. Direct excitation of hypocretin/orexin cells by extracellular ATP at P2X receptors. J Neurophysiol. 2005;94:2195–206.
146. Fuxe K, Rivera A, Jacobsen KX, Hoistad M, Leo G, Horvath TL, Staines W, De la Calle A, Agnati LF. Dynamics of volume transmission in the brain. Focus on catecholamine and opioid peptide communication and the role of uncoupling protein 2. J Neural Transm. 2005;112:65–76.
147. Andrews ZB, Diano S, Horvath TL. Mitochondrial uncoupling proteins in the central nervous system: in support of function and survival. Nat Rev Neurosci. 2005;6(11):829–40.
148. Ibrahim N, et al. Hypothalamic proopiomelanocortin neurons are glucose responsive and express K(ATP) channels. Endocrinology. 2003;144:1331–40.
149. Wang R, et al. The regulation of glucose-excited neurons in the hypothalamic arcuate nucleus by glucose and feeding-relevant peptides. Diabetes. 2004;53:1959–65.
150. Parton LE, et al. Glucose sensing by POMC neurons regulates glucose homeostasis and is impaired in obesity. Nature. 2007;449:228–32.
151. Fioramonti X, et al. Characterization of glcosensing neuron subpopulations in the arcuate ncleus: integration in neuropeptideY and pro-opio-melanocortin networks? Diabetes. 2007;56:1219–27.
152. Dietrich M, Andrews ZB, Horvath TL. Exercise-induced synaptogenesis in the hippocampus is dependent on UCP2-dependent mitochondrial adaptation. J Neurosci. 2008;28:10766–71.
153. Andrews AB, Horvath TL. Uncoupling Protein 2 regulates lifespan in mice. Am J Physiol Endo Met. 2009;296(4):E621–7. Epub 2009 Jan 13.

The Neurogenetics of Energy Balance

6

Martin G. Myers Jr.

Abstract

Energy balance is tightly controlled in humans and other animals: perturbations that increase or decrease energy (fat) stores provoke countervening responses to restore adiposity to baseline. Changes in energy expenditure, whether at the level of basal metabolic rate or activity, generally impact energy balance only modestly due to compensatory changes in feeding. Thus, alterations in systems that modulate food intake relative to energy expenditure, rather than systems that regulate energy utilization alone, dominate in the control of adiposity. Genetic perturbations that alter body weight, both in animal models and in human patients and populations have aided in the identification and study of the systems that control energy balance. The genetic causes of obesity for which we have a rudimentary understanding of underlying mechanism fall within the CNS, and many of these lesions lie in genes important to the function of crucial hypothalamic circuits that control feeding. Others encode more general modulators of neurophysiology that impact feeding systems in addition to other neural circuits. From the standpoint of neural mechanism, however, a great many genetic lesions associated with altered energy balance remain incompletely understood. Understanding the neural systems impacted by these genetic alterations will not only reveal the physiologic underpinnings of energy homeostasis, but may also identify targets for therapeutic intervention.

Introduction

Energy Balance Is Regulated in Mammals

Over the long term, body weight and adiposity are regulated within the relatively narrow limits in humans and other animals [1, 2]. Animals defend against changes in body weight and adiposity via a variety of mechanisms: negative

M.G. Myers Jr., MD, PhD (✉)
Department of Internal Medicine,
University of Michigan, 1000 Wall Street,
Ann Arbor, MI 48105, USA

Department of Molecular and Integrative Physiology,
University of Michigan, 1000 Wall Street,
Ann Arbor, MI 48105, USA
e-mail: mgmyers@umich.edu

P.J. Shiromani et al. (eds.), *Sleep Loss and Obesity: Intersecting Epidemics*,
DOI 10.1007/978-1-4614-3492-4_6, © Springer Science+Business Media, LLC 2012

energy balance due to increased energy expenditure (e.g., increased exercise) or decreased food intake (e.g., dieting) provokes compensatory processes that tend to restore adipose tissue mass to its previous state—hunger increases and energy expenditure is restrained. In the absence of artificial food restriction (dieting), feeding increases to compensate for increased energy expenditure. Diet-induced weight loss results in decreased energy expenditure and is difficult to maintain.

While it is true that the response to negative energy balance (weight loss) is more robust than the response to positive energy balance, involuntary increases in body weight and adiposity, as by overfeeding via a tube in the stomach, also decrease hunger and promote increased energy expenditure. During ad libitum feeding followed forced overfeeding, food intake decreases to below baseline, until the previous levels of weight and adiposity are restored [3]. Thus, homeostatic systems exist to modulate food intake relative to energy expenditure and adiposity, maintaining body energy stores within a relatively narrow range over the long term.

That is not to say that all individuals should be expected to maintain the same adiposity or food intake relative to energy expenditure at any particular body weight, however. A variety of environmental and genetic factors interact to determine the level of adiposity for any individual. The impact of the environment is evident by comparing rates of adiposity and obesity today to those from even 20 years ago: around 1990, 23% of US adults were classified as obese, while today over 35% meet this definition [4]. It would be difficult to argue that natural selection, occurring within a single generation, could underlie this shift in adiposity. Rather, palatable, high-calorie foods have become increasingly cheap and available during this time, and the overconsumption of these foods represents the most plausible explanation for the increased rates of obesity in industrialized societies. Other environmental factors, such as the increased feeding promoted by stress and/or altered circadian rhythms, discussed elsewhere in this text, also contribute [5, 6]. Yet, even within our current environment, some individuals remain lean into old age, while others rapidly progress to morbid obesity during their youth. A substantial portion of this variability in body weight and adiposity within human populations in a given environment is attributable to the genetic makeup of each person [7]. In this chapter, we consider genetic contributors to energy balance, both in laboratory animals (including spontaneous mutations and molecular genetic models) and in human populations. We also discuss what the genetics of adiposity has taught us thus far, as well as the substantial amount that we have yet to completely understand about genetic determinants of feeding and body weight.

Food Intake Is the Controlled Variable in Energy Balance

While obesity (or leanness) is often attributed to a slow (or fast) metabolism, the systems that control energy balance dictate that metabolic rate is not the primary driver of body weight, since the level of energy expenditure in any individual will modulate food intake to achieve a certain level of adiposity for that individual within a given environment. Not only does the increased drive to feed in humans who exercise a great deal serve to maintain adiposity in the face of acutely increased energy expenditure, but genetic alterations that solely alter energy expenditure minimally alter adiposity in animal models, except under special circumstances: for instance, the decreased energy expenditure conferred by deletion of the mitochondrial uncoupling protein, UCP1, from metabolically active brown adipose tissue in mice produces little change in body weight and adiposity under most conditions [8]. While alterations in adrenergic signaling, as by deletion of all β-adrenergic receptors or a regulatory subunit of protein kinase A, promotes obesity, these alterations impact CNS function, not just peripheral metabolic control [9, 10]. As will become clear below, however, there is ample evidence to suggest that genetic changes that alter CNS processes (especially those that control food intake) underlie most identified dramatic and/or monogenic causes of obesity, as well as more subtle contributors to energy balance identified in genome-wide association studies (GWAS).

Genetic Causes of Obesity

Many single gene lesions that promote obesity in rodents and humans define neural systems integral to the control of feeding and energy balance.

Many insights into the neural mechanisms that control energy balance have emerged from the study of rodent models; the importance of many of these affected genes and systems has subsequently been verified in humans. In general, these rodent models and human subjects exhibit obesity, rather than leanness. This is presumably attributable both to the multiple redundant pathways that protect against decreased energy stores, and to the fact that, while hyperphagia produces obesity and tends not to limit fecundity, animals with severe hypophagia are unlikely to survive, let alone reproduce.

Leptin and Its Receptor

The most dramatic mouse models of obesity were first detected as spontaneously arising mutants within the breeding colony at Jackson Labs; these are the *obese* (*ob/ob*) and *diabetes* (*db/db*) mice [11]. These animals exhibit similarly extreme hyperphagia, decreased metabolism, and a neuroendocrine phenotype reminiscent of the stereotypical response to starvation [11–13]. Early studies revealed that the *ob* and *db* loci were distinct, and subsequent elegant parabiosis studies suggested that the gene product affected by *ob* was a hormone, while *db* disrupted the receptor for this hormone [11, 14]. Indeed, molecular cloning revealed that the *ob* mutation disrupts an adipocyte-specific hormone (termed leptin) [15]; soon thereafter, *db* was shown to disrupt the leptin receptor (LepRb) [16, 17].

Leptin is produced by adipocytes in approximate proportion to their triglyceride (energy) stores; leptin acts via LepRb to convey the adequacy (or by its absence, the inadequacy) of these fat stores, and modulates appetite, energy expenditure, and a variety of complex behaviors in concert with energy balance [12, 18, 19]. During caloric restriction, for instance, leptin levels drop—provoking increased hunger and decreased energy utilization; leptin administration reverses these effects [12, 20]. LepRb is expressed predominantly in brain areas known to be involved in the control of (or response to) feeding and energy balance, and leptin action via LepRb in the brain mediates most leptin action [21–24].

The discovery of leptin and LepRb also enabled the screening of humans for alterations in the genes that encode them; while no genetic alterations in these loci have been shown to contribute to common forms of human obesity (indeed, obese humans generally have increased leptin commensurate with their adipose mass), a few very rare individuals with homozygous null mutations in these genes have been identified in cohorts of individuals with extreme early onset obesity [25, 26]. The clinical syndrome exhibited by these patients is similar to the phenotype exhibited by $Lep^{ob/ob}$ and $Lepr^{db/db}$ mice, including hyperphagic obesity, neuroendocrine dysfunction, and hypoactivity. Thus, rodent and human genetics demonstrate that leptin and its receptor represent crucial controllers of feeding and energy balance.

The LepRb Signaling Pathway

The cellular mechanisms of LepRb signaling and the contributions of these signals to leptin action have been extensively studied in cultured cells and in rodents [27]. Mice mutant for a number of specific LepRb signals display hyperphagia and obesity, including animals disrupted for LepRb signaling via the signal transducer and activator of transcription-3 (STAT3) pathway—a major downstream signaling pathway that is directly activated by LepRb [28].

SH2B1 is an adaptor protein that interacts with the LepRb signaling complex to increase receptor signaling and recruit some downstream signaling proteins into the complex [29, 30]. Animals bearing inactivating *Sh2b1* mutations or

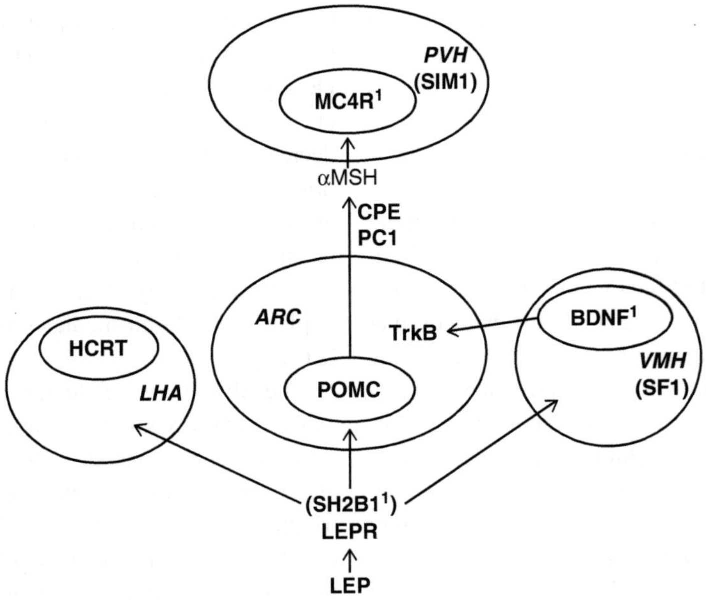

Fig. 6.1 Neural circuits and pathways corresponding to human obesity genes of understood function. Large ovals correspond to hypothalamic nuclei (*names in italics*); smaller ovals correspond to neurons within those nuclei that contain the indicated gene product. [1]Mutations of the genes encoding the indicated gene products not only produce obesity in rare null syndromes in humans, but have been implicated as obesity genes by GWAS as well

neuronal *Sh2b1* deletions exhibit hyperphagia extreme enough to promote obesity in the face of (presumably leptin-independent) increased energy expenditure [31, 32]. Interestingly, the *SH2B1* locus lies near the affected chromosomal region in a group of patients with severe obesity syndromes secondary to a deletion of chromosome 16 [33, 34]. Furthermore, recent GWAS have suggested a potential role for *SH2B1* variants as risk alleles for common obesity [35, 36]. Thus, *SH2B1* represents an important controller of energy balance in humans as well as rodents, presumably due to its role in leptin signaling.

The Melanocortin System

Contemporaneous with the explosion of knowledge about the leptin signaling system came the genetic dissection of the leptin-regulated hypothalamic melanocortin system. Based upon an enormous amount of work from numerous laboratories, we now understand that a group of neurons that express proopiomelanocortin (POMC) and that lie within the arcuate nucleus of the hypothalamus (ARC) play a central role in the control of energy balance (Fig. 6.1) [37, 38]. While *POMC* is also expressed in the anterior pituitary, where it is processed to adrenocorticotropic hormone (ACTH), processing in the ARC produces a number of other peptides, including α-melanocyte-stimulating hormone (αMSH) [39]. ARC POMC neurons project to a number of other hypothalamic regions involved in the control of food intake and energy expenditure, including the paraventricular hypothalamic nucleus (PVH) [40]; αMSH release at these target sites activates the melanocortin 4 receptor (MC4R), as well as MC3R, on target neurons to diminish feeding [41–43].

The molecular genetic dissection of the obese *agouti* yellow mouse (A^y; which ubiquitously overproduces agouti, a melanocortin receptor antagonist) led to the discovery of agouti-related protein (AgRP, an endogenous inverse agonist of MC3R and MC4R), and contributed importantly to the understanding of the melanocortin system and its role in energy balance [44–46].

ARC AgRP neurons act in opposition to the ARC POMC neurons by secreting AgRP to attenuate αMSH signaling when energy stores are low [37, 38]. Leptin is an important regulator of the ARC melanocortin system, as many POMC and AgRP neurons express LepRb, and leptin potentiates signaling by POMC neurons while inhibiting AgRP neurons.

Mutant mouse lines have also been produced to examine most components of this system: while animals null for *Agrp* have little phenotype [47], animals and humans containing mutations that interfere with melanocortin action are obese [44–46, 48–51]. Mutations in *Pomc* itself promote hyperphagic obesity, as do mutations in the *Mc3r* and *Mc4r*, although the phenotype of animals null for *Mc4r* is much stronger than that of *Mc3r* nulls. Furthermore, humans carrying mutations in *POMC* and *MC4R* exhibit early onset obesity. Indeed, not only does the existence of these patients confirm the underlying importance of the melanocortin system for human physiology and energy balance, but mutations affecting *MC4R* underlie a substantial portion (between 2 and 5%) of extreme early onset obesity in a number of cohorts. Additionally, GWAS have repeatedly identified polymorphisms associated with the *MC4R* locus as contributors to obesity in large populations of humans [35, 52]. Thus, the melanocortin system represents an important neural controller of energy balance; genetic alteration of this system produces obesity in rodent genetic models, underlies a significant amount of extreme obesity in humans, and also contributes to the genetic predisposition of many humans to increased adiposity.

The Peptide Processing Machinery and the Control of Energy Balance

Like many other peptide hormones and neurotransmitters, melanocortin peptides must be proteolytically processed to generate active peptides, such as αMSH, in order to act on their receptors and mediate their biological functions [39]. POMC itself is cleaved by prohormone convertase 1 (PC1, also called PCSK1 or PC1/3) and PC2 (aka PCSK2) to generate a precursor peptide that is further cleaved by carboxypeptidase E (CPE) to generate active αMSH; acetylation by unknown enzymes may also modify αMSH activity. Following its release, αMSH is inactivated by prolylcarboxypeptidase (PRCP) [39, 53].

As one might expect, genetic changes in this pathway modify adiposity—animals null for *Pcsk1* are obese, as are the very few identified human patients with mutations for this enzyme [54, 55]. Surprisingly, animals null for *Pcsk2* are not obese, although they exhibit phenotypes consistent with other defects in peptide processing; this lack of obesity may result from the partial activity of the POMC PC1 product on MC3/4R even in the absence of PC2, combined with other hormonal and neural changes in these animals [56]. No humans defective in PCSK2 function have been identified.

The phenotype of the spontaneously obese *Fat* mouse results from altered CPE function, and mice with targeted disruption of *Cpe* are obese [57]. Conversely, mice null for *Prcp*, which inactivates αMSH, are lean [53]. Humans mutant for *CPE* or *PRCP* have not been identified, however.

Genes Controlling Important Hypothalamic Structures

While not entirely specific to the melanocortin system, disruption of the genes encoding the transcription factors, SIM1 and SF1, alter the organization and function of the PVH and ventromedial hypothalamic nucleus (VMH), respectively [58, 59]. The PVH is a major output pathway for leptin and the melanocortins [60]. Indeed, expression of MC4R in SIM1-expressing neurons reverses much of the phenotype of *Mc4r*-null mice [61]. Humans containing *SIM1* mutations have also been identified and are extremely obese, as are mice with targeted mutation of *Sim1* [62, 63]. The VMH, which plays an important role in feeding and energy balance, interacts with the ARC in general and with the melanocortin pathway specifically, and, while no human patients deficient for SF1 have been described, mice lacking SF1 are hyperphagic and obese [64].

Other Neural Systems That Control Energy Balance

Recent work in the neurobiology of energy balance, aided by the use of molecular genetic techniques, has identified other somewhat interrelated, neural systems that contribute to the control of energy balance [37, 65]. Both leptin and the melanocortins have been implicated in the control of two important sets of neurons that lie within the lateral hypothalamic area (LHA). One such set of neurons contains the neuropeptide melanin concentrating hormone (MCH; not related to POMC or any of its peptides) [66, 67]. MCH promotes feeding, and animals null for the gene encoding MCH (or its receptor) are lean [66, 68]. No humans mutant for the genes encoding MCH or its receptor have been discovered, consistent with the difficulty in identifying spontaneous genetic alterations conferring leanness.

A distinct set of LHA neurons express the neuropeptide hypocretin (HCRT, also known as orexin) [69–71]. Based upon early acute pharmacologic studies, HCRT was originally conceived of as an orexigen; subsequent work has revealed animals null for *Hcrt* or one of its receptors to be mildly obese, however [70, 72–74]. Indeed, narcolepsy, which results from the loss of HCRT action in mice and humans, is associated with increased adiposity [74, 75] (see Chap. 3 for further details regarding hypocretin and arousal).

The brain serotonin (5-HT) system also contributes to the control of energy balance, and 5-HT agonists (such as the formerly prescribed diet drug d-phenfluramine) diminish feeding and body weight [76]. This anorectic action of brain 5-HT is mediated via the 5HT2c receptor, and animals null for this receptor are obese [77]. The stimulation of ARC POMC neurons represents a major output pathway for 5HT2cR [78, 79]. No human mutations or risk alleles have been identified in this pathway, however. (See also Chap. 3 for details on the serotonin receptor.)

Obesity Syndromes and the Broad Alteration of Neural Function

Molecular Genetic Models and Human Obesity Syndromes

With the advent of gene targeting (i.e., knockout mice) has come the discovery that mutations in many genes produce unexpected changes in adiposity. Included in these are mutations in genes associated with the immune system, cell membrane and nuclear receptors, basic cellular metabolism, and, perhaps most prominently, neuronal and CNS function. The genes whose disruption promotes obesity but which were not previously associated with CNS function presumably possess unrecognized roles in the neural circuits that contribute to the control of feeding.

As one might expect, given the known and rather diverse functions of many genes whose disruption causes increased adiposity, many of these mutant mouse lines exhibit obesity within the context of additional phenotypes (a "syndrome"), rather than presenting with a discrete lesion in energy balance. Indeed, both mouse models and humans null for the more pleiotropic players listed for the leptin and melanocortin pathways, above, display a complex phenotype, of which obesity represents only one feature. Since leptin controls energy utilization via a plethora of pathways, disruption of leptin action produces hypothalamic hypogonadism, hypothyroidism, and immune defects, among other phenotypes [25, 26, 11–13]. Disruption of *POMC*, which produces ACTH to control adrenal function as well as peptides that control pigmentation, results in hypopigmentation and adrenal insufficiency [80]. Similarly, mice and humans mutant for PC1, which control the processing and subsequent activity of numerous peptides in addition to POMC, demonstrate hypogonadism, decreased growth, adrenal insufficiency, and hypoglycemia [54, 55].

Determining the mechanisms underlying the energy homeostasis phenotypes of mouse knockout lines with unanticipated obesity or of human with pleiotropic genetic syndromes that include obesity presents an enormous challenge. The general response to this challenge has been to look for the mechanism "under the streetlamp" by examining potential roles for known mediators of energy homeostasis, such as leptin and the melanocortin axis. While a few enlightening observations have been made in this manner, most of the diagnoses of "leptin resistance" or observed changes in (e.g., gene expression) within the melanocortin system revealed in mouse models have fallen rather short of mechanistic certitude. Consequently, most mouse models with complex phenotypes that include unexpected changes in adiposity have not been subjected to long-term follow-up. Mouse models that mirror known human conditions represent an important exception to this rule, however. Although it may not be practical to understand the molecular mechanisms underlying all obese knockout models in the near term, all such models, whether they mirror a human condition or not, are potentially informative, since the propensity to obesity in most affected humans likely represents the cumulative effects of small changes in multiple genes [7], potentially including those genes that produce large and pleiotropic changes when completely disrupted.

The analysis of human obesity syndromes and the rodent models that mirror these syndromes has begun to make mechanistic inroads, although a great deal of work remains for most of these. Certainly, the syndromes whose biology is best understood at present have in common the underlying alteration of neurophysiology, consistent with the causality of CNS processes that control feeding relative to energy stores in the predisposition to obesity.

BDNF and Its Receptor, TRKB

Brain-derived neurotrophic factor (BDNF) is widely expressed in the nervous system during development, as well as being expressed within several brain regions important for energy homeostasis in adults [81]. It acts via its receptor, TrkB, to control a variety of basic neural processes, including proliferation, survival, and plasticity. Given its many important roles in the CNS, alterations in BDNF expression (or that of its receptor, TrkB) would be predicted to interfere with multiple processes. Indeed, humans haploinsufficient for *BDNF* display impaired cognitive function and hyperactivity in addition to hyperphagic obesity, as do mice bearing similar mutations [82–84]. Mutations in *TRKB* produce similar hyperphagia and obesity in rare human patients, along with impaired cognitive function and nociception [85].

Studies in rodent models have revealed the regulation of *Bdnf* expression within the VMH by nutrition (feeding vs fasting), and have suggested a potential role for hypothalamic BDNF in the control of energy balance, perhaps via action on TrkB in the ARC [86]. Thus, BDNF action in the mediobasal hypothalamus may modulate food intake in response to physiologic changes in adult animals, in addition to controlling the development of many brain areas.

Interestingly, a coding polymorphism in *BDNF* (Val66Met) is associated both with obesity and with binge eating disorders in humans, consistent with the role for BDNF/TrkB signaling in energy balance, and suggesting a broader role for this system in the genetic determination of adiposity in humans [87]. Indeed, recent GWAS data have demonstrated an association between *BDNF*-linked single nucleotide polymorphisms (SNPs) and obesity [35].

Ciliopathies

A subset of mutations causing defects in primary cilia also promote obesity syndromes [88, 89]. The primary cilium is found on most cells; while structurally related to motile cilia (such as flagella), the primary cilium is immotile and does not participate in propulsion. The primary cilium plays a crucial sensory role in cells, including cell-specific sensing, such as olfaction in sensory epithelium, photoreception in retinal cells, mechanical transduction in kidney cells, and signaling via a variety of cell surface receptors, including many GPCRs. A broad group of disease-causing human mutations (and induced mutations in mouse models) have now been recognized to result from mutations in genes affecting ciliary functions (the "ciliopathies"). The clinical presentation of these diseases variably includes anosmia, retinal degeneration, kidney malformations, and a variety of developmental and neural defects, many of which are idiosyncratic to the particular gene that is mutated. A number of these mutations produce obesity in addition to the other phenotypes noted above, both in mice and in humans. Included in these obesity-causing ciliopathies are Bardet–Beidel Syndrome (BBS), McKusic–Kaufman Syndrome, Alström Syndrome, and Joubert Syndrome.

In most of these diseases, no structural defect in the primary cilium is observed. Rather, the BBS proteins, many of which associate in a "BBS-some" complex associated with the base of the primary cilium/basal body, may participate in the trafficking of proteins to and within the cilium. Indeed, mutation of *Ift88*, which encodes a protein specific for trafficking within the cilium, in mice results in an obesity phenotype similar to that produced by mutations in BBS genes [90]. While the particular protein(s) whose impaired trafficking may underlie this obesity is not yet clear, the primary cilium is crucial for signaling via a variety of receptor signaling pathways, including the Wnt and Shh pathways, tyrosine kinases such as the receptor for PDGF, and numerous GPCRs. Such alterations could impair the development or function of a variety of neural circuits important for the regulation of energy balance. The receptor for MCH is also trafficked to the primary cilium [91]. Thus, the mechanisms underlying the obesity associated with mutations in BBS proteins and associated proteins likely stem from alterations in the function of multiple receptors in more than one type of neuron important for energy balance.

Prader–Willi Syndrome

Prader–Willi syndrome (PWS) presents in infancy with low birth weight, hypotonia and feeding difficulties, with a progressive transition to hyperphagia and obesity starting around age 2 [92, 93]. Additional features include short stature (correctible with growth hormone therapy), hypogonadism, characteristic behaviors (especially around feeding), and often mental retardation. Most cases result from a 5 to 7 Mb deletion of an imprinted region (PWS region) on the paternal chromosome 15 (15q11q13). Within this deletion lie a number of genetic elements, including the genes encoding MAGEL2 and NECDIN, which are thought to be involved in neural development and function, and a complex noncoding locus. This noncoding locus consists of a transcribed noncoding gene (*SNURF-SNRPN*) that hosts a multitude of C/D box small nucleolar (sno-) RNA genes, including *SNORD116*; the RNA products of these *SNORD* genes are thought to be involved in RNA editing, perhaps of specific mRNA species.

A few patients with variant PWS apparently resulting from smaller genetic changes with the larger PWS region have been described [92]; some of these patients have demonstrated obesity and developmental delay in the absence of many of the other features of PWS, and affect primarily *SNORD* loci, prominently including *SNORD116*, suggesting an important role for this *SNORD* in the obesity of PWS. The *Snord116* locus has been deleted from mouse models, which display a growth defect and behavioral abnormalities, including a relative hyperphagia that develops after weaning, but which is balanced by increased activity [94]. Thus, the effects of *SNORD116* likely contribute to the PWS, but cannot account for all of the phenotypes. The functions of *Necdin* and *Magel2* have also been examined in genetically targeted mouse models: *Magel2*-null mice display early growth retardation with late-onset obesity, and *Necdin*-null mice display early postnatal respiratory failure along with a subset of PWS-associated behaviors [95–97]. Thus, the full PWS likely results from the combined effects of multiple genes; multiple of the loci within the PWS region also likely contribute to the full obesity phenotype. It is not yet clear how each of the loci within the PWS alters neurophysiology and/or which neurons they might specifically affect to alter energy balance.

Other Syndromes

A variety of other human obesity syndromes have been described, including Carpenter's syndrome. Carpenter's syndrome patients suffer from a pleiotropic set of musculoskeletal deformities, learning disabilities, and obesity; this syndrome may be attributable to defects in *RAB23*, which encodes a member of the Rab family of small GTPases that control intracellular trafficking [98, 99]. Interestingly, RAB23 is involved in the formation of the primary cilium, as well as playing roles in Shh signaling [89].

Variably sized deletions or alterations in chromosome 16p11.2 are also highly associated with obesity in the setting of a variety of neural/cognitive disorders and other variable phenotypes that may correlate with the size of the deletion [33, 34]. While the deletion locus may encompass or lie near *SH2B1*, the essential gene(s) of the approximately 30 genes affected in subjects with this deletion syndrome remains to be determined.

Other Risk Alleles Identified in Humans

The sequencing of multiple human genomes has permitted the detection and cataloguing of common SNPs throughout the human genome [7, 35]. This information has, in turn, enabled large-scale studies for linkage between huge numbers of SNPs and a variety of disease-associated phenotypes, such as obesity, in GWAS. Over 40 loci linked to increased measures of adiposity have been identified at this time, and are more likely to be identified as cohorts become larger, analysis becomes more sophisticated, and especially once whole-genome sequencing becomes the standard for such studies [35]. Several of the loci shown to associate with obesity in GWAS fit into the systems discussed above: roles for *MC4R*, *SH2B1*, and *BDNF* loci have been detected and replicated in numerous cohorts of people. Furthermore, the products of many of the most highly replicated loci are clearly associated with neuronal function: TMEM18 is a transmembrane protein that is highly expressed in neural tissues and has been implicated in neuronal migration [100]; NEGR1 is a GPI-linked cell adhesion molecule that controls neuronal growth [101, 102]; SEC16B is an ER membrane protein that controls the transport of vesicles important for the function of some neurons [103]; FAIM2 is a fas inhibitor that plays an anti-apoptotic role in the CNS, and has a role in neural protection during ischemia [104].

The function of many additional genes linked to adiposity remains unclear, however. Highly replicated loci include *ETV5*, which encodes an ETS family transcription factor that controls the expression of matrix metalloproteinases, and *GNPDA2*, which encodes a glucosamine-6-phosphate deamidase [35]. Indeed, the locus with the strongest linkage to adiposity and which contributes the largest amount to the genetic component of polygenic human obesity, *FTO*, remains mysterious in terms of its physiologic function (it appears to be a DNA demethylase) [105, 106]. Indeed, the *FTO* locus overlaps substantially with *FTM*, which encodes a protein involved in primary cilium function [107, 108]. *FTO* is expressed in the brain, including in hypothalamic feeding centers, however, where its expression is modulated by leptin and feeding status [107, 108]. Furthermore, its role in controlling food intake and body weight is suggested not only by GWAS results, but also from the phenotype of *FTO*-null mice, which are lean [109]. In contrast, mice ubiquitously overexpressing *FTO* or overexpressing *FTO* in the brain demonstrate increased food intake and adiposity [110]. Thus, while a great deal remains to be learned about how the *FTO*-associated SNPs may contribute to brain function and the control of body weight, the evidence suggests that it is important and has much to teach us about the systems that control energy balance.

Conclusion

Stemming from the study of spontaneous single-gene mutations that cause obesity (both in rodents and in humans), the results of mouse molecular genetic studies, and new human genetic association data, our understanding of the systems that control energy balance has burgeoned over the past few years. While we still have a great deal to learn, roles for leptin and melanocortin signaling, along with modifying systems and genes acting upstream and downstream of ARC melanocortin neurons, have been revealed. Similarly, contributions for BDNF signaling, along with those of numerous other mediators of neural function discovered by rodent and human genetics, demonstrate the primacy of CNS systems for the control of energy balance—consistent with our understanding of the underlying physiology. Given the primitive nature of our conception of these systems and the plethora of new, poorly understood, loci associated with altered energy balance, the next decade can only be more enlightening than the last.

References

1. Schwartz MW, Woods SC, Seeley RJ, Barsh GS, Baskin DG, Leibel RL. Is the energy homeostasis system inherently biased toward weight gain? Diabetes. 2003;52(2):232–8.
2. Spiegelman BM, Flier JS. Obesity and the regulation of energy balance. Cell. 2001;104(4):531–43.
3. Tremblay A, Despres JP, Theriault G, Fournier G, Bouchard C. Overfeeding and energy expenditure in humans. Am J Clin Nutr. 1992;56(5):857–62.
4. Control CfD. Prevalence of overweight, obesity and extreme obesity among adults: United States, trends 1960–62 through 2005–2006. 2011; http://www.cdc.gov/nchs/data/hestat/overweight/overweight_adult.htm.
5. Spiegel K, Tasali E, Leproult R, Van Cauter E. Effects of poor and short sleep on glucose metabolism and obesity risk. Nat Rev Endocrinol. 2009;5(5):253–61.
6. Bass J, Takahashi JS. Circadian integration of metabolism and energetics. Science. 2010;330(6009):1349–54.
7. Hebebrand J, Volckmar AL, Knoll N, Hinney A. Chipping away the 'missing heritability': GIANT steps forward in the molecular elucidation of obesity - but still lots to go. Obes Facts. 2010;3(5):294–303.
8. Feldmann HM, Golozoubova V, Cannon B, Nedergaard J. UCP1 ablation induces obesity and abolishes diet-induced thermogenesis in mice exempt from thermal stress by living at thermoneutrality. Cell Metab. 2009;9(2):203–9.
9. Bachman ES, Dhillon H, Zhang CY, et al. BetaAR signaling required for diet-induced thermogenesis and obesity resistance. Science. 2002;297(5582):843–5.
10. Cummings DE, Brandon EP, Planas JV, Motamed K, Idzerda RL, McKnight GS. Genetically lean mice result from targeted disruption of the RII beta subunit of protein kinase A. Nature. 1996;382(6592):622–6.
11. Coleman DL. Obese and diabetes: two mutant genes causing diabetes-obesity syndromes in mice. Diabetologia. 1978;14(3):141–8.
12. Ahima RS, Prabakaran D, Mantzoros C, et al. Role of leptin in the neuroendocrine response to fasting. Nature. 1996;382(6588):250–2.
13. Leibel RL, Chung WK, Chua Jr SC. The molecular genetics of rodent single gene obesities. J Biol Chem. 1997;272(51):31937–40.
14. Coleman DL. Effects of parabiosis of obese with diabetic and normal mice. Diabetologia. 1973;4:294–8.
15. Zhang Y, Proenca R, Maffei M, Barone M, Leopold L, Friedman JM. Positional cloning of the mouse obese gene and its human homologue. Nature. 1994;372(6505):425–32.
16. Chua Jr SC, Chung WK, Wu-Peng XS, et al. Phenotypes of mouse diabetes and rat fatty due to mutations in the OB (leptin) receptor. Science. 1996;271(5251):994–6.
17. Lee GH, Proenca R, Montez JM, et al. Abnormal splicing of the leptin receptor in diabetic mice. Nature. 1996;379:632–5.
18. Friedman JM, Halaas JL. Leptin and the regulation of body weight in mammals. Nature. 1998;395(6704):763–70.
19. Schwartz MW, Woods SC, Porte Jr D, Seeley RJ, Baskin DG. Central nervous system control of food intake. Nature. 2000;404(6778):661–71.
20. Rosenbaum M, Goldsmith R, Bloomfield D, et al. Low-dose leptin reverses skeletal muscle, autonomic, and neuroendocrine adaptations to maintenance of reduced weight. J Clin Invest. 2005;115(12):3579–86.
21. de Luca C, Kowalski TJ, Zhang Y, et al. Complete rescue of obesity, diabetes, and infertility in db/db mice by neuron-specific LEPR-B transgenes. J Clin Invest. 2005;115(12):3484–93.
22. Cohen P, Zhao C, Cai X, et al. Selective deletion of leptin receptor in neurons leads to obesity. J Clin Invest. 2001;108(8):1113–21.
23. Scott MM, Lachey JL, Sternson SM, et al. Leptin targets in the mouse brain. J Comp Neurol. 2009;514(5):518–32.
24. Patterson CM, Leshan RL, Jones JC, Myers Jr MG. Molecular mapping of mouse brain regions innervated by leptin receptor-expressing cells. Brain Res. 2011;1378:18–28.
25. Farooqi IS, Jebb SA, Langmack G, et al. Effects of recombinant leptin therapy in a child with congenital leptin deficiency. N Engl J Med. 1999;341(12):879–84.

26. Vaisse C, Lahlou N, Cabrol S, et al. A mutation in the human leptin receptor gene causes obesity and pituitary dysfunction. Nature. 1998;392(6674):398–401.
27. Robertson SA, Leinninger GM, Myers Jr MG. Molecular and neural mediators of leptin action. Physiol Behav. 2008;94(5):637–42.
28. Bates SH, Stearns WH, Schubert M, et al. STAT3 signaling is required for leptin regulation of energy balance but not reproduction. Nature. 2003;421:856–9.
29. Li Z, Zhou Y, Carter-Su C, Myers Jr MG, Rui L. SH2B1 enhances leptin signaling by both Janus kinase 2 Tyr813 phosphorylation-dependent and -independent mechanisms. Mol Endocrinol. 2007;21(9):2270–81.
30. Maures TJ, Kurzer JH, Carter-Su C. SH2B1 (SH2-B) and JAK2: a multifunctional adaptor protein and kinase made for each other. Trends Endocrinol Metab. 2007;18(1):38–45.
31. Ren D, Li M, Duan C, Rui L. Identification of SH2-B as a key regulator of leptin sensitivity, energy balance, and body weight in mice. Cell Metab. 2005;2(2):95–104.
32. Ren D, Zhou Y, Morris D, Li M, Li Z, Rui L. Neuronal SH2B1 is essential for controlling energy and glucose homeostasis. J Clin Invest. 2007;117(2):397–406.
33 Walters RG, Jacquemont S, Valsesia A, et al. A new highly penetrant form of obesity due to deletions on chromosome 16p11.2. Nature. 2010;463(7281):671–5.
34. Bachmann-Gagescu R, Mefford HC, Cowan C, et al. Recurrent 200-kb deletions of 16p11.2 that include the SH2B1 gene are associated with developmental delay and obesity. Genet Med. 2010;12(10):641–7.
35. Speliotes EK, Willer CJ, Berndt SI, et al. Association analyses of 249,796 individuals reveal 18 new loci associated with body mass index. Nat Genet. 2010;42(11):937–48.
36. Willer CJ, Speliotes EK, Loos RJ, et al. Six new loci associated with body mass index highlight a neuronal influence on body weight regulation. Nat Genet. 2009;41(1):25–34.
37. Morton GJ, Cummings DE, Baskin DG, Barsh GS, Schwartz MW. Central nervous system control of food intake and body weight. Nature. 2006;443(7109):289–95.
38. Elmquist JK, Coppari R, Balthasar N, Ichinose M, Lowell BB. Identifying hypothalamic pathways controlling food intake, body weight, and glucose homeostasis. J Comp Neurol. 2005;493(1):63–71.
39. Wardlaw SL. Hypothalamic proopiomelanocortin processing and the regulation of energy balance. Eur J Pharmacol. 2011;660(1):213–9.
40. Simerly RB. Hypothalamic substrates of metabolic imprinting. Physiol Behav. 2008;94(1):79–89.
41. Kishi T, Aschkenasi CJ, Lee CE, Mountjoy KG, Saper CB, Elmquist JK. Expression of melanocortin 4 receptor mRNA in the central nervous system of the rat. J Comp Neurol. 2003;457(3):213–35.
42. Huszar D, Lynch CA, Fairchild-Huntress V, et al. Targeted disruption of the melanocortin-4 receptor results in obesity in mice. Cell. 1997;88:131–41.
43. Butler AA, Kesterson RA, Khong K, et al. A unique metabolic syndrome causes obesity in the melanocortin-3 receptor-deficient mouse. Endocrinology. 2000;141(9):3518–21.
44. Fan W, Boston BA, Kesterson RA, Hruby VJ, Cone RD. Role of melanocortinergic neurons in feeding and the agouti obesity syndrome. Nature. 1997;385:165–8.
45. Duhl DM, Vrieling H, Miller KA, Wolff GL, Barsh GS. Neomorphic agouti mutations in obese yellow mice. Nat Genet. 1994;8(1):59–65.
46. Ollmann MM, Wilson BD, Yang YK, et al. Antagonism of central melanocortin receptors in vitro and in vivo by agouti-related protein. Science. 1997;278(5335):135–8.
47. Qian S, Chen H, Weingarth D, et al. Neither agouti-related protein nor neuropeptide Y is critically required for the regulation of energy homeostasis in mice. Mol Cell Biol. 2002;22(14):5027–35.
48. Ellacott KL, Cone RD. The role of the central melanocortin system in the regulation of food intake and energy homeostasis: lessons from mouse models. Philos Trans R Soc Lond B Biol Sci. 2006;361(1471):1265–74.
49. Yeo GS, Farooqi IS, Aminian S, Halsall DJ, Stanhope RG, O'Rahilly S. A frameshift mutation in MC4R associated with dominantly inherited human obesity. Nat Genet. 1998;20(2):111–2.
50. Vaisse C, Clement K, Durand E, Hercberg S, Guy-Grand B, Froguel P. Melanocortin-4 receptor mutations are a frequent and heterogeneous cause of morbid obesity. J Clin Invest. 2000;106(2):253–62.
51. Challis BG, Pritchard LE, Creemers JW, et al. A missense mutation disrupting a dibasic prohormone processing site in pro-opiomelanocortin (POMC) increases susceptibility to early-onset obesity through a novel molecular mechanism. Hum Mol Genet. 2002;11(17):1997–2004.
52. Loos RJ, Lindgren CM, Li S, et al. Common variants near MC4R are associated with fat mass, weight and risk of obesity. Nat Genet. 2008;40(6):768–75.
53. Wallingford N, Perroud B, Gao Q, et al. Prolylcarboxypeptidase regulates food intake by inactivating alpha-MSH in rodents. J Clin Invest. 2009;119(8):2291–303.
54. Jackson RS, Creemers JW, Farooqi IS, et al. Small-intestinal dysfunction accompanies the complex endocrinopathy of human proprotein convertase 1 deficiency. J Clin Invest. 2003;112(10):1550–60.
55. Farooqi IS, Volders K, Stanhope R, et al. Hyperphagia and early-onset obesity due to a novel homozygous missense mutation in prohormone convertase 1/3. J Clin Endocrinol Metab. 2007;92(9):3369–73.

56. Furuta M, Yano H, Zhou A, et al. Defective prohormone processing and altered pancreatic islet morphology in mice lacking active SPC2. Proc Natl Acad Sci USA. 1997;94(13):6646–51.
57. Fricker LD, Berman YL, Leiter EH, Devi LA. Carboxypeptidase E activity is deficient in mice with the fat mutation. Effect on peptide processing. J Biol Chem. 1996;271(48):30619–24.
58. Michaud JL, Rosenquist T, May NR, Fan CM. Development of neuroendocrine lineages requires the bHLH-PAS transcription factor SIM1. Genes Dev. 1998;12(20):3264–75.
59. Dhillon H, Zigman JM, Ye C, et al. Leptin directly activates SF1 neurons in the VMH, and this action by leptin is required for normal body-weight homeostasis. Neuron. 2006;49(2):191–203.
60. Simerly RB. Wired on hormones: endocrine regulation of hypothalamic development. Curr Opin Neurobiol. 2005;15(1):81–5.
61. Balthasar N, Dalgaard LT, Lee CE, et al. Divergence of melanocortin pathways in the control of food intake and energy expenditure. Cell. 2005;123(3): 493–505.
62. Michaud JL, Boucher F, Melnyk A, et al. Sim1 haploinsufficiency causes hyperphagia, obesity and reduction of the paraventricular nucleus of the hypothalamus. Hum Mol Genet. 2001;10(14):1465–73.
63. Holder Jr JL, Butte NF, Zinn AR. Profound obesity associated with a balanced translocation that disrupts the SIM1 gene. Hum Mol Genet. 2000;9(1):101–8.
64. Kim KW, Sohn JW, Kohno D, Xu Y, Williams K, Elmquist JK. SF-1 in the ventral medial hypothalamic nucleus: a key regulator of homeostasis. Mol Cell Endocrinol. 2011;336(1–2):219–23.
65. Berthoud HR. Interactions between the "cognitive" and "metabolic" brain in the control of food intake. Physiol Behav. 2007;91(5):486–98.
66. Qu D, Ludwig DS, Gammeltoft S, et al. A role for melanin-concentrating hormone in the central regulation of feeding behavior. Nature. 1996;380(6571): 243–7.
67. Ludwig DS, Mountjoy KG, Tatro JB, et al. Melanin-concentrating hormone: a functional melanocortin antagonist in the hypothalamus. Am J Physiol. 1998;37:E627–33.
68. Shimada M, Tritos NA, Lowell BB, Flier JS, Maratos-Flier E. Mice lacking melanin-concentrating hormone are hypophagic and lean. Nature. 1998;396(6712):670–4.
69. Tritos NA, Mastaitis JW, Kokkotou E, Maratos-Flier E. Characterization of melanin concentrating hormone and preproorexin expression in the murine hypothalamus. Brain Res. 2001;895(1–2):160–6.
70. Sakurai T, Amemiya A, Ishii M, et al. Orexins and orexin receptors: a family of hypothalamic neuropeptides and G protein-coupled receptors that regulate feeding behavior. Cell. 1998;92(4):573–85.
71. Broberger C, de Lecea L, Sutcliffe JG, Hokfelt T. Hypocretin/orexin- and melanin-concentrating hormone-expressing cells form distinct populations in the rodent lateral hypothalamus: relationship to the neuropeptide Y and agouti gene-related protein systems. J Comp Neurol. 1998;402(4):460–74.
72. Chemelli RM, Willie JT, Sinton CM, et al. Narcolepsy in orexin knockout mice: molecular genetics of sleep regulation. Cell. 1999;98(4):437–51.
73. Funato H, Tsai AL, Willie JT, et al. Enhanced orexin receptor-2 signaling prevents diet-induced obesity and improves leptin sensitivity. Cell Metab. 2009;9(1):64–76.
74. Hara J, Yanagisawa M, Sakurai T. Difference in obesity phenotype between orexin-knockout mice and orexin neuron-deficient mice with same genetic background and environmental conditions. Neurosci Lett. 2005;380(3):239–42.
75. Schuld A, Hebebrand J, Geller F, Pollmacher T. Increased body-mass index in patients with narcolepsy. Lancet. 2000;355(9211):1274–5.
76. Heisler LK, Cowley MA, Kishi T, et al. Central serotonin and melanocortin pathways regulating energy homeostasis. Ann N Y Acad Sci. 2003;994:169–74.
77. Nonogaki K, Strack AM, Dallman MF, Tecott LH. Leptin-independent hyperphagia and type 2 diabetes in mice with a mutated serotonin 5-HT2C receptor gene. Nat Med. 1998;4(10):1152–6.
78. Xu Y, Jones JE, Lauzon DA, et al. A serotonin and melanocortin circuit mediates D-fenfluramine anorexia. J Neurosci. 2010;30(44):14630–4.
79. Xu Y, Jones JE, Kohno D, et al. 5-HT2CRs expressed by pro-opiomelanocortin neurons regulate energy homeostasis. Neuron. 2008;60(4):582–9.
80. Smart JL, Tolle V, Low MJ. Glucocorticoids exacerbate obesity and insulin resistance in neuron-specific proopiomelanocortin-deficient mice. J Clin Invest. 2006;116(2):495–505.
81. Bernd P. The role of neurotrophins during early development. Gene Expr. 2008;14(4):241–50.
82. Gray J, Yeo GS, Cox JJ, et al. Hyperphagia, severe obesity, impaired cognitive function, and hyperactivity associated with functional loss of one copy of the brain-derived neurotrophic factor (BDNF) gene. Diabetes. 2006;55(12):3366–71.
83. Han JC, Liu QR, Jones M, et al. Brain-derived neurotrophic factor and obesity in the WAGR syndrome. N Engl J Med. 2008;359(9):918–27.
84. Kernie SG, Liebl DJ, Parada LF. BDNF regulates eating behavior and locomotor activity in mice. EMBO J. 2000;19(6):1290–300.
85. Yeo GS, Connie Hung CC, Rochford J, et al. A de novo mutation affecting human TrkB associated with severe obesity and developmental delay. Nat Neurosci. 2004;7(11):1187–89.
86. Xu B, Goulding EH, Zang K, et al. Brain-derived neurotrophic factor regulates energy balance downstream of melanocortin-4 receptor. Nat Neurosci. 2003;6(7):736–42.
87. Friedel S, Horro FF, Wermter AK, et al. Mutation screen of the brain derived neurotrophic factor gene (BDNF): identification of several genetic variants and association studies in patients with obesity,

eating disorders, and attention-deficit/hyperactivity disorder. Am J Med Genet B Neuropsychiatr Genet. 2005;132B(1):96–9.
88. Zaghloul NA, Katsanis N. Mechanistic insights into Bardet-Biedl syndrome, a model ciliopathy. J Clin Invest. 2009;119(3):428–37.
89. Sharma N, Berbari NF, Yoder BK. Ciliary dysfunction in developmental abnormalities and diseases. Curr Top Dev Biol. 2008;85:371–427.
90. Davenport JR, Watts AJ, Roper VC, et al. Disruption of intraflagellar transport in adult mice leads to obesity and slow-onset cystic kidney disease. Curr Biol. 2007;17(18):1586–94.
91. Berbari NF, O'Connor AK, Haycraft CJ, Yoder BK. The primary cilium as a complex signaling center. Curr Biol. 2009;19(13):R526–35.
92. Buiting K. Prader-Willi syndrome and Angelman syndrome. Am J Med Genet C Semin Med Genet. 2010;154C(3):365–76.
93. Whittington J, Holland A. Neurobehavioral phenotype in Prader-Willi syndrome. Am J Med Genet C Semin Med Genet. 2010;154C(4):438–47.
94. Ding F, Li HH, Zhang S, et al. SnoRNA Snord116 (Pwcr1/MBII-85) deletion causes growth deficiency and hyperphagia in mice. PLoS One. 2008;3(3):e1709.
95. Bischof JM, Stewart CL, Wevrick R. Inactivation of the mouse Magel2 gene results in growth abnormalities similar to Prader-Willi syndrome. Hum Mol Genet. 2007;16(22):2713–9.
96. Mercer RE, Kwolek EM, Bischof JM, van Eede M, Henkelman RM, Wevrick R. Regionally reduced brain volume, altered serotonin neurochemistry, and abnormal behavior in mice null for the circadian rhythm output gene Magel2. Am J Med Genet B Neuropsychiatr Genet. 2009;150B(8):1085–99.
97. Muscatelli F, Abrous DN, Massacrier A, et al. Disruption of the mouse Necdin gene results in hypothalamic and behavioral alterations reminiscent of the human Prader-Willi syndrome. Hum Mol Genet. 2000;9(20):3101–10.
98. Jenkins D, Seelow D, Jehee FS, et al. RAB23 mutations in Carpenter syndrome imply an unexpected role for hedgehog signaling in cranial-suture development and obesity. Am J Hum Genet. 2007;80(6):1162–70.
99. Jenkins D, Baynam G, De Catte L, et al. Carpenter syndrome: extended RAB23 mutation spectrum and analysis of nonsense-mediated mRNA decay. Hum Mutat. 2011;32(4):E2069–78.
100. Almen MS, Jacobsson JA, Shaik JH, et al. The obesity gene, TMEM18, is of ancient origin, found in majority of neuronal cells in all major brain regions and associated with obesity in severely obese children. BMC Med Genet. 2010;11:58.
101. Funatsu N, Miyata S, Kumanogoh H, et al. Characterization of a novel rat brain glycosylphosphatidylinositol-anchored protein (Kilon), a member of the IgLON cell adhesion molecule family. J Biol Chem. 1999;274(12):8224–30.
102. Miyata S, Funatsu N, Matsunaga W, Kiyohara T, Sokawa Y, Maekawa S. Expression of the IgLON cell adhesion molecules Kilon and OBCAM in hypothalamic magnocellular neurons. J Comp Neurol. 2000;424(1):74–85.
103. Miller EA, Barlowe C. Regulation of coat assembly–sorting things out at the ER. Curr Opin Cell Biol. 2010;22(4):447–53.
104. Somia NV, Schmitt MJ, Vetter DE, Van Antwerp D, Heinemann SF, Verma IM. LFG: an anti-apoptotic gene that provides protection from Fas-mediated cell death. Proc Natl Acad Sci USA. 1999;96(22):12667–72.
105. Frayling TM, Timpson NJ, Weedon MN, et al. A common variant in the FTO gene is associated with body mass index and predisposes to childhood and adult obesity. Science. 2007;316(5826):889–94.
106. Gerken T, Girard CA, Tung YC, et al. The obesity-associated FTO gene encodes a 2-oxoglutarate-dependent nucleic acid demethylase. Science. 2007;318(5855):1469–72.
107. Stratigopoulos G, Padilla SL, LeDuc CA, et al. Regulation of Fto/Ftm gene expression in mice and humans. Am J Physiol Regul Integr Comp Physiol. 2008;294(4):R1185–96.
108. Stratigopoulos G, LeDuc CA, Cremona ML, Chung WK, Leibel RL. Cut-like homeobox 1 (CUX1) regulates expression of the fat mass and obesity-associated and retinitis pigmentosa GTPase regulator-interacting protein-1-like (RPGRIP1L) genes and coordinates leptin receptor signaling. J Biol Chem. 2011;286(3):2155–70.
109. Fischer J, Koch L, Emmerling C, et al. Inactivation of the Fto gene protects from obesity. Nature. 2009;458(7240):894–8.
110. Church C, Moir L, McMurray F, et al. Overexpression of Fto leads to increased food intake and results in obesity. Nat Genet. 2010;42(12):1086–92.

Short Sleep and Obesity Risk in Children

7

Chantelle N. Hart, Elizabeth S. Kuhl, and Elissa Jelalian

Abstract

Accumulating evidence suggests that short sleep is associated with both concurrent and prospective obesity risk in children. Although recent studies point to potential mediators and moderators of the sleep weight association in children, it is unclear whether enhancing children's sleep duration could significantly impact weight status. The present chapter reviews the existing literature on sleep and obesity risk in children, and highlights areas in need of additional work in an effort to determine whether enhancing children's sleep will help to combat the current obesity epidemic.

Introduction

Pediatric obesity has reached epidemic proportions. The most recent estimates indicate that 15% of U.S. children 2–19 years old are overweight (BMI ≥ 85th percentile, but less than the 95th percentile for age and gender) and an additional 17% are obese (BMI ≥ 95th percentile) [1]. Although trends for obesity appear to have leveled off over the past 10 years [1], current estimates represent a threefold increase in pediatric obesity over the past 30–40 years [1–3]. The most recent estimates show that groups at greatest risk of being obese include those from ethnic minority backgrounds and older children and adolescents [1].

The negative health and psychosocial consequences associated with being obese as a child and adolescent are clear. Health risks include impaired glucose tolerance [4], insulin resistance [5], development of type 2 diabetes [6–8] and metabolic syndrome [9], high diastolic and systolic blood pressure, left ventricular hypertrophy, atherosclerosis [7, 10], asthma [11], and nonalcoholic fatty liver disease [7]. Psychosocial risks include decreased health-related quality of life [12, 13], decreased physical self-worth, higher body dissatisfaction [14, 15], and increased risk of teasing and/or peer victimization [16, 17]. Collectively, the evidence highlights the urgency of developing effective prevention and intervention

C.N. Hart, PhD (✉) • E.S. Kuhl, PhD
Psychiatry & Human Behavior/Weight Control & Diabetes Research Center, Alpert Medical School of Brown University/The Miriam Hospital, 196 Richmond Street, Providence, RI 02903, USA
e-mail: Chantelle_Hart@brown.edu

E. Jelalian, PhD
Psychiatry & Human Behavior/Bradley Hasbro Children's Research Center, Alpert Medical School of Brown University/The Miriam Hospital, 196 Richmond Street, Providence, RI 02903, USA

P.J. Shiromani et al. (eds.), *Sleep Loss and Obesity: Intersecting Epidemics*, DOI 10.1007/978-1-4614-3492-4_7, © Springer Science+Business Media, LLC 2012

programs to change weight trajectories in children and adolescents.

A number of prevention and intervention approaches targeted at healthier eating and activity behaviors and weight status have been evaluated. To date, prevention approaches that combine modification of diet and physical activity in school settings have demonstrated promise [18]. However, as a whole, there has been considerable variability in terms of their effectiveness for weight-related outcomes [19]. In contrast, family-based pediatric intervention approaches that combine dietary and physical activity prescriptions with the use of behavioral strategies are considered empirically supported and efficacious [20] with some evidence for long-term changes in weight status [21]. Furthermore, at least for certain adolescent populations, there is some evidence supporting the effectiveness of both medication and surgery [20, 22].

Although there is empirical support for a number of strategies outlined above, it is important to note that most children remain overweight and obese despite best efforts. This has led to recent interest in identifying novel targets-beyond eating and activity behaviors—that may play a role in this growing epidemic and thus could be targeted by future interventions [23]. One factor that has received considerable attention over the past few years is children's sleep duration. A growing body of research suggests that sleep duration is associated with children's obesity risk, and thus may represent a novel treatment target for future intervention. This chapter will review this evidence, and highlight areas for future research to determine whether enhancing children's sleep would help to change weight trajectories and thus produce benefits for downstream health and psychosocial outcomes.

Increasing Rates of Obesity and Decreasing Sleep Duration: Co-occurring Phenomena

At the most basic level, as the number of children in this country who are overweight and obese has risen, reported nocturnal sleep duration has decreased. A recent review of international studies using data from 1905 through 2008 found that sleep in children 5–18 years old has declined [24]. Specifically, findings demonstrated that nightly sleep decreased by 0.75 nightly min per year over the 103-year period with the greatest decreases in sleep length found in the United States, Canada, Europe, and Asia. Additional studies demonstrated decreases in children's sleep over the past 20–30 years [25, 26]—the period of time during which the proportion of children who are overweight and obese has increased threefold [1–3]. Currently, children are reported to sleep about 9.50 h each night [27–29]. While these estimates are based on parent- and self-report, they are largely supported by objective measures of sleep. One large cohort study of school-age children that used home polysomnography showed that children 6–11 years old obtained 8.45–9.25 h of nightly sleep [30]. It is important to note that all of these estimates of average sleep fall below expert recommendations (i.e., 10–11 h of sleep/night) [31, 32], suggesting that on average children may not be obtaining sufficient sleep each night.

It has been argued that changes in sleep duration over time likely reflect changes in lifestyle rather than a biological need for less sleep [33, 34]. Lifestyle changes that are often implicated in changes in sleep length include increased use of technology, busier lifestyles, and changes in parental attitudes toward children's sleep needs [33, 24, 34]. Taken together with findings documenting high levels of self-reported daytime sleepiness in school-age children [35] as well as discrepancies in children's sleep length on weekdays and weekends (when they can sleep more) [29, 36], it appears that children may need more sleep than they are obtaining. These findings combined with additional evidence documenting that children who are reported to sleep less are identified as having greater cognitive difficulties than their more rested counterparts, including problems with attention and memory, and decreased processing speed [37] make a compelling case against biology and in favor of other factors such as lifestyle changes.

Pediatric Epidemiological Studies Assessing Short Sleep and Weight Status

Beyond these noted temporal changes in sleep and weight status, a growing body of evidence is available to support an association between sleep duration and obesity risk. While studies contributing to this literature have not been designed to assess causality, the consistency among findings is striking. Two recent meta-analyses suggest that across reviewed studies, children with shorter sleep were 56–89% more likely to be obese than their more rested counterparts [38, 39]. Findings are particularly notable given that studies were conducted in 16 different countries with children from birth to 19 years of age, using both cross-sectional and prospective research designs, and controlling for a number of potential confounding factors [40]. Cross-sectional and prospective studies that have assessed the association between sleep duration and obesity risk are reviewed below. For a more detailed review of individual studies, the reader is referred to a number of recent reviews and meta-analyses [38–41].

Cross-Sectional Studies

One of the first studies to assess the association between children's sleep duration and obesity risk was conducted almost 20 years ago [42]. That study found that 5-year-old children sleeping less than 11 h/night were at a 40% increased risk for being obese than children sleeping 11 h or more/night. Subsequent studies have largely confirmed these early findings [38–41].

For example, in a sample of 4-year-old children, Anderson and Whitaker [43] found that compared to those who slept less than 10.5 h/night, those who slept more had a 14% reduced risk of being obese-even after accounting for a number of factors such as child age and gender, maternal BMI and education, screen time, and poverty status. Although most cross-sectional findings are limited by self-reported (or caretaker-reported) sleep length, two studies that used actigraphy to measure sleep length found consistent results. Using one night of actigraphy, Nixon and colleagues [44] found that for 7-year-olds, those who slept less than 9 h/night had over a threefold increased risk of obesity than those who slept 9 h or more. Similarly, a study with children 11–16 years of age found that for each additional hour of sleep (as measured by one night of actigraphy), the risk of obesity decreased by 80% [45].

In addition to risk for overweight and obesity, more recent studies assessed the association between sleep duration and other indices of body composition, including more precise measures of body fat. Findings regarding the effect of sleep duration on measured body fat are generally consistent with those that assessed BMI. Two studies using bioimpedance assay (BIA) found that 5- to 7-year-old children who slept less had higher percentages of body fat than those who slept more [44, 46]. An additional study which used dual-energy X-ray absorptiometry (DEXA) with adolescent twins found that females who slept less had higher total and truncal body fat and less lean mass [47]. However, these findings were not consistent in males. Studies in regards to waist circumference have been less consistent than those using BMI or percent body fat as outcomes, although there is some indication that shorter sleep is associated with higher waist circumference [47–50].

Prospective Studies

Prospective studies are consistent with cross-sectional findings. For example, one study with children 3–18 years old found that for each additional hour of sleep reported at the first assessment, there was a 0.75 kg/m^2 decrease in BMI at follow-up [28]. Furthermore, a study that assessed sleep patterns across the toddler and preschool years (2.5–6 years old) found that children who were persistently short sleepers had almost a threefold increased risk of obesity when they were 6 years old [51]. An additional study demonstrated that the association between short sleep and obesity risk may be particularly pronounced in younger children. Bell and Zimmerman [52] found that for children birth–4 years, sleeping less at night was associated with increased risk of overweight and obesity 5 years later. Interestingly, they found no influence for daytime sleep on later obesity risk. They also found no association between sleep and obesity for the older cohort of children 5–13 years of age. Additional prospective studies have demonstrated temporal associations between childhood sleep and obesity risk in adolescence and adulthood. For example, Seegers and colleagues [53] found that each hour less of nightly sleep at 10 years of age was associated with an increased risk of overweight and obesity at 13 years of age. Even more striking, Landhuis and colleagues [54] found that nightly sleep during childhood (from 5 to 11 years of age) was associated with obesity risk at 32 years of age. Thus, prospective studies have documented that sleep duration is not only associated with concurrent, but future risk of obesity as well.

There is additional evidence that discrepancies in sleep length on weekdays and weekends may be associated with obesity status. In a community cohort study of children 4–10 years old, obese children not only slept less, but more variably on weekends than on weekdays [61]. Children with shorter and more variable sleep also had greater metabolic dysfunction in this study, including higher C-reactive protein levels and altered insulin and low density lipoprotein [61]. An additional study showed that children who persistently slept 8 h or less on weekdays and weekends or on weekdays and during holidays were at greater risk for obesity than children who compensated for short sleep on weekends or holidays [62]. Finally, one recent study found that females (but not males) in high school who slept in on weekends had lower body fat and a healthier weight status than those who did not [63]. Thus, in addition to sleep duration, it may be important to assess variability in sleep length as an additional risk for obesity.

The Association of Other Sleep Parameters with Obesity Risk

Most cross-sectional and prospective pediatric studies to date have focused on the potential role of duration of sleep on obesity risk. However, there is ample evidence from animal models suggesting that circadian clocks play an important role in metabolism and obesity [55]. Thus, more recent studies have begun to assess the association between additional sleep parameters and obesity risk, including the timing and variability of sleep [56–59]. With one notable exception [57], studies have found that later bedtimes (as opposed to rise times) are associated with increased obesity risk [28, 56, 58, 59]. Sleep irregularity may also be associated with obesity risk. One study found that children who were reported by parents to have irregular sleeping habits between 2 and 4 years of age had a higher risk of obesity in young adulthood [60]. Findings persisted even after adjustment for a number of potential confounding variables such as child sex, maternal BMI, dietary patterns during adolescence, and television viewing.

Pathways Through Which Sleep May Influence Obesity Risk

Studies reviewed above provide compelling evidence for a potential role of sleep, particularly sleep duration, on obesity risk. With this path now well established, more recent studies have begun to assess the potential pathways through which sleep may influence weight status. A number of pathways have been proposed [40, 41, 64]. Because of their role in energy balance, eating and activity behaviors have been variables of high interest. Driven by key findings from experimental studies with adults, the influence of sleep duration on appetite regulation (due to metabolic changes in appetite regulating hormones) and eating behaviors have been of particular interest. In addition to mediating pathways, there may be shared underlying causes for both short sleep and obesity in children [65]. Two that have been postulated are low socioeconomic status (SES) and stressors associated with low SES [66], and less effective parenting styles [65]. Each of these areas will be discussed briefly below.

Physiological Changes That Could Affect Eating Behaviors

Evidence from experimental studies with adults point toward plausible physiological pathways through which sleep could influence eating behaviors and obesity risk. Although to date, we are unaware of similar published studies with children (although studies are ongoing at present), tentative inferences can be made to help provide a framework from which to understand the sleep–weight relationship in children. A review of this experimental work with adults is beyond the scope of this chapter (please see Chap. 10); however, it is relevant to note that the secretion of a number of hormones is influenced by sleep [67]. In particular, two hormones involved in appetite regulation [68–70]—leptin and ghrelin—have been implicated as possible mediators of the association between sleep and obesity status. Studies have documented that levels of circulating leptin decrease [71, 72] and levels of ghrelin increase [71, 73] upon sleep restriction (compared to sleep extension). Although experimental studies with children have yet to be published, extant correlational research with children demonstrate that short sleep is associated with a number of physiological and metabolic changes such as lower fasting C-peptide [74], hyperglycemia [75], and higher fasting insulin, peak insulin, and insulin resistance [76]. Thus, findings from adult studies may be of relevance for our understanding of how sleep influences obesity risk in children as well.

Eating Behaviors

Beyond metabolic changes, recent pediatric studies have begun to focus on the association between sleep and eating behaviors. In a large sample of school-aged children, shorter sleep was associated with greater consumption of energy-dense foods (e.g., pizza and refined sugars), particularly in boys and on weekdays [77]. An additional study with children found that when parent report of child overeating (at 6 years old) was removed from the statistical model to predict BMI, there was a small increase in the odds for children classified as "short persistent" sleepers to be obese, which suggests that eating pathways may mediate the association between sleep and obesity risk [51]. At least two additional studies have been conducted with adolescents. In the first, adolescents with more daytime/less nocturnal sleep reported greater food cravings [78]. A second study found that adolescents with shorter weekday sleep duration (as measured with actigraphy) reported consuming a higher percentage of calories from fat and lower percentage from carbohydrates [79]. Thus, evidence to date provides some support for a potential role of eating behaviors in the sleep–obesity relationship. However, these findings can only speak to the associative links between sleep and eating behaviors, and cannot demonstrate that enhancing sleep in children will influence what they eat.

Activity Behaviors

Most studies to date have focused on eating pathways. However, there is some evidence supporting an association between sleep and both sedentary and physical activities. Several studies demonstrate that television (TV) viewing and having a TV in the bedroom is negatively associated with good sleep hygiene. For example, one study with school-age children found that having a TV in the bedroom, watching TV around bedtime, and daily TV viewing were associated with bedtime resistance and sleep onset delay as well as short sleep duration [80]. A second study with toddlers found that increased TV viewing was associated with irregular nap and bedtime schedules [81]. Studies with adolescents are consistent. Having a TV in the bedroom and watching more TV was associated with later bedtimes on weekdays and weekends, later rise times on weekends, and less time spent in bed [82]. Similarly, watching more television during adolescence was associated with an increased risk of developing sleep problems during young adulthood [83]. When considered in light of data demonstrating an association between TV viewing and obesity risk [84–86] and the effectiveness of decreasing TV viewing to enhance weight loss [87–89], sedentary behaviors such as TV viewing seem like important factors that may link sleep duration with children's weight status. However, it is unclear whether sedentary behaviors mediate associations between short sleep and obesity risk or represent a common underlying cause of both of these conditions.

Although several studies have assessed the impact of technology and sedentary behaviors on children's sleep, less is known regarding the association between sleep and engagement in physical activity, particularly how sleep duration affects children's subsequent engagement in physical activity. One study with adolescents found that sleep duration was positively associated with engagement in exercise [90]. However, at least one additional study found no association between sleep duration and physical activity [45]. There is some evidence from experimental studies with adults that sleep restriction may affect subsequent engagement in physical activity [91, 92], although not all studies have found this effect [93]. Clearly, more work needs to be done to determine whether changing children's sleep—either through experimental manipulation or intervention approaches—leads to subsequent changes in sedentary and physical activities.

Shared Underlying Causes of Short Sleep and Obesity Risk

Although most research has focused on short sleep leading to obesity risk, it is also possible that other, third variables may influence both short sleep and obesity in children [65]. van Cauter and Spiegel [66] proposed that sleep may mediate the association between socioeconomic status and health (including obesity). In fact, one pediatric study found that economic strain (i.e., presence of a number of economic difficulties within the past 12 months) was associated with later bedtimes in children [94]. However, this finding was not significant in adolescents and was not associated with the total number of hours that children or adolescents were reported to sleep [94]. Additional studies demonstrate that individuals from lower SES backgrounds are at increased risk for being obese [95, 96]. Thus, socioeconomic factors may underlie problems with both sleep and obesity. However, additional studies are needed to demonstrate exactly how SES may influence both sleep and weight status.

In addition to socioeconomic conditions, parenting style may play an important role in determining both risk of short sleep and obesity in children. There is some evidence that parenting factors, including parenting style, perceived warmth, and the number of family rules are associated with sleep quantity and quality in children and adolescents [97]. For example, one study found that, for school-age children, greater parental warmth was associated with more weekday sleep, and that for adolescents, stricter household rules were associated with greater weekday sleep, which was likely due to implementation of earlier bedtimes [94]. Similarly, emergent data suggest that an authoritative parenting style may be protective against obesity risk [98], and that use of positive reinforcement and monitoring are positively associated with healthier eating and activity habits in children [99]. Beyond parenting style per se, it is interesting to note that effective intervention strategies for both short sleep and obesity include the use of behavioral strategies that enhance the appropriate use of parental limit setting, and promote the use of positive reinforcement and monitoring of key behaviors [100]. Thus, further exploration of the role of parenting behaviors in both short sleep and obesity risk, particularly given their relevance for effective behavioral interventions, seems warranted.

Summary and Review of Study Limitations

In sum, accumulating evidence provides strong support for an association between short sleep and obesity risk in children with meta-analyses demonstrating a significant effect of short sleep on pediatric obesity risk [38, 39]. Recent studies have provided additional support for this association by demonstrating that sleep duration is associated with hypothesized mediators of the sleep–weight relationship, and by pointing toward additional factors which may moderate the influence of sleep on obesity risk (and thus may account for some discrepancies in study findings to date). In addition, there is evidence that common underlying factors may also link short sleep with obesity risk in children. While research demonstrating possible pathways through which sleep may influence obesity risk are provocative, considerable work remains prior to determining whether enhancing sleep will help curb the current obesity epidemic.

In an effort to inform future research, it is important to note limitations of previous work. Perhaps, the two key methodological limitations of previous studies are the reliance upon epidemiological work to support a link between sleep and obesity risk in children, and the almost exclusive reliance on parent- or self-report of sleep [38, 40, 65]. Although several large studies have assessed prospective associations between sleep and obesity and have controlled for confounding factors that could influence the association between these variables over time, they cannot demonstrate that less sleep causes increases in weight status, and hence obesity. Clearly, experimental and intervention studies with children in which sleep length is carefully manipulated (or lengthened with effective treatment approaches) are needed. More studies using objective measures of sleep such as actigraphy and polysomnography are also needed to strengthen confidence in the reliability and validity of previous findings.

Beyond these two key limitations, it is also important to note some inconsistencies across studies. First, not all studies found a dose–response association between sleep duration and obesity risk. For example, one study with children 2–4 years old found that compared to sleeping greater than 11 h each night, sleeping 9–10 h each night was associated with a 34% increased risk of obesity [57]. However, sleeping less than 9 h per night was not associated with increased risk. In addition, one study found that how sleep was measured (i.e., self-report or time diary self-report) influenced whether a significant association was observed between sleep duration and obesity [101]. Furthermore, an additional recent study found that despite significant bivariate associations between sleep and obesity status, multivariate results showed that sleep duration was not independently associated with a child's risk of being obese [102]. Although the preponderance of data suggests a significant association between sleep and children's weight status, the above-noted limitations and inconsistencies speak to the need to think critically about the design of future work, including consideration of key factors that may influence study findings.

Areas for Future Research

Based on the previous work, a number of areas in need of future research seem warranted. Perhaps, most importantly, future studies need to move beyond epidemiological work by using experimental designs and interventions to actively manipulate children's sleep duration [40, 65]. Experimental and intervention studies need to focus not only on the main outcome of obesity, but also on a number of other relevant factors, including mediators and moderators of the sleep–weight relationship. Future work also needs to consider other underlying mechanisms, which may influence both sleep and weight status. Finally, consideration of other sleep parameters such as variability in sleep–wake schedules could broaden our understanding of how sleep influences children's obesity risk. Each of these areas will be reviewed briefly below.

Need to Better Understand the Role of Sleep Timing

Findings from recent work suggesting that consistency in sleep across weekdays and weekend days may confer protection against obesity risk are interesting. The premise that variability in sleep could influence weight status is supported by research on the influence of the circadian system on appetite regulation [55]. Experimental studies may be particularly useful for better understanding how variability versus duration (or their combination) may influence eating, activity, and weight status in children. Furthermore, the idea that "catching up" on sleep on weekends and holidays can confer protection against obesity risk in children [62] is one that merits future study. In particular, it will be important to better understand the potential benefit of "catch-up" sleep for obesity risk versus the importance of consistency in sleep–wake habits (a common component of sleep hygiene recommendations) in the promotion of healthy sleep in children [40].

Exploration of Mediators, Moderators, and Underlying Common Mechanisms

Research studies to date have provided some interesting findings regarding possible mediators and moderators of the sleep–weight association. First and foremost, future experimental and intervention studies need to explore how changes in sleep influence subsequent eating and activity behaviors. Thus far pediatric findings linking sleep with these potential mediators have been based on cross-sectional, observational studies. However, mounting experimental work with adults suggests that short sleep may influence both subsequent eating and engagement in activity [91–93, 103, 104]. It will be important to demonstrate whether similar effects (including metabolic changes) are evident in pediatric populations.

In addition, it will be important to determine whether certain factors may moderate the association between short sleep and obesity risk. This is particularly important as it can speak to specific populations who may benefit the most from a sleep intervention targeted at decreasing weight status. Studies to date suggest two potential variables to consider in future work: child age and sex. Reviews of studies across the lifespan demonstrate that associations between sleep and obesity risk are stronger in children than in adults, and in younger as opposed to older adults [38, 41]. Within pediatric studies, a dose–response association between sleep and obesity risk was found only in children younger than 10 years of age (although additional analyses did not demonstrate a stronger association between sleep and obesity risk in younger children) [39]. Moreover, subsequent work found that the association between sleep duration and obesity risk was stronger in younger children [52, 105]. Thus, whether interventions to enhance sleep should target younger children is a plausible question in need of future work.

In addition to age, gender may moderate the association between sleep length and obesity risk. A recent meta-analysis found that the association between sleep and obesity risk was stronger in boys than in girls [39]. A subsequent study supported this finding [48] while two additional studies that were published following the meta-analysis found the opposite effect [49, 106]. Given the potential influence of age and gender as well as the role of sleep in hormone regulation, it is possible that some of the effects evidenced for both gender and age may be related to developmental maturation and pubertal status. However, the role of pubertal development has largely remained unexplored.

Associations between short sleep and obesity risk may also be explained by common underlying factors [65]. Although studies to date have primarily focused on how lengthening sleep can improve weight status, cross-sectional associations between sleep and weight status also support the potential utility of addressing both short sleep and obesity as comorbid conditions in need of treatment [40, 100]. As noted previously in this chapter, a number of factors point to the potential utility of targeting parenting behaviors to address both sleep duration and obesity risk. Behavioral interventions for sleep disturbance and obesity share the use of key parenting strategies, including positive reinforcement, appropriate limit setting, and self-monitoring [40, 100]. Interventions designed for young children may benefit the most from parenting interventions. Young children are particularly dependent upon parents to have their needs met, and in this way parent behaviors are highly influential in shaping lifestyle behavioral habits and routines. Taken together with data suggesting that young children may be at higher risk for the effects of short sleep on obesity risk [39, 52, 105] as well as recommendations for less intensive obesity interventions with younger children [107], targeting the use of key parenting behaviors that could influence both sleep and obesity risk may be particularly relevant to this age group. In fact, it is notable that a recent, effective multicomponent intervention designed to decrease obesity risk in infants included parenting and sleep components [108].

Conclusion

Shortened sleep duration is associated with both concurrent and future risk of obesity in children. Although recent studies have documented potential mediators and moderators of the sleep–weight association in children, it remains unclear whether enhancing children's nightly sleep will help to curb the current obesity epidemic. Experimental and intervention studies that change children's sleep length and assess how these changes influence eating, activity, and weight status are greatly needed. Studies that also aim to identify individuals at greatest risk of the influence of short sleep on weight status will also help to focus work at those in greatest need.

Acknowledgments This work was supported in part by Grant Nos. 1-09-JF-22 from the American Diabetes Association, and R01HL092910 from the National Heart Lung and Blood Institute.

References

1. Ogden CL, Carroll MD, Curtin LR, Lamb MM, Flegal KM. Prevalence of high body mass index in US children and adolescents, 2007–2008. JAMA. 2010;303(3):242–9.
2. Troiano RP, Flegal KM. Overweight children and adolescents: description, epidemiology, and demographics. Pediatrics. 1998;101(3 Pt 2):497–504.
3. Ogden CL, Flegal KM, Carroll MD, Johnson CL. Prevalence and trends in overweight among US children and adolescents, 1999–2000. J Am Med Assoc. 2002;288:1728–32.
4. Sinha R, Fisch G, Teague B, et al. Prevalence of impaired glucose tolerance among children and adolescents with marked obesity. N Engl J Med. 2002;346(11):802–10.
5. Valente AM, Strong W, Sinaiko AR. Obesity and insulin resistance in young people. Am Heart J. 2001;142(3):440–4.
6. Cruz ML, Shaibi GQ, Weigensberg MJ, Spruijt-Metz D, Ball GD, Goran MI. Pediatric obesity and insulin resistance: chronic disease risk and implications for treatment and prevention beyond body weight modification. Annu Rev Nutr. 2005;25:435–68.
7. Daniels SR. The consequences of childhood overweight and obesity. Future Child. 2006;16(1):47–67.
8. Hannon TS, Rao G, Arslanian SA. Childhood obesity and type 2 diabetes mellitus. Pediatrics. 2005; 116(2):473–80.
9. Nathan BM, Moran A. Metabolic complications of obesity in childhood and adolescence: more than just diabetes. Curr Opin Endocrinol Diabetes Obes. 2008;15(1):21–9.
10. Freedman DS, Dietz WH, Srinivasan SR, Berenson GS. The relation of overweight to cardiovascular risk factors among children and adolescents: the Bogalusa heart study. Pediatrics. 1999;103(6 Pt 1): 1175–82.
11. Story RE. Asthma and obesity in children. Curr Opin Pediatr. 2007;19(6):680–4.
12. Williams J, Wake M, Hesketh K, Maher E, Waters E. Health-related quality of life of overweight and obese children. JAMA. 2005;293(1):70–6.
13. Schwimmer JB, Burwinkle TM, Varni JW. Health-related quality of life of severely obese children and adolescents. JAMA. 2003;289(14):1813–9.
14. Wardle J, Cooke L. The impact of obesity on psychological well-being. Best Pract Res Clin Endocrinol Metab. 2005;19(3):421–40.
15. Thompson JK, Shroff H, Herbozo S, Cafri G, Rodriguez J, Rodriguez M. Relations among multiple peer influences, body dissatisfaction, eating disturbance, and self-esteem: a comparison of average weight, at risk of overweight, and overweight adolescent girls. J Pediatr Psychol. 2007;32(1):24–9.
16. Pearce MJ, Boergers J, Prinstein MJ. Adolescent obesity, overt and relational peer victimization, and romantic relationships. Obes Res. 2002;10(5): 386–93.
17. Neumark-Sztainer D, Falkner N, Story M, Perry C, Hannan PJ, Mulert S. Weight-teasing among adolescents: correlations with weight status and disordered eating behaviors. Int J Obes Relat Metab Disord. 2002;26(1):123–31.
18. Brown T, Summerbell C. Systematic review of school-based interventions that focus on changing dietary intake and physical activity levels to prevent childhood obesity: an update to the obesity guidance produced by the national institute for health and clinical excellence. Obes Rev. 2009;10(1):110–41.
19. Summerbell CD, Waters E, Edmunds LD, Kelly S, Brown T, Campbell KJ. Interventions for preventing obesity in children. Cochrane Database Syst Rev. 2005;(3):CD001871.
20. Oude Luttikhuis H, Baur L, Jansen H, et al. Interventions for treating obesity in children. Cochrane Database Syst Rev. 2009;(1):CD001872.
21. Epstein LH, Valoski A, Wing RR, McCurley J. Ten-year outcomes of behavioral family-based treatment

for childhood obesity. Health Psychol. 1994;13(5): 373–83.
22. Pratt JS, Lenders CM, Dionne EA, et al. Best practice updates for pediatric/adolescent weight loss surgery. Obesity (Silver Spring). 2009;17(5):901–10.
23. Keith SW, Redden DT, Katzmarzyk PT, et al. Putative contributors to the secular increase in obesity: exploring the roads less traveled. Int J Obes (Lond). 2006;30(11): 1585–94.
24. Matricciani L, Olds T, Petkov J. In search of lost sleep: secular trends in the sleep time of school-aged children and adolescents. Sleep Med Rev. 2011.
25. Dollman J, Ridley K, Olds T, Lowe E. Trends in the duration of school-day sleep among 10- to 15-year-old South Australians between 1985 and 2004. Acta Paediatr. 2007;96(7):1011–4.
26. Iglowstein I, Jenni OG, Molinari L, Largo RH. Sleep duration from infancy to adolescence: reference values and generational trends. Pediatrics. 2003;111(2): 302–7.
27. Foundation NS. Final report: 2004 sleep in America poll; 2004.
28. Snell EK, Adam EK, Duncan GJ. Sleep and the body mass index and overweight status of children and adolescents. Child Dev. 2007;78(1):309–23.
29. Spilsbury JC, Storfer-Isser A, Drotar D, et al. Sleep behavior in an urban US sample of school-aged children. Arch Pediatr Adolesc Med. 2004;158(10): 988–94.
30. Quan SF, Goodwin JL, Babar SI, et al. Sleep architecture in normal Caucasian and Hispanic children aged 6–11 years recorded during unattended home polysomnography: experience from the Tucson Children's Assessment of Sleep Apnea Study (TuCASA). Sleep Med. 2003;4(1):13–9.
31. Mindell JA, Owens J. A clinical guide to pediatric sleep: diagnosis and management of sleep problems. Philadelphia: Lippincott Williams & Wilkins; 2003.
32. Ferber R. Solve your child's sleep problems: new, revised and expanded edition. NY: Fireside; 1996.
33. Guilleminault C, Powell NB, Martinez S, et al. Preliminary observations on the effects of sleep time in a sleep restriction paradigm. Sleep Med. 2003; 4(3):177–84.
34. Broman JE, Lundh LG, Hetta J. Insufficient sleep in the general population. Neurophysiol Clin. 1996;26(1): 30–9.
35. Amschler DH, McKenzie JF. Elementary students' sleep habits and teacher observations of sleep-related problems. J Sch Health. 2005;75(2):50–6.
36. Thorleifsdottir B, Bjornsson JK, Benediktsdottir B, Gislason T, Kristbjarnarson H. Sleep and sleep habits from childhood to young adulthood over a 10-year period. J Psychosom Res. 2002;53(1):529–37.
37. Fallone G, Acebo C, Seifer R, Carskadon MA. Experimental restriction of sleep opportunity in children: effects on teacher ratings. Sleep. 2005;28(12): 1561–7.
38. Cappuccio FP, Taggart FM, Kandala NB, et al. Meta-analysis of short sleep duration and obesity in children and adults. Sleep. 2008;31(5):619–26.
39. Chen X, Beydoun MA, Wang Y. Is sleep duration associated with childhood obesity? A systematic review and meta-analysis. Obesity (Silver Spring). 2008;16(2):265–74.
40. Hart CN, Cairns A, Jelalian E. Sleep and obesity in children and adolescents. Pediatr Clin North Am. 2011;58(3):715–33.
41. Patel SR, Hu FB. Short sleep duration and weight gain: a systematic review. Obesity (Silver Spring). 2008;16(3):643–53.
42. Locard E, Mamelle N, Billette A, Miginiac M, Munoz F, Rey S. Risk factors of obesity in a five year old population. Parental versus environmental factors. Int J Obes Relat Metab Disord. 1992;16(10): 721–9.
43. Anderson SE, Whitaker RC. Household routines and obesity in US preschool-aged children. Pediatrics. 2010;125(3):420–8.
44. Nixon GM, Thompson JM, Han DY, et al. Short sleep duration in middle childhood: risk factors and consequences. Sleep. 2008;31(1):71–8.
45. Gupta NK, Mueller WH, Chan W, Meininger JC. Is obesity associated with poor sleep quality in adolescents? Am J Hum Biol. 2002;14(6):762–8.
46. von Kries R, Toschke AM, Wurmser H, Sauerwald T, Koletzko B. Reduced risk for overweight and obesity in 5- and 6-y-old children by duration of sleep – a cross-sectional study. Int J Obes. 2002;26(5): 710–6.
47. Yu Y, Lu BS, Wang B, et al. Short sleep duration and adiposity in Chinese adolescents. Sleep. 2007;30(12): 1688–97.
48. Ozturk A, Mazicioglu M, Poyrazoglu S, Cicek B, Gunay O, Kurtoglu S. The relationship between sleep duration and obesity in Turkish children and adolescents. Acta Paediatr. 2009;98(4):699–702.
49. Hitze B, Bosy-Westphal A, Bielfeldt F, et al. Determinants and impact of sleep duration in children and adolescents: data of the Kiel obesity prevention study. Eur J Clin Nutr. 2009;63(6):739–46.
50. Eisenmann JC, Ekkekakis P, Holmes M. Sleep duration and overweight among Australian children and adolescents. Acta Paediatr. 2006;95(8):956–63.
51. Touchette E, Petit D, Tremblay RE, et al. Associations between sleep duration patterns and overweight/obesity at age 6. Sleep. 2008;31(11):1507–14.
52. Bell JF, Zimmerman FJ. Shortened nighttime sleep duration in early life and subsequent childhood obesity. Arch Pediatr Adolesc Med. 2010;164(9):840–5.
53. Seegers V, Petit D, Falissard B, et al. Short sleep duration and body mass index: a prospective longitudinal study in preadolescence. Am J Epidemiol. 2011;173(6):621–9.
54. Landhuis CE, Poulton R, Welch D, Hancox RJ. Childhood sleep time and long-term risk for obesity: a 32-year prospective birth cohort study. Pediatrics. 2008;122(5):955–60.
55. Bray MS, Young ME. Circadian rhythms in the development of obesity: potential role for the circadian clock within the adipocyte. Obes Rev. 2007; 8(2):169–81.

56. Maddah M, Rashidi A, Mohammadpour B, Vafa R, Karandish M. In-school snacking, breakfast consumption, and sleeping patterns of normal and overweight Iranian high school girls: a study in urban and rural areas in Guilan, Iran. J Nutr Educ Behav. 2009;41(1):27–31.
57. Sekine M, Yamagami T, Hamanishi S, et al. Parental obesity, lifestyle factors and obesity in preschool children: results of the Toyama birth cohort study. J Epidemiol. 2002;12(1):33–9.
58. Sekine M, Yamagami T, Handa K, et al. A dose–response relationship between short sleeping hours and childhood obesity: results of the Toyama birth cohort study. Child Care Health Dev. 2002; 28(2):163–70.
59. Olds T, Blunden S, Dollman J, Maher CA. Day type and the relationship between weight status and sleep duration in children and adolescents. Aust N Z J Public Health. 34(2):165–171.
60. Al Mamun A, Lawlor DA, Cramb S, O'Callaghan M, Williams G, Najman J. Do childhood sleeping problems predict obesity in young adulthood? Evidence from a prospective birth cohort study. Am J Epidemiol. 2007;166(12):1368–73.
61. Spruyt K, Molfese DL, Gozal D. Sleep duration, sleep regularity, body weight, and metabolic homeostasis in school-aged children. Pediatrics. 2011;127(2):e345–52.
62. Wing YK, Li SX, Li AM, Zhang J, Kong AP. The effect of weekend and holiday sleep compensation on childhood overweight and obesity. Pediatrics. 2009;124(5):e994–1000.
63. Lytle LA, Pasch KE, Farbakhsh K. The relationship between sleep and weight in a sample of adolescents. Obesity (Silver Spring). 2011;19(2):324–31.
64. Knutson KL, Van Cauter E. Associations between sleep loss and increased risk of obesity and diabetes. Ann N Y Acad Sci. 2008;1129:287–304.
65. Nielsen LS, Danielsen KV, Sorensen TI. Short sleep duration as a possible cause of obesity: critical analysis of the epidemiological evidence. Obes Rev. 2010;12(2):78–92.
66. Van Cauter E, Spiegel K. Sleep as a mediator of the relationship between socioeconomic status and health: a hypothesis. Ann N Y Acad Sci. 1999;896: 254–61.
67. Copinschi G. Metabolic and endocrine effects of sleep deprivation. Essent Psychopharmacol. 2005;6(6): 341–7.
68. Wren AM, Seal LJ, Cohen MA, et al. Ghrelin enhances appetite and increases food intake in humans. J Clin Endocrinol Metab. 2001;86(12): 5992.
69. Levin F, Edholm T, Schmidt PT, et al. Ghrelin stimulates gastric emptying and hunger in normal-weight humans. J Clin Endocrinol Metab. 2006;91(9): 3296–302.
70. Mars M, de Graaf C, de Groot CP, van Rossum CT, Kok FJ. Fasting leptin and appetite responses induced by a 4-day 65%-energy-restricted diet. Int J Obes (Lond). 2006;30(1):122–8.
71. Spiegel K, Tasali E, Penev P, Van Cauter E. Brief communication: sleep curtailment in healthy young men is associated with decreased leptin levels, elevated ghrelin levels, and increased hunger and appetite. Ann Intern Med. 2004;141(11):846–50.
72. Spiegel K, Leproult R, L'Hermite-Baleriaux M, Copinschi G, Penev PD, Van Cauter E. Leptin levels are dependent on sleep duration: relationships with sympathovagal balance, carbohydrate regulation, cortisol, and thyrotropin. J Clin Endocrinol Metab. 2004;89(11):5762–71.
73. Schmid SM, Hallschmid M, Jauch-Chara K, Born J, Schultes B. A single night of sleep deprivation increases ghrelin levels and feelings of hunger in normal-weight healthy men. J Sleep Res. 2008; 17(3):331–4.
74. Verhulst SL, Schrauwen N, Haentjens D, et al. Sleep duration and metabolic dysregulation in overweight children and adolescents. Arch Dis Child. 2008;93(1): 89–90.
75. Tian Z, Ye T, Zhang X, et al. Sleep duration and hyperglycemia among obese and nonobese children aged 3 to 6 years. Arch Pediatr Adolesc Med. 2010; 164(1):46–52.
76. Flint J, Kothare SV, Zihlif M, et al. Association between inadequate sleep and insulin resistance in obese children. J Pediatr. 2007;150(4):364–9.
77. Westerlund L, Ray C, Roos E. Associations between sleeping habits and food consumption patterns among 10-11-year-old children in Finland. Br J Nutr. 2009;102(10):1531–7.
78. Landis AM, Parker KP, Dunbar SB. Sleep, hunger, satiety, food cravings, and caloric intake in adolescents. J Nurs Scholarsh. 2009;41(2):115–23.
79. Weiss A, Xu F, Storfer-Isser A, Thomas A, Ievers-Landis CE, Redline S. The association of sleep duration with adolescents' fat and carbohydrate consumption. Sleep. 2010;33(9):1201–9.
80. Owens J, Maxim R, McGuinn M, Nobile C, Msall M, Alario A. Television-viewing habits and sleep disturbance in school children. Pediatrics. 1999; 104(3):e27.
81. Thompson DA, Christakis DA. The association between television viewing and irregular sleep schedules among children less than 3 years of age. Pediatrics. 2005;116(4):851–6.
82. Van den Bulck J. Television viewing, computer game playing, and internet use and self-reported time to bed and time out of bed in secondary-school children. Sleep. 2004;27(1):101–4.
83. Johnson JG, Cohen P, Kasen S, First MB, Brook JS. Association between television viewing and sleep problems during adolescence and early adulthood. Arch Pediatr Adolesc Med. 2004;158(6): 562–8.
84. Andersen RE, Crespo CJ, Bartlett SJ, Cheskin LJ, Pratt M. Relationship of physical activity and television watching with body weight and level of fatness among children: results from the third national health and nutrition examination survey. JAMA. 1998; 279(12):938–42.

85. Dennison BA, Erb TA, Jenkins PL. Television viewing and television in bedroom associated with overweight risk among low-income preschool children. Pediatrics. 2002;109(6):1028–35.
86. Robinson TN. Television viewing and childhood obesity. Pediatr Clin North Am. 2001;48(4):1017–25.
87. Epstein LH, Paluch RA, Gordy CC, Dorn J. Decreasing sedentary behaviors in treating pediatric obesity. Arch Pediatr Adolesc Med. 2000;154(3): 220–6.
88. Epstein LH, Roemmich JN, Robinson JL, et al. A randomized trial of the effects of reducing television viewing and computer use on body mass index in young children. Arch Pediatr Adolesc Med. 2008;162(3):239–45.
89. Robinson TN. Reducing children's television viewing to prevent obesity: a randomized controlled trial. JAMA. 1999;282(16):1561–7.
90. Chen MY, Wang EK, Jeng YJ. Adequate sleep among adolescents is positively associated with health status and health-related behaviors. BMC Public Health. 2006;6:59.
91. Benedict C, Hallschmid M, Lassen A, et al. Acute sleep deprivation reduces energy expenditure in healthy men. Am J Clin Nutr. 2011;93(6):1229–36.
92. Schmid SM, Hallschmid M, Jauch-Chara K, et al. Short-term sleep loss decreases physical activity under free-living conditions but does not increase food intake under time-deprived laboratory conditions in healthy men. Am J Clin Nutr. 2009;90(6):1476–82.
93. St-Onge MP, Roberts AL, Chen J, et al. Short sleep duration increases energy intakes but does not change energy expenditure in normal-weight individuals. Am J Clin Nutr. 2011;94(2):410–6.
94. Adam EK, Snell EK, Pendry P. Sleep timing and quantity in ecological and family context: a nationally representative time-diary study. J Fam Psychol. 2007;21(1):4–19.
95. Shrewsbury V, Wardle J. Socioeconomic status and adiposity in childhood: a systematic review of cross-sectional studies 1990–2005. Obesity (Silver Spring). 2008;16(2):275–84.
96. Grow HM, Cook AJ, Arterburn DE, Saelens BE, Drewnowski A, Lozano P. Child obesity associated with social disadvantage of children's neighborhoods. Soc Sci Med. 2010;71(3):584–91.
97. Meltzer LJ, Montgomery-Downs HE. Sleep in the family. Pediatr Clin North Am. 2011;58(3):765–74.
98. Sleddens EF, Gerards SM, Thijs C, de Vries NK, Kremers SP. General parenting, childhood overweight and obesity-inducing behaviors: a review. Int J Pediatr Obes. 2011;6(2–2):e12–27.
99. Arredondo EM, Elder JP, Ayala GX, Campbell N, Baquero B, Duerksen S. Is parenting style related to children's healthy eating and physical activity in Latino families? Health Educ Res. 2006;21(6): 862–71.
100. Hart CN, Jelalian E. Shortened sleep duration is associated with pediatric overweight. Behav Sleep Med. 2008;6(4):251–67.
101. Knutson KL, Lauderdale DS. Sleep duration and overweight in adolescents: self-reported sleep hours versus time diaries. Pediatrics. 2007;119(5):e1056–62.
102. Hassan F, Davis MM, Chervin RD. No independent association between insufficient sleep and childhood obesity in the National Survey of Children's Health. J Clin Sleep Med. 2011;7(2):153–57.
103. Nedeltcheva AV, Kilkus JM, Imperial J, Kasza K, Schoeller DA, Penev PD. Sleep curtailment is accompanied by increased intake of calories from snacks. Am J Clin Nutr. 2009;89(1):126–33.
104. Brondel L, Romer MA, Nougues PM, Touyarou P, Davenne D. Acute partial sleep deprivation increases food intake in healthy men. Am J Clin Nutr. 2010; 91(6):1550–9.
105. Shi Z, Taylor AW, Gill TK, Tuckerman J, Adams R, Martin J. Short sleep duration and obesity among Australian children. BMC Public Health. 2010; 10:609.
106. Sun Y, Sekine M, Kagamimori S. Lifestyle and overweight among Japanese adolescents: the Toyama birth cohort study. J Epidemiol. 2009;19(6):303–10.
107. Spear BA, Barlow SE, Ervin C, et al. Recommendations for treatment of child and adolescent overweight and obesity. Pediatrics. 2007;120 Suppl 4:S254–88.
108. Paul IM, Savage JS, Anzman SL, et al. Preventing obesity during infancy: a pilot study. Obesity (Silver Spring). 2011;19(2):353–61.

Circadian Misalignment and Sleep Disruption in Shift Work: Implications for Fatigue and Risk of Weight Gain and Obesity *

8

Rachel R. Markwald and Kenneth P. Wright Jr.

Abstract

The interplay of circadian timing and metabolic physiology represents a new frontier in biomedical research. Emerging evidence from animal models indicates that circadian physiology impacts weight gain, including the observation of obesity in clock gene mutants and most recently the finding that food intake restricted to the habitual sleep time of mice leads to weight gain as compared to the same amount of food intake during the normal wake episode. Eating at night is common in work schedules with long work hours and with work operations during the nighttime hours (e.g., health care, emergency response, security personnel) and in circadian sleep disorders including, but not limited to, shift work disorder. Shift work and shift work disorder are associated with circadian misalignment, sleep disruption, and fatigue, all of which may contribute to weight gain and obesity via the modification of feeding hormones and perhaps total daily energy expenditure. Future research is needed to explore the impact of circadian misalignment/sleep disruption and the resulting fatigue on metabolic physiology in shift workers, the mechanisms underlying this association and to develop effective countermeasures to promote shift worker health and well-being.

Introduction

Emerging evidence from animal models indicates a fundamental interplay between circadian and metabolic physiology that may have important implications in understanding metabolic health and disease [1, 2]. The role of the circadian timekeeping system in coordinating physiological and behavioral events so that they occur at an appropriate environmental time of day is widely recognized (e.g., sleeping during the solar day and activity at night for nocturnal species).

* Preparation of this manuscript was supported in part by NIH HL085705 and HL081761.

R.R. Markwald, PhD • K.P. Wright Jr., PhD (✉)
Department of Integrative Physiology,
University of Colorado at Boulder,
1725 Pleasant Street, Clare Small 114,
Boulder, CO 80309-0354, USA
e-mail: kenneth.wright@colorado.edu

P.J. Shiromani et al. (eds.), *Sleep Loss and Obesity: Intersecting Epidemics*,
DOI 10.1007/978-1-4614-3492-4_8,

Our understanding of how the circadian timing of physiological events is critical for health represents an important emerging field of study in biomedical research. Circadian clock genes are now recognized to exist in central and peripheral tissues outside of the core circadian clock of the hypothalamic suprachiasmatic nucleus (SCN). For example, clock genes are present in brain regions regulating endocrine physiology and in peripheral metabolic tissues including adipose, liver, muscle, and pancreas [3–8]. Based on the findings from animal models, we are beginning to understand how circadian misalignment and disruption of circadian clock genes lead to impaired tissue function, risk of obesity, metabolic syndrome, and diabetes [1, 2, 9]. Shift workers are at a higher risk of developing obesity and diabetes [10–14] compared to their day-working counterparts. Night eating syndrome is a disorder associated with obesity where the proportion of calories eaten at night after dinner is high (e.g., >50% after 7 pm) [15–17]. Circadian misalignment is associated with disturbed sleep, and sleep loss is recognized as a risk factor for weight gain, obesity, and diabetes. This chapter provides an overview of the interplay between circadian, sleep, and metabolic physiology from experimental animals to humans and provides a working model for how circadian disruption may contribute to weight gain and obesity primarily focusing on circadian misalignment during shift work.

Circadian Clock and Sleep–Wakefulness Homeostasis Interact to Regulate Daily Activity Patterns

Arousal systems located in the brain stem, basal forebrain, and hypothalamus promote wakefulness, while inhibition of these systems promotes sleep (see Chap. 3 for details). These arousal systems are under homeostatic [18–21] and circadian control [22, 23] (see Chaps. 1 and 2 for details regarding the biological clock). In general, sleep homeostasis builds up with increased duration of wakefulness. It has been reported that the metabolic byproduct of energy utilization, adenosine, is a sleep factor that reflects homeostatic sleep drive, rising with increased time awake, dissipating during sleep, and influencing brain regulatory wakefulness/sleep centers [21, 24, 25]. Thus, sleep and metabolic systems are integrated at the molecular level. When daily sleep need is not met, homeostatic sleep drive is increased resulting in fatigue and impaired alertness and performance. In addition, the SCN regulates near-24-h rhythms or circadian rhythms in brain arousal [26, 27]. Rhythms of the pineal hormone melatonin and body temperature are the most commonly used marker rhythms driven by the circadian clock in humans. Low melatonin and high body temperature levels represent the biological day, whereas high melatonin and low body temperature levels represent the biological night (Fig. 8.1). During the biological day, the clock system promotes wakefulness and associated functions (e.g., activity, energy intake) and during the biological night, the clock system promotes sleep and associated functions (e.g., rest, energy conservation) [28]. Misalignment between sleep and wakefulness schedules and the internal circadian timing system occurs in many occupational schedules in today's 24-h society. This misalignment occurs because the internal biological time keeping system of most shift workers does not appear to easily adapt to working at night [29, 30]. Rotating and permanent shift workers and travelers who rapidly cross multiple time zones are commonly awake during the biological night when levels of the sleep-promoting hormone melatonin are high.

Circadian Misalignment

Circadian misalignment can be defined by an altered relationship between sleep and wakefulness timing relative to internal circadian timing. Such circadian misalignment is common in work operations with long work hours and work operations during the nighttime hours (e.g., health care, emergency response, security personnel) and in circadian sleep disorders. For shift workers in particular, inappropriately timed exposure to

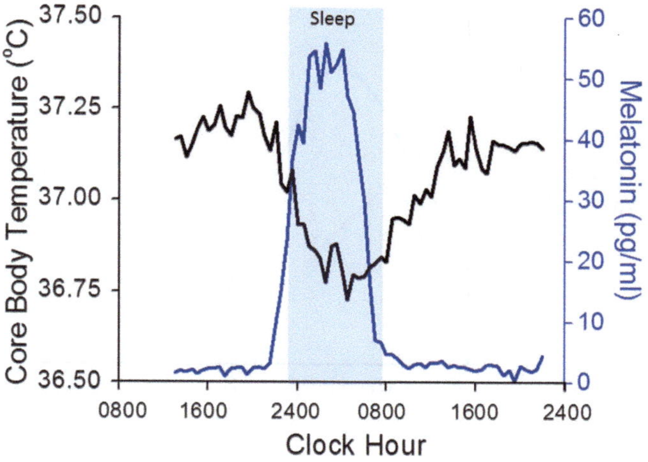

Fig. 8.1 *High circadian-driven melatonin levels and low core body temperature levels represent the biological night*. Circadian rhythms in melatonin and body temperature are two commonly used markers of the phase of the internal circadian clock. The biological night is associated with high levels of circulating melatonin and low core body temperature levels. Conversely, the biological day is associated with low levels of circulating melatonin and higher core body temperature levels

natural and artificial environmental light are hypothesized to be primary factors that impede adaptation of the human circadian clock to shift work schedules [31]. During circadian misalignment, sleep is often attempted during the biological day when levels of the clock-driven sleep-promoting hormone melatonin are low and wakefulness occurs during the biological night when endogenous melatonin levels are high. We and others have shown that circadian misalignment leads to impaired cognition [32, 33], disturbed sleep [32, 34–36], and altered feeding hormones [37, 38].

Findings from animal studies also indicate that circadian misalignment can occur as a result of a change in the relative timing of circadian-driven molecular rhythms in peripheral tissues (e.g., clock gene rhythms in the liver take longer to adjust to simulated jet lag than do clock gene rhythms in the SCN [39]). Additionally, it has been shown that the timing of food intake can modulate peripheral clocks and lead to desynchronization from the central pacemaker. Damiola and colleagues showed that food restriction resets the phase of clock gene rhythm expression in tissues such as the liver and kidney [40]. The latter study was one of the first to show that entrainment of peripheral oscillators could occur by metabolic signals and temporarily uncouple them from the master clock in the SCN. The potential health implications of eating at an inappropriate circadian time for physiology are talked about later in this chapter.

Shift Work and Disturbed Sleep

Wakefulness and sleep schedules associated with shift work go against human biology. Humans evolved to be awake, intake food, and be metabolically active during the solar day and to be asleep, not take in food, and be less metabolically active during the solar night. Thus, the biological day in humans is optimized to occur during the solar day and the biological night is optimized to occur during the solar night. Circadian misalignment is considered to be a primary mechanism underlying many of the health and safety concerns associated with shift work. As noted, the circadian system of night shift workers is typically misaligned such that the biological night occurs during the geophysical night, even though night shift workers are awake and working [29, 30, 41, 42]. Furthermore, shift workers sleep after their night shift during the biological day, when melatonin levels are low and the circadian clock

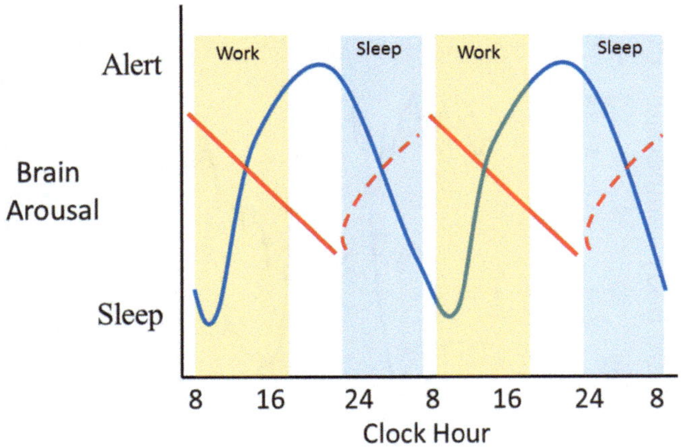

Fig. 8.2 *Opponent process interaction between sleep homeostasis and circadian phase during entrainment.* Homeostatic sleep drive (*solid red line*) decreases brain arousal during wakefulness and is dissipated during sleep (*dashed red line in the blue shading*). Circadian rhythm in brain arousal (*blue line*) oscillates with a near-24-h pattern such that brain arousal is lowest in the morning hours after habitual wake time and brain arousal is highest in the evening hours prior to habitual bedtime. These two central nervous system processes interact such that the circadian arousal signal counters the buildup of homeostatic sleep drive. The result is relatively stable levels of brain arousal or alertness across a typical day especially across the hours of the work day (*orange shading*). This schematic represents a simplified opponent process model as sleep homeostasis and circadian processes interact in a nonlinear manner such that the amplitude of the circadian drive is larger when homeostatic sleep drive is high

is promoting wakefulness. Because the circadian clock is promoting wakefulness during the biological day, sleep is of inadequate duration and quality. The average sleep duration of shift workers is reported to be approximately 5–6 h per 24 h [43–46].

Disturbed sleep is a common health problem reported by shift workers [45–49]. Sleep electroencephalographic (EEG) recordings of permanent night shift and rotating shift workers indicates that daytime sleep is composed of reduced stage 2 and rapid eye movement (REM) sleep and increased wakefulness after sleep onset (WASO). Detailed reviews of sleep disruption associated with various shift work schedules is provided elsewhere [43, 45]. Figures 8.2 and 8.3 illustrate a wakefulness–sleep model showing the typical interaction between sleep homeostasis for a day worker versus a night shift worker. The model represents an adaptation of the classic two process model of sleep regulation [50, 51] and opponent process models of sleep–wakefulness regulation [52–54]. Such models include a homeostatic sleep component and an oscillating circadian component that interact to promote the daily pattern of sleep and wakefulness. The homeostatic drive for sleep can be likened to other homeostatically regulated processes such that sleep drive builds with time awake and is influenced by prior sleep history. The circadian system modulates brain arousal such that wakefulness is promoted during the biological day and sleep is promoted during the biological night. Daytime workers awaken from nighttime sleep with a relatively low homeostatic drive for sleep assuming that the prior sleep was of sufficient duration and quality (Fig. 8.2). The homeostatic drive for sleep subsequently builds up across the day with increasing time awake and this drive promotes sleep. Homeostatic sleep drive is then dissipated during sleep. As the wakefulness episode progresses, the circadian system promotes wakefulness to counteract the buildup of homeostatic sleep drive in such a way that brain arousal from a circadian perspective is highest shortly prior to habitual bedtime. Near the onset of endogenous melatonin secretion, which demarcates the beginning of the biological night, the circadian clock begins to promote sleep and continues to promote sleep until the early morning hours. From

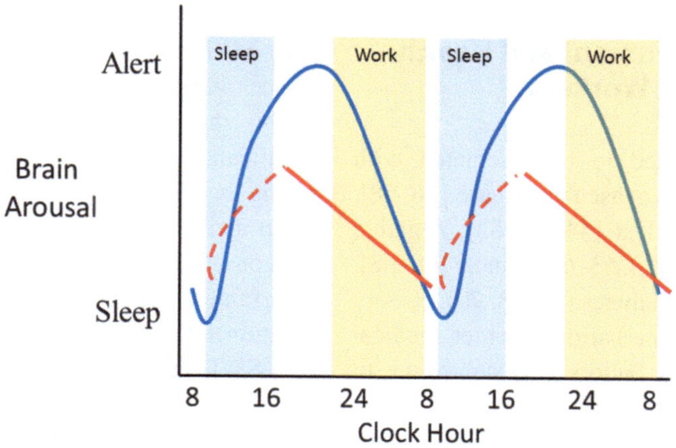

Fig. 8.3 *Interaction between sleep homeostasis and circadian phase during circadian misalignment typical of night shift work.* Circadian misalignment is when sleep and wakefulness occur at inappropriate circadian times. This example shows the altered relationship between sleep homeostasis and circadian phase for a typical night shift worker. Sleep during the daytime (*blue shading*) is shorter and is disturbed compared to sleep at night (Fig. 8.2) because the circadian arousal signal is promoting brain arousal. Following awakening, homeostatic sleep drive (*solid red line*) is larger than following nighttime sleep due to the daytime sleep disruption and thus the smaller dissipation of sleep drive (*dashed red line*) as compared to nighttime sleep. Brain arousal is relatively high after awakening because the circadian arousal signal is promoting wakefulness (*blue line*) and because some sleep drive has been dissipated. During the timing of the nighttime work shift (*orange shading*), both sleep homeostasis and circadian phase promote sleep. Thus, the loss of the typical opponent process relationship between sleep homeostasis and circadian phase leads to high levels of sleepiness and fatigue and to performance impairment when working at night

a circadian perspective, brain arousal is lowest shortly after habitual wake time, a time when the homeostatic drive for sleep is relatively low. Thus, the sleep homeostatic and circadian systems work together in an opponent process manner to promote the typical pattern of consolidated wakefulness during the daytime and sleep at night. With respect to the timing of work, the day worker is required to perform at a time when brain arousal is relatively high. Figure 8.3 illustrates the interaction between the sleep homeostatic and circadian systems during circadian misalignment in a night shift worker. Failure to adapt the circadian system to the imposed shift work schedule results in the circadian system promoting brain arousal during the daytime sleep episode. This leads to disturbed sleep and short sleep duration. The homeostatic drive for sleep upon awakening from sleep in the shift worker is relatively higher (lower brain arousal) than that in the daytime worker because of the reduced sleep duration. However, the arousal signal from the circadian clock initially helps to counter the higher homeostatic sleep drive. Subsequently, the sleep homeostatic and circadian systems no longer work together in an opponent process manner and instead, as the wakefulness episode progresses both circadian and sleep homeostatic systems promote sleep. Thus, with respect to the timing of work, the night shift worker is required to perform at a time when brain arousal is relatively low.

In addition to the biological drives noted above, sleep in the night shift worker is also compromised by a number of environmental factors including family responsibilities (curtailing sleep to provide child care), noisy environment (e.g., family members awake, kids playing, phone calls during the daytime), drugs (e.g., caffeine taken to promote wakefulness during the night shift), light exposure (e.g., bedrooms with inadequate shades that let light in), and bedroom temperature (too hot or too cold) [43, 55].

Circadian Misalignment and Health Problems in Shift Workers

Shift work is reported to be associated with increased risk of cardiovascular disease [56–58], gastrointestinal disorders [44, 58–62], negative pregnancy outcomes [58, 63–65], cancer [66–68], depression [44, 69], diabetes [11, 58, 70], obesity [10–12, 71], and exacerbation of other medical disorders [72]. Many factors may contribute to health problems in shift workers including: stress, sleep disruption, circadian misalignment, socio-economic status, smoking, and unhealthy food habits [73]. Shift work models in nonhumans provide further evidence for the negative effects of shift work schedules. For example, cardiomyopathic Syrian hamsters exposed to a weekly 12-h phase shift of the light dark cycle showed a decrease in survival time by 11% [74]. Sleep during the biological day can be improved with melatonin, melatonin analogues, and sleep medications [35, 75, 76]. However, it is unknown whether promoting daytime sleep reduces the burden of the negative health outcomes associated with shift work.

Circadian Sleep Disorders

Circadian sleep disorders are characterized by circadian misalignment. A common symptom observed in patients with circadian sleep disorders is that patients cannot sleep when sleep is desired and that sleep is disturbed and inadequate. These disorders often cause excessive daytime sleepiness and fatigue and have negative consequences for quality of life. The most common circadian sleep disorders are delayed sleep phase type, advanced sleep phase type, jet lag, non-24-h sleep wake type, and shift work disorder (SWD) [77, 78].

Shift Work Disorder

SWD is thought to be caused in part by a greater vulnerability to the circadian misalignment due to sleep and work schedules occurring at inappropriate biological times of day. Patients with SWD have clinically significant performance impairments when work is required during the biological night and/or clinically significant sleep disruption when sleep occurs during the biological day [36, 44, 45, 55, 79, 80]. Sleep and performance problems are less common on days off when sleep occurs at night and wakefulness occurs during the day (e.g., during vacation). During the work schedule, patients with SWD are reported to show disturbed sleep that is as bad as that seen in patients with severe insomnia [36, 44, 79]; furthermore, patients with SWD are as sleepy during the nighttime as are patients with disorders of hypersomnolence, such as narcolepsy and sleep apnea during the daytime [36, 79]. The wakefulness promoting medications modafinil and r-modafinil are currently approved by the FDA to treat excessive sleepiness in patients with SWD [36, 79]. No clinical trials have tested the effectiveness of sleep-promoting countermeasures in patients with SWD.

In comparison with other shift workers, patients with SWD have higher rates of ulcers, depression, excessive sleepiness on the night shift, sleepiness-related accidents, reduced satisfaction for the work schedule, and increased absenteeism from work [36, 43]. Whether these and other negative health outcomes in SWD are associated with the degree or duration of circadian misalignment is unknown.

Circadian Misalignment, Sleep Disruption, Energy Metabolism, and Weight Gain

An important function of the circadian system is the regulation of metabolic machinery in preparation for temporal variations in the abundance of nutrients. Therefore, it is not surprising that various humoral factors associated with metabolic control show wakefulness–sleep patterns or circadian rhythms (e.g., glucose, fatty acids and triglycerides, but also glucocorticoids, insulin and catecholamines). The finding that many metabolic factors fluctuate with wakefulness-sleep patterns and/or circadian rhythms suggests that misalignment between circadian clocks, sleep–wakefulness and metabolic physiology and behavior may contribute to or be casually related to the development of metabolic disorders [38, 81–84]. Related, genetic variations in human clock genes (e.g., gene polymorphisms) have been reported to be associated with metabolic disorders such as obesity and the metabolic syndrome [85] as well as type II diabetes [86].

Clock Genes and Energy Metabolism

At the cellular level, regulation of mammalian clock proteins includes posttranslational modifications (PTMs). These PTMs provide an entry route for metabolic signals or environmental cues. To explore PTMs and nutrient signals, Lamia and colleagues used mass spectrometry and bioinformatics to locate sites of regulated phosphorylation in the mouse clock protein cryptochrome (CRY). They found that phosphorylation of the energy sensing molecule, adenosine monophosphate (AMP) activated protein kinase (AMPK) enables CRY to transduce nutrient signals to circadian clocks in mammalian peripheral organs [87]. In this way, peripheral clocks may be modulated by information from metabolic signals. Reciprocally, output genes, under the control of the circadian clock, have been shown to be involved in metabolic processes. Evidence for direct coupling between the circadian clock and the regulation of metabolism comes from cultured adipocytes, where it has been shown that overexpression of the clock gene Brain and Muscle ARNT (Arylhydrocarbon Receptor Nuclear Translocator)-Like Protein (BMAL1) increases lipid synthesis perhaps through induction of the peroxisome proliferator-activated receptor gamma (PPARγ), adipocyte protein 2 (AP_2), and other binding proteins [88]. In mice, rhythmic transcriptional activation by CLOCK/BMAL1 is a key regulator of lipid metabolic enzymes (e.g., acyl-CoA oxidase), [83]. Additionally, both CLOCK mutant and BMAL-deficient mice show impaired gluconeogenesis, glucose tolerance, and insulin sensitivity [82]. In animals deficient of liver-specific BMAL, glucose balance is compromised and reflected by a loss of rhythmicity in the expression of genes important for glucose mobilization such as glucose-6-phosphate translocase, phosphoenolpyruvate carboxykinase 2, adenylate kinase 4, and glucose transporter 2 [89].

Energy Expenditure and Weight Gain

There are three primary components that contribute to 24 h energy expenditure: (1) ~60–70% basal metabolic rate (BMR), which is energy needed to sustain resting cellular, mechanical, and chemical processes, including heart rate, neural activity, respiration, and body temperature (sleeping metabolic rate is lower than BMR); (2) ~10% thermic effect of food [90], which is the energy expended to digest, metabolize and store ingested macronutrients; and (3) ~20–30% activity thermogenesis, which is the energy expended during physical activity. Whole room indirect calorimetry is the gold standard method of estimating 24 h energy expenditure and substrate oxidation [91]. When total daily energy expenditure is lower than total daily energy intake, weight gain ensues [92].

Sleep Reduces Energy Expenditure and Disturbed Sleep Increases Energy Expenditure

Since circadian misalignment is physiologically associated with sleep disruption, we briefly discuss how alterations in sleep may contribute to alterations in energy metabolism in humans. Energy expenditure is lower during sleep when compared to wakefulness [93–95]. We examined the influence of total sleep deprivation on 24 h energy expenditure in healthy subjects who lived in a whole room calorimeter under bed rest and controlled food intake conditions [93]. As seen in Fig. 8.4, sleep deprivation per se increased 24 h energy expenditure. This finding is contrary to what would be expected based on the hypothesis of a role for sleep deprivation in weight gain, but it is consistent with the hypothesis that sleep serves to conserve energy. The finding that sleep deprivation increases energy expenditure in humans is also consistent with findings from

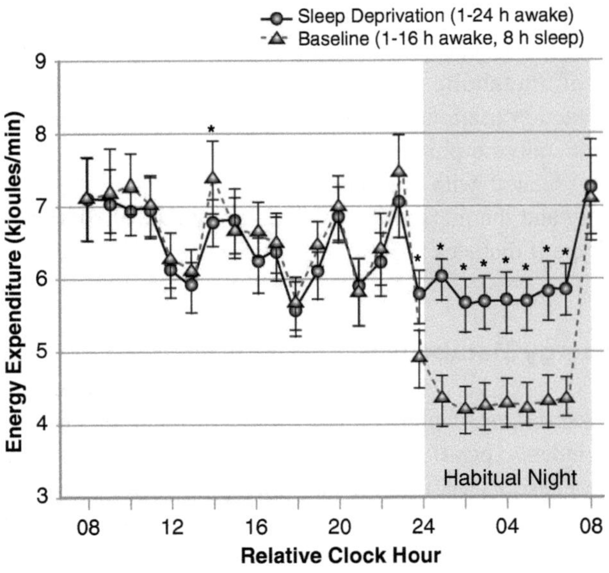

Fig. 8.4 *Sleep deprivation increases energy expenditure.* Average hourly energy expenditure during a typical day of 16 h scheduled wakefulness and 8 h scheduled sleep versus 24 h of total sleep deprivation. Subjects were studied in a whole room calorimeter under modified constant routine conditions that limited physical activity during bed rest and exposed subjects to dim ambient light and thermoneutral ambient temperature. Meals (breakfast, lunch, dinner, and a snack) of the exact same content were provided at the same time each day, and no food intake occurred during the habitual night. One night of sleep deprivation increased energy expenditure providing support for the hypothesis that sleep conserves energy in humans and that sleep loss has a metabolic cost. Modified with permission from [93]

nonhuman models showing sleep deprivation increases 24 h energy expenditure [96, 97]. Since sleep disruption increases energy expenditure, it would be hypothesized that disturbed sleep due to circadian misalignment will also be associated with higher energy expenditure during a daytime sleep episode. To date however, no study has examined the influence of circadian misalignment or of chronic sleep restriction on energy expenditure in humans.

Circadian Variation of Energy Expenditure

A circadian rhythm in energy expenditure has been reported by Spengler et al. [98] and Krauchi et al. [99], but the circadian pattern was inconsistent. Specifically, Spengler et al. [98] reported CO_2 production to be lowest during the early evening between 180 and 270 circadian degrees with the core body temperature rhythm minimum designated as zero degrees. These circadian times correspond to clock hours of 1800–2400 h in an individual with a wake time of 0800 h. Krauchi et al. [99] reported a circadian rhythm in energy expenditure (kcal/day) with low levels between 2400 and 0600 h. The reason for the discrepancy in the timing of low circadian-driven energy expenditure is unclear as controlled constant routine conditions were used in both studies. Therefore, additional research is needed to better understand the circadian contribution to energy expenditure in humans.

Hypothalamic and Peripheral Energy Balance Hormones That Contribute to the Regulation of Food Intake Are Altered by Sleep Disruption and Circadian Misalignment

Within the hypothalamus, the arcuate nucleus has two opposing sets of neuronal circuitry, appetite

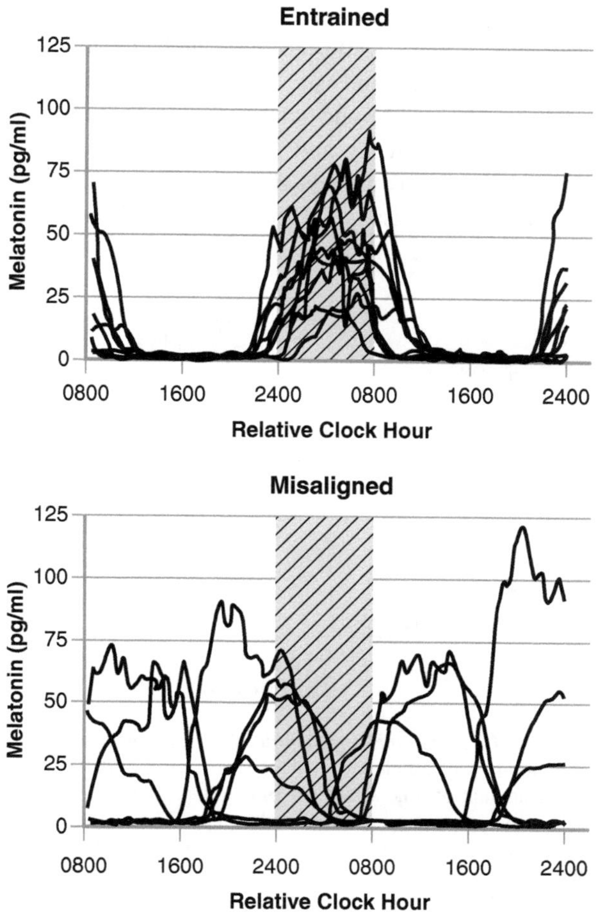

Fig. 8.5 *Sleep occurs during the biological day when endogenous melatonin levels are low during circadian misalignment.* Melatonin rhythms for individual subjects who lived in the laboratory and who were scheduled to 24 h or non-24-h day lengths for more than 1 month. Subjects whose circadian clock entrained to the scheduled day length slept (*dashed blue box*) during the biological night when melatonin levels were high. Subjects whose circadian clock failed to entrain to the scheduled day length (misaligned) were awake during the biological night when melatonin levels were high. Modified with permission from [32]

stimulating (orexigenic) and appetite inhibiting (anorexigenic). These structures receive neural and peripheral inputs that serve to regulate energy balance, predominantly by modifying food intake. The SCN projects to hypocretin/orexin neurons in the hypothalamus and hypocretin/orexin neurons influence wakefulness and feeding [100–103]. Peripheral hormonal inputs include anorexigenic (e.g., PYY, insulin, and leptin) and orexigenic (e.g., ghrelin) mediators. The SCN also contains leptin receptors [104], and thus the circadian clock in addition to feeding centers in the hypothalamus may respond to energy balance hormones. Sympathetic projections from the SCN to adipose tissue [105] suggest circadian modulation of energy balance hormones such as leptin.

Leptin is an anorexigenic hormone produced by adipocytes. Circulating plasma levels are thought to represent total body lipid stores. In humans, acute caloric restriction decreases leptin levels whereas acute caloric excess increases leptin levels [106]. Leptin levels also change across the 24 h day. A daily variation in plasma leptin levels is reported to be related to a daily variation in satiety [107, 108] with high leptin levels at night [109–111].

Circadian misalignment has been shown to reduce the energy balance hormone leptin in humans [37, 38], a physiological signal that should promote food intake. Figures 8.5–8.8 show findings from a 55-day inpatient study on circadian misalignment [32]. Half of the subjects maintained a normal relationship between sleep–wakefulness and internal circadian time (entrained group), while the other half did not (circadian misalignment), (Fig. 8.5). Sleep efficiency (Fig. 8.6) and total sleep time were reduced, sleep latency and REM latency were shortened, and wakefulness after sleep onset was increased in the misaligned group [32]. We found

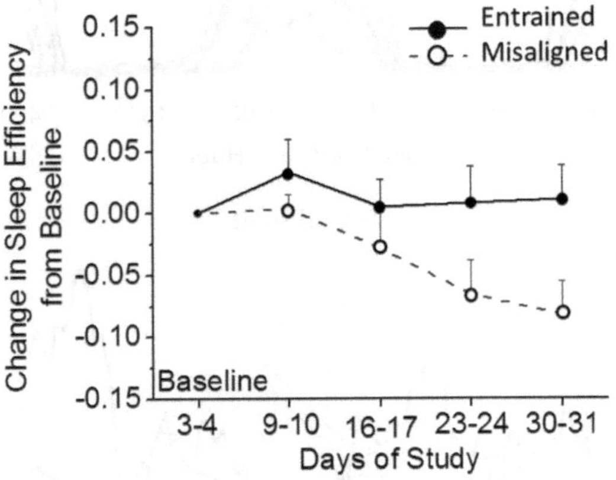

Fig. 8.6 *Circadian misalignment leads to chronic sleep disruption.* Sleep efficiency is a marker of sleep continuity or the amount of time asleep per time in bed. Low sleep efficiency can represent disturbed sleep. When sleep occurs during the biological day, at a time when the circadian clock is promoting wakefulness (Fig. 8.3), sleep is disturbed. Sleep efficiency became progressively lower in subjects who failed to entrain to the scheduled day length (misaligned). Modified with permission from [32]

that circadian misalignment reduced daytime plasma leptin levels by ~10% [37]. The circadian misalignment that occurred in this study was akin to that which commonly occurs in non-24-h disorder/free running disorder, a circadian sleep disorder where the clock fails to entrain to the 24 h day [78]. Failure to entrain to the light–dark cycle also occurs aboard submarines [112] and during spaceflight [113] when crewmembers are scheduled to non-24-h day lengths. Scheer et al. [38] also showed that circadian misalignment associated with a rapid delay of bedtime, 4 h per day using a 28 h forced desynchrony protocol, decreased leptin levels. The decrease in leptin during circadian misalignment remained even when controlling for sleep disruption. The circadian misalignment in the study by Scheer is akin to that which occurs during rapid jet travel across multiple time zones. Thus, both chronic and transient circadian misalignment alters leptin in such a way that would be hypothesized to promote food intake.

A small circadian rhythm in leptin levels has been reported with an amplitude of approximately 10% of the total daily minimum to maximum values [81]. Furthermore, there was no circadian variation in the soluble leptin receptor (sOB-r), which indirectly determines the amount of biologically active free leptin at the

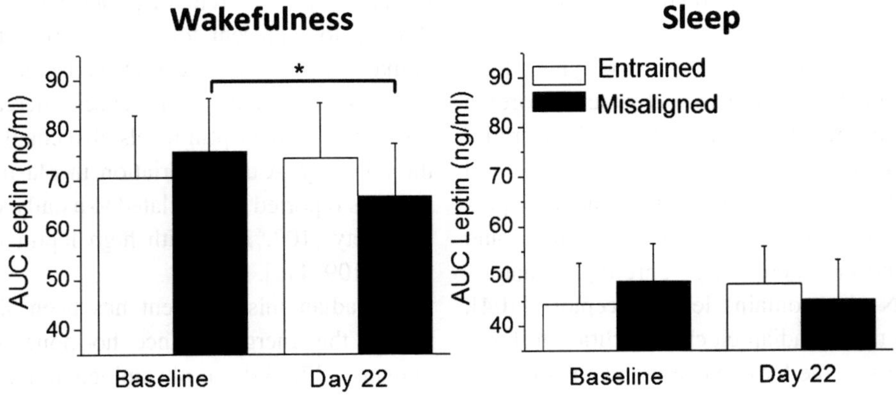

Fig. 8.7 *Circadian misalignment reduces leptin levels predominantly during wakefulness.* Average area under the curve leptin levels during scheduled wakefulness and scheduled sleep assessed every hour across 24 h. Leptin levels were lower during scheduled wakefulness on day 22 of the study when subjects were awake during the biological night (circadian misalignment) as compared to baseline when wakefulness occurred during the biological day. Modified from [37]. Copyright 2010, with permission from Dove Medical Press Ltd.

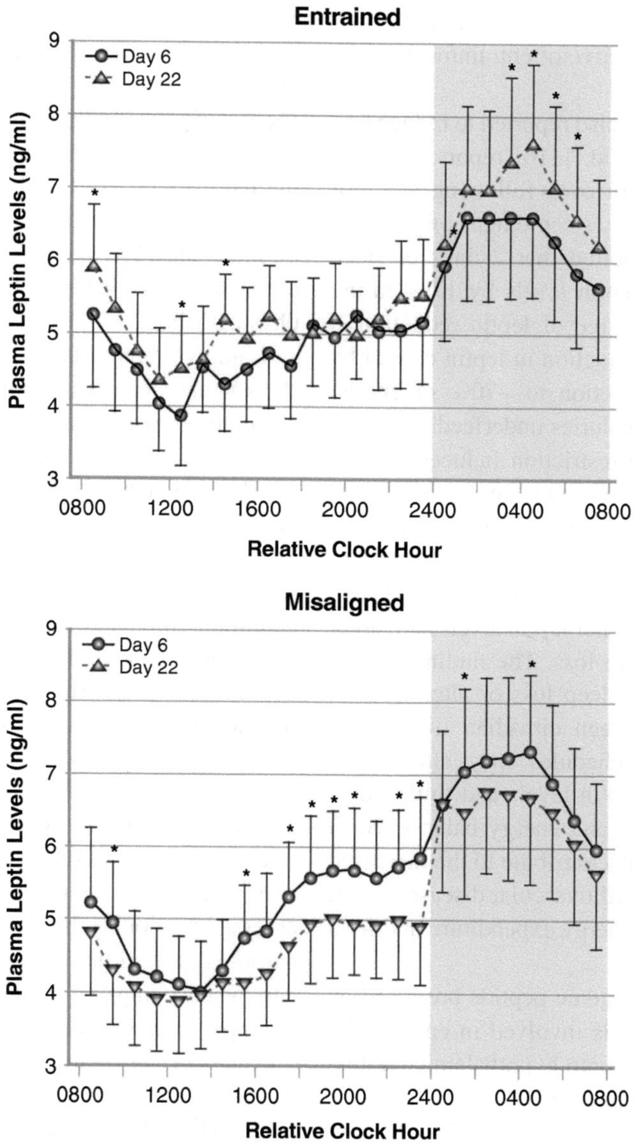

Fig. 8.8 *The pattern of low daytime and high nighttime leptin levels is maintained during circadian misalignment.* Average hourly leptin levels assessed every hour across 24 h at baseline (day 6) and after living on a 24 h or non-24-h day lengths for several weeks. Leptin levels were significantly lower predominantly during the late afternoon and evening hours during circadian misalignment. Unlike melatonin that shows a robust circadian rhythm (Fig. 8.5), leptin levels show a robust wakefulness–sleep pattern with high leptin levels during sleep even during circadian misalignment. Modified from [37]. Copyright 2010, with permission from Dove Medical Press Ltd

hypothalamus [114, 115]. The lack of circadian variation in the sOB-r suggests that the varied levels of leptin that are available at different times of day are biologically active. Findings from the 55-day study reported above provide further support to the findings that the circadian rhythm in leptin is small. Specifically, Figs. 8.5 and 8.8 show that the circadian-driven hormone melatonin is misaligned with the imposed sleep–wakefulness schedule, whereas the daily pattern in leptin shows low levels across the day and high levels at night irrespective of entrainment or circadian misalignment.

Sleep disruption is also reported to reduce leptin levels. Spiegel et al. [116] reported a 19% reduction in 24 h leptin levels following 6 days of sleep restriction versus sleep extension. Spiegel et al. [117] also reported that 2 days of sleep restriction reduced leptin levels by 18% during the daytime. This degree of leptin reduction is similar to the 22% reduction in leptin caused by 3 days of food restriction to ~70% of energy requirements (~900 calories underfeeding) [106] indicating that sleep restriction induces physiologically meaningful changes in leptin. Taheri et al. [118] also reported that habitual short sleep duration was associated with reduced leptin levels. In the original study by Spiegel et al. [116], those subjects with higher leptin levels had larger decreases during sleep loss. The findings above suggest that chronic sleep loss or alterations in the relationship between circadian timing and sleep–wakefulness schedules may alter leptin levels. Changes in leptin levels and timing may influence food intake and energy balance. Such changes in leptin may contribute to the increased risk of obesity and cardiovascular disease in shift workers by altering energy expenditure and food intake.

Ghrelin is an orexigenic peptide produced by the stomach. Ghrelin is involved in energy balance by feeding back onto hypothalamic feeding centers to stimulate appetite [119]. Ghrelin levels rise between meals and decrease rapidly following food intake and are high at night [120]. Taheri et al. [118] reported that habitual short sleep duration was associated with elevated ghrelin and higher BMI levels. Spiegel et al. [117] reported that 2 days of sleep restriction (4 h per night) compared to 2 days of sleep extension (10 h per night) increased ghrelin levels by 28%. The finding that changes in ghrelin occurred within 2 days suggests that energy balance hormones can respond quickly to sleep loss. Further, using a moderate bout of exercise in the evening followed by a simulated night shift condition, Morris and colleagues found that the post-exercise decrease in acylated ghrelin levels, observed during the biological day, does not occur during the biological night. In fact, plasma acylated ghrelin levels were ~17% higher following the evening exercise bout than compared to the no exercise control condition [121]. These findings may have important implications for eating behaviors in night shift work that involves physical labor.

Findings from studies in humans have also shown that sleep restriction in healthy subjects promotes increased appetite, especially for calorie dense foods with high carbohydrate content [117]. Nedeltcheva et al. restricted sleep in healthy young subjects to 5.5 h per night for 14 days and found an increase in food intake from snacks (1,087 kcals/day) in the sleep restriction condition compared to the 8.5 h per night sleep condition (866 kcal/day). This was accompanied by an increase in carbohydrate consumption particularly during 1900 to 0700 hours [122]. In another study by this same group, the combined exposure of overweight middle-aged adults to 2 weeks of caloric (90% of resting metabolic rate) and sleep restriction (5.5 h per night) increased hunger, reduced energy expenditure as assessed by doubly labeled water and increased carbohydrate oxidation assessed with a metabolic cart when compared to the same subjects in a 8.5 h per night sleep condition [123].

Eating Patterns in Shift Workers

The eating patterns and behavior in shift workers may contribute to weight gain and obesity. Night shift workers are reported to consume more sugar when compared to day workers, and this has been hypothesized to be an attempt to overcome fatigue in the workplace [124, 125]. Research findings from a nurse cohort working the night shift indicated that many displayed abnormal eating behaviors as identified by one or more categories on the Dutch Eating Behavior Questionnaire(DEBQ) [126]. These categories include: (1) emotional eating in response to fear, anger, or anxiety; (2) external in response to sight and smell of food; and (3) restrained eating which refers to overeating when the cognitive resolve to diet is abandoned after a period of slimming [127]. Comparisons of total daily caloric intake between night and day shift workers have found no difference; however, there was a redistribution of the major timing of food intake [128, 129] that may impact metabolic physiology. Similar findings have been reported for patients with night eating syndrome, where more than 50% of the calories are consumed at night [15–17]. Biological factors (e.g., levels of ghrelin/leptin) that are affected by circadian misalignment and/or sleep loss may in turn impact hunger, food intake, and macronutrient preference in night shift workers. Additionally, many environmental factors (e.g., limited food choices on shift, unpredictable working conditions, exhaustion, and stress) may also contribute to eating patterns and behavior in night shift workers.

Circadian Clock Gene Mutation Alters Energy Balance

In a study by Turek et al. [130], the influence of a clock gene mutation was examined on metabolic physiology. Average caloric intake was assessed over a 10-week period in wild-type (WT) and Clock mutant (CL) mice fed a regular or high-fat diet. CL mice showed a significant increase in energy intake and body weight. The average metabolic rate during the light episode was higher in CL and the average metabolic rate was lower in CL mice during the dark episode indicating an alteration in metabolism across the 24 h day. Overall, energy expenditure was decreased by 10% in CL mice. Serum triglyceride, cholesterol, glucose, and leptin levels were higher in CL mice suggesting that circadian disruption through the Clock mutation alters metabolic factors involved in obesity. Note that the changes in leptin observed in CL mice are opposite to those observed in response to sleep loss in humans. The key difference between prior chronic sleep loss studies in humans and Clock mutant mice may be that the mice studies examined chronic circadian and sleep disruption during adlib food intake. It is likely that the metabolic changes and positive energy balance are related to this increased food intake. It is also possible that the Clock mutation altered functioning of other systems involved in energy balance. Specifically, animal models show that circadian clock gene disruption, abnormal light–dark cycles, and imposed circadian misalignment (eating during the time normally reserved for sleep) all lead to weight gain [130–132].

Working Model for Effects of Circadian Misalignment on Energy Balance

Figure 8.9 shows a working model that predicts circadian misalignment in shift workers will disrupt sleep and alter levels of feeding hormones, which are expected to increase appetite. The model predicts that increased food intake and unhealthy food choices will promote positive energy balance. It is unknown how circadian misalignment will influence total daily energy expenditure. Circadian misalignment could lead to fatigue and reduced physical activity. Circadian misalignment could also increase total daily energy expenditure due to sleep disruption. The latter finding would be consistent with findings of increased energy expenditure during sleep deprivation [93]. Future research is needed to improve our understanding of the mechanisms by which circadian misalignment/sleep disruption and disruption of circadian clock genes contribute to weight gain and obesity. Understanding of such mechanisms will provide new avenues for the development of treatments for obesity and will provide critical information needed for the education of the public on the importance of sleep and circadian physiology for health and well-being.

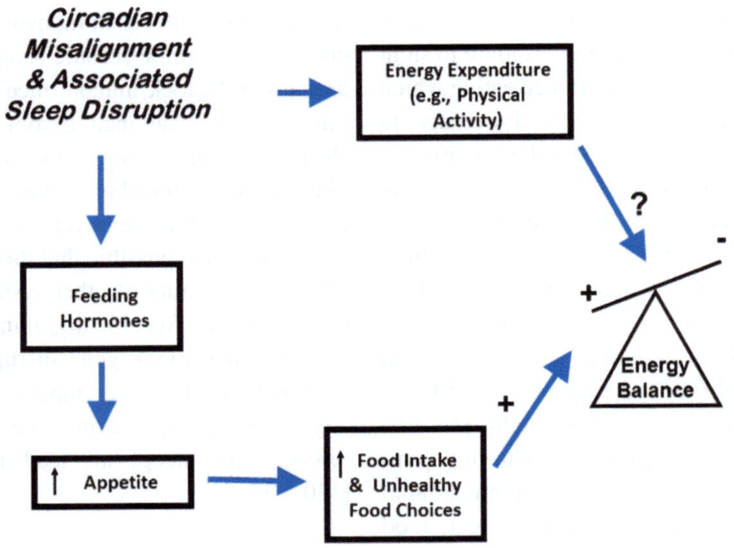

Fig. 8.9 Working model on mechanisms that may contribute to weight gain and obesity during circadian misalignment. Circadian misalignment leads to food intake at an internal biological time when the circadian clock is less prepared to intake nutrients. Circadian misalignment also leads to sleep disruption and shorter sleep duration. Both circadian misalignment and sleep disruption lead to alterations in feeding hormones that promote appetite and this could lead to increased food intake. Longer episodes of wakefulness provide more opportunity for food intake. Unhealthy food choices due to biological (cravings for sugary/starchy foods) or environmental factors (e.g., food availability) promote positive energy balance. It is unknown how circadian misalignment influences energy expenditure. If food intake during circadian misalignment is higher than energy expended, then energy storage and weight gain ensue

Acknowledgement This work was supported by NIH/NCRR Colorado CTSI Grant Number UL1 RR025780.

References

1. Laposky AD, Bass J, Kohsaka A, Turek FW. Sleep and circadian rhythms: key components in the regulation of energy metabolism. FEBS Lett. 2008;582(1):142–51.
2. Laposky AD, Turek FW. Physiologic and health consequences of circadian disruption (in animal models). Sleep Med Clin. 2009;4:127–42.
3. Zvonic S, Ptitsyn AA, Conrad SA, Scott LK, Floyd ZE, Kilroy G, et al. Characterization of peripheral circadian clocks in adipose tissues. Diabetes. 2006;55(4):962–70.
4. Storch KF, Lipan O, Leykin I, Viswanathan N, Davis FC, Wong WH, et al. Extensive and divergent circadian gene expression in liver and heart. Nature. 2002; 417(6884):78–83.
5. Davidson AJ, London B, Block GD, Menaker M. Cardiovascular tissues contain independent circadian clocks. Clin Exp Hypertens. 2005;27(2–3): 307–11.
6. Davidson AJ, Castanon-Cervantes O, Stephan FK. Daily oscillations in liver function: diurnal vs circadian rhythmicity. Liver Int. 2004;24(3):179–86.
7. Yamazaki S, Straume M, Tei H, Sakaki Y, Menaker M, Block GD. Effects of aging on central and peripheral mammalian clocks. Proc Natl Acad Sci USA. 2002;99(16):10801–6.
8. Nystrom T, Nygren A, Sjoholm A. Persistent endothelial dysfunction is related to elevated C-reactive protein (CRP) levels in Type II diabetic patients after acute myocardial infarction. Clin Sci (Lond). 2005;108(2):121–8.
9. Bray MS, Young ME. Circadian rhythms in the development of obesity: potential role for the circadian clock within the adipocyte. Obes Rev. 2007;8(2):169–81.
10. Ishizaki M, Morikawa Y, Nakagawa H, Honda R, Kawakami N, Haratani T, et al. The influence of work characteristics on body mass index and waist to hip ratio in Japanese employees. Ind Health. 2004;42(1):41–9.
11. Karlsson B, Knutsson A, Lindahl B. Is there an association between shift work and having a metabolic syndrome? Results from a population based study of 27,485 people. Occup Environ Med. 2001;58(11): 747–52.
12. Di Lorenzo L, De Pergola G, Zocchetti C, L'Abbate N, Basso A, Pannacciulli N, et al. Effect of shift work on body mass index: results of a study performed in 319 glucose-tolerant men working in a Southern Italian industry. Int J Obes Relat Metab Disord. 2003;27(11):1353–8.

13. Esquirol Y, Bongard V, Mabile L, Jonnier B, Soulat JM, Perret B. Shift work and metabolic syndrome: respective impacts of job strain, physical activity, and dietary rhythms. Chronobiol Int. 2009;26(3):544–59.
14. Ayas NT, White DP, Al Delaimy WK, Manson JE, Stampfer MJ, Speizer FE, et al. A prospective study of self-reported sleep duration and incident diabetes in women. Diabet Care. 2003;26(2):380–4.
15. Rogers NL, Dinges DF, Allison KC, Maislin G, Martino N, O'Reardon JP, et al. Assessment of sleep in women with night eating syndrome. Sleep. 2006;29(6):814–9.
16. Goel N, Stunkard AJ, Rogers NL, Van Dongen HP, Allison KC, O'Reardon JP, et al. Circadian rhythm profiles in women with night eating syndrome. J Biol Rhythms. 2009;24(1):85–94.
17. Howell MJ, Schenck CH, Crow SJ. A review of nighttime eating disorders. Sleep Med Rev. 2009;13(1):23–34.
18. Basheer R, Strecker RE, Thakkar MM, McCarley RW. Adenosine and sleep-wake regulation. Prog Neurobiol. 2004;73(6):379–96.
19. Chou TC, Bjorkum AA, Gaus SE, Lu J, Scammell TE, Saper CB. Afferents to the ventrolateral preoptic nucleus. J Neurosci. 2002;22(3):977–90.
20. Saper CB, Chou TC, Scammell TE. The sleep switch: hypothalamic control of sleep and wakefulness. Trends Neurosci. 2001;24(12):726–31.
21. Porkka-Heiskanen T, Strecker RE, Thakkar M, Bjorkum AA, Greene RW, McCarley RW. Adenosine: a mediator of the sleep-inducing effects of prolonged wakefulness. Science. 1997;276(5316):1265–8.
22. Deurveilher S, Semba K. Indirect projections from the suprachiasmatic nucleus to major arousal-promoting cell groups in rat: implications for the circadian control of behavioural state. Neuroscience. 2005;130(1):165–83.
23. Aston-Jones G, Chen S, Zhu Y, Oshinsky ML. A neural circuit for circadian regulation of arousal. Nat Neurosci. 2001;4(7):732–8.
24. Rainnie DG, Grunze HC, McCarley RW, Greene RW. Adenosine inhibition of mesopontine cholinergic neurons: implications for EEG arousal. Science. 1994;263(5147):689–92.
25. Chamberlin NL, Arrigoni E, Chou TC, Scammell TE, Greene RW, Saper CB. Effects of adenosine on gabaergic synaptic inputs to identified ventrolateral preoptic neurons. Neuroscience. 2003;119(4):913–8.
26. Moore RY, Eichler VB. Loss of a circadian adrenal corticosterone rhythm following suprachiasmatic lesions in the rat. Brain Res. 1972;42:201–6.
27. Morse D, Sassone-Corsi P. Time after time: inputs to and outputs from the mammalian circadian oscillators. Trends Neurosci. 2002;25(12):632–7.
28. Berger RJ, Phillips NH. Comparative physiology of sleep, thermoregulation and metabolism from the perspective of energy conservation. Prog Clin Biol Res. 1990;345:41–50.
29. Sack RL, Blood ML, Lewy AJ. Melatonin rhythms in night shift workers. Sleep. 1992;15(5):434–41.
30. Dumont M, Benhaberou-Brun D, Paquet J. Profile of 24-h light exposure and circadian phase of melatonin secretion in night workers. J Biol Rhythms. 2001;16(5):502–11.
31. Smith MR, Cullnan EE, Eastman CI. Shaping the light/dark pattern for circadian adaptation to night shift work. Physiol Behav. 2008;95(3):449–56.
32. Wright Jr KP, Hull JT, Hughes RJ, Ronda JM, Czeisler CA. Sleep and wakefulness out of phase with internal biological time impairs learning in humans. J Cogn Neurosci. 2006;18:508–21.
33. Wright Jr KP, Hull JT, Czeisler CA. Relationship between alertness, performance, and body temperature in humans. Am J Physiol Regul Integr Comp Physiol. 2002;283(6):R1370–7.
34. Grady S, Aeschbach D, Wright Jr KP, Czeisler CA. Effect of modafinil on impairments in neurobehavioral performance and learning associated with extended wakefulness and circadian misalignment. Neuropsychopharmacology. 2010;35(9):1910–20.
35. Markwald RR, Lee-Chiong TL, Burke TM, Snider JA, Wright Jr KP. Effects of the melatonin MT-1/MT-2 agonist ramelteon on daytime body temperature and sleep. Sleep. 2010;33(6):825–31.
36. Czeisler CA, Walsh JK, Roth T, Hughes RJ, Wright KP, Kingsbury L, et al. Modafinil for excessive sleepiness associated with shift-work sleep disorder. N Engl J Med. 2005;353(5):476–86.
37. Nguyen J, Wright Jr KP. Influence of weeks of circadian misalignment on leptin levels. Nat Sci Sleep. 2010;2:9–18.
38. Scheer FAJL, Hilton MF, Mantzoros CS, Shea SA. Adverse metabolic and cardiovascular consequences of circadian misalignment. Proc Natl Acad Sci USA. 2009;106(11):4453–8.
39. Davidson AJ, Yamazaki S, Arble DM, Menaker M, Block GD. Resetting of central and peripheral circadian oscillators in aged rats. Neurobiol Aging. 2008;29(3):471–7.
40. Damiola F, Le Minh N, Preitner N, Kornmann B, Fleury-Olela F, Schibler U. Restricted feeding uncouples circadian oscillators in peripheral tissues from the central pacemaker in the suprachiasmatic nucleus. Genes Dev. 2000;14(23):2950–61.
41. Waterhouse J, Minors D, Redfern P. Some comments on the measurement of circadian rhythms after time-zone transitions and during night work. Chronobiol Int. 1997;14(2):125–32.
42. Boivin DB, James FO. Circadian adaptation to night-shift work by judicious light and darkness exposure. J Biol Rhythms. 2002;17(6):556–67.
43. Drake CL, Wright KP Jr. Shift work, shift work disorder, and jet lag. In: Kryger MH, Roth T, Dement WC, editors. Principles and practice of sleep medicine. Saunders;2011. p. 784–98.
44. Drake CL, Roehrs T, Richardson G, Walsh JK, Roth T. Shift work sleep disorder: prevalence and conse-

quences beyond that of symptomatic day workers. Sleep. 2004;27(8):1453–62.
45. Åkerstedt TA, Wright Jr KP. Sleep loss and fatigue in shift work and shift work disorder. Sleep Med Clin. 2009;4:257–71.
46. Gold DR, Rogacz S, Bock N, Tosteson TD, Baum TM, Speizer FE, et al. Rotating shift work, sleep, and accidents related to sleepiness in hospital nurses. Am J Public Health. 1992;82(7):1011–4.
47. Axelsson J, Akerstedt T, Kecklund G, Lowden A. Tolerance to shift work-how does it relate to sleep and wakefulness? Int Arch Occup Environ Health. 2004;77(2):121–9.
48. Akerstedt T, Ingre M, Broman JE, Kecklund G. Disturbed sleep in shift workers, day workers, and insomniacs. Chronobiol Int. 2008;25(2):333–48.
49. Barger LK, Lockley SW, Rajaratnam SM, Landrigan CP. Neurobehavioral, health, and safety consequences associated with shift work in safety-sensitive professions. Curr Neurol Neurosci Rep. 2009; 9(2):155–64.
50. Borbely AA. A two process model of sleep regulation. Hum Neurobiol. 1982;1(3):195–204.
51. Achermann P, Borbely AA. Mathematical models of sleep regulation. Front Biosci. 2003;8:s683–93.
52. Dijk D-J, Edgar DM. Circadian and homeostatic control of wakefulness and sleep. In: Turek FW, Zee PC, editors. Regulation of sleep and wakefulness. New York: Marcel Dekker; 1999. p. 111–47.
53. Czeisler CA, Dijk D-J, Duffy JF. Entrained phase of the circadian pacemaker serves to stabilize alertness and performance throughout the habitual waking day. In: Ogilvie RD, Harsh JR, editors. Sleep onset: normal and abnormal processes. Washington, DC: American Psychological Association; 1994. p. 89–110.
54. Dijk DJ, Czeisler CA. Contribution of the circadian pacemaker and the sleep homeostat to sleep propensity, sleep structure, electroencephalographic slow waves, and sleep spindle activity in humans. J Neurosci. 1995;15(5 Pt 1):3526–38.
55. Zee PC, Wright KP Jr. Primary care update: shift work disorder. Elsevier Inc. 2010. p. 5–24.
56. Knutsson A, Hallquist J, Reuterwall C, Theorell T, Akerstedt T. Shiftwork and myocardial infarction: a case-control study. Occup Environ Med. 1999; 56(1):46–50.
57. Ayas NT, White DP, Manson JE, Stampfer MJ, Speizer FE, Malhotra A, et al. A prospective study of sleep duration and coronary heart disease in women. Arch Intern Med. 2003;163(2):205–9.
58. Knutsson A. Health disorders of shift workers. Occup Med (Lond). 2003;53(2):103–8.
59. Harrington JM. Health effects of shift work and extended hours of work. Occup Environ Med. 2001;58:68–72.
60. Ha M, Park J. Shiftwork and metabolic risk factors of cardiovascular disease. J Occup Health. 2005; 47(2):89–95.
61. Harrington JM. Shift work and health–a critical review of the literature on working hours. Ann Acad Med Singapore. 1994;23(5):699–705.
62. Segawa K, Nakazawa S, Tsukamoto Y, Kurita Y, Goto H, Fukui A, et al. Peptic ulcer is prevalent among shift workers. Dig Dis Sci. 1987;32(5): 449–53.
63. Mozurkewich EL, Luke B, Avni M, Wolf FM. Working conditions and adverse pregnancy outcome: a meta-analysis. Obstet Gynecol. 2000;95(4): 623–35.
64. Zhu JL, Hjollund NH, Andersen AM, Olsen J. Shift work, job stress, and late fetal loss: the national birth cohort in Denmark. J Occup Environ Med. 2004; 46(11):1144–9.
65. Zhu JL, Hjollund NH, Olsen J. Shift work, duration of pregnancy, and birth weight: the national birth cohort in Denmark. Am J Obstet Gynecol. 2004; 191(1):285–91.
66. Megdal SP, Kroenke CH, Laden F, Pukkala E, Schernhammer ES. Night work and breast cancer risk: a systematic review and meta-analysis. Eur J Cancer. 2005;41(13):2023–32.
67. Schernhammer ES, Laden F, Speizer FE, Willett WC, Hunter DJ, Kawachi I, et al. Night-shift work and risk of colorectal cancer in the nurses' health study. J Natl Cancer Inst. 2003;95(11):825–8.
68. Schernhammer ES, Laden F, Speizer FE, Willett WC, Hunter DJ, Kawachi I, et al. Rotating night shifts and risk of breast cancer in women participating in the nurses' health study. J Natl Cancer Inst. 2001;93(20):1563–8.
69. Skipper Jr JK, Jung FD, Coffey LC. Nurses and shiftwork: effects on physical health and mental depression. J Adv Nurs. 1990;15(7):835–42.
70. Nagaya T, Yoshida H, Takahashi H, Kawai M. Markers of insulin resistance in day and shift workers aged 30-59 years. Int Arch Occup Environ Health. 2002;75(8):562–8.
71. van Amelsvoort LG, Schouten EG, Kok FJ. Duration of shiftwork related to body mass index and waist to hip ratio. Int J Obes Relat Metab Disord. 1999; 23(9):973–8.
72. Scott AJ, LaDou J. Shiftwork: effects on sleep and health with recommendations for medical surveillance and screening. Occup Med. 1990;5(2): 273–99.
73. Knutsson A, Boggild H. Shiftwork and cardiovascular disease: review of disease mechanisms. Rev Environ Health. 2000;15(4):359–72.
74. Penev PD, Kolker DE, Zee PC, Turek FW. Chronic circadian desynchronization decreases the survival of animals with cardiomyopathic heart disease. Am J Physiol. 1998;275(6 Pt 2):H2334–7.
75. Porcu S, Bellatreccia A, Ferrara M, Casagrande M. Performance, ability to stay awake, and tendency to fall asleep during the night after a diurnal sleep with temazepam or placebo. Sleep. 1997;20(7):535–41.
76. Walsh JK, Schweitzer PK, Anch AM, Muehlbach MJ, Jenkins NA, Dickins QS. Sleepiness alertness on a simulated night-shift following sleep at home with Triazolam. Sleep. 1991;14(2):140–6.
77. Sack RL, Auckley D, Auger R, Carskadon MA, Wright Jr KP, Vitiello MV, et al. Circadian rhythm

sleep disorders: part I, basic principles, shift work and jet lag disorders: an American Academy of Sleep Medicine review. Sleep. 2007;30(11):1456–79.
78. Sack RL, Auckley D, Auger R, Carskadon MA, Wright Jr KP, Vitiello MV, et al. Circadian rhythm sleep disorders: Part II, advanced sleep phase disorder, delayed sleep phase disorder, free-running disorder, and irregular sleep-wake rhythm: an American Academy of Sleep Medicine review. Sleep. 2007;30(11):1480–97.
79. Czeisler CA, Walsh JK, Wesnes KA, Arora S, Roth T. Armodafinil for treatment of excessive sleepiness associated with shift work disorder: a randomized controlled study. Mayo Clin Proc. 2009;84(11):958–72.
80. Leger D, Scheuermaier K, Roger M. The relationship between alertness and sleep in a population of 769 elderly insomniacs with and without treatment with zolpidem. Arch Gerontol Geriatr. 1999;29(2): 165–73.
81. Shea SA, Hilton MF, Orlova C, Ayers RT, Mantzoros CS. Independent circadian and sleep/wake regulation of adipokines and glucose in humans. J Clin Endocrinol Metab. 2005;90(5):2537–44.
82. Rudic RD, McNamara P, Curtis AM, Boston RC, Panda S, Hogenesch JB, et al. BMAL1 and CLOCK, two essential components of the circadian clock, are involved in glucose homeostasis. PLoS Biol. 2004; 2(11):e377.
83. Inoue I, Shinoda Y, Ikeda M, Hayashi K, Kanazawa K, Nomura M, et al. CLOCK/BMAL1 is involved in lipid metabolism via transactivation of the peroxisome proliferator-activated receptor (PPAR) response element. J Atheroscler Thromb. 2005;12(3):169–74.
84. Marcheva B, Ramsey KM, Affinati A, Bass J. Clock genes and metabolic disease. J Appl Physiol. 2009; 107(5):1638–46.
85. Scott EM, Carter AM, Grant PJ. Association between polymorphisms in the Clock gene, obesity and the metabolic syndrome in man. Int J Obes (Lond). 2007;32(4):658–62.
86. Woon PY, Kaisaki PJ, Braganca J, Bihoreau MT, Levy JC, Farrall M, et al. Aryl hydrocarbon receptor nuclear translocator-like (BMAL1) is associated with susceptibility to hypertension and type 2 diabetes. Proc Natl Acad Sci USA. 2007;104(36):14412–7.
87. Lamia KA, Sachdeva UM, DiTacchio L, Williams EC, Alvarez JG, Egan DF, et al. AMPK regulates the circadian clock by cryptochrome phosphorylation and degradation. Science. 2009;326(5951):437–40.
88. Shimba S, Ishii N, Ohta Y, Ohno T, Watabe Y, Hayashi M, et al. Brain and muscle Arnt-like protein-1 (BMAL1), a component of the molecular clock, regulates adipogenesis. Proc Natl Acad Sci USA. 2005;102(34):12071–6.
89. Lamia KA, Storch KF, Weitz CJ. Physiological significance of a peripheral tissue circadian clock. Proc Natl Acad Sci USA. 2008;105(39):15172–7.
90. Miller DS, Mumford P. Gluttony. 1. An experimental study of overeating low- or high-protein diets. Am J Clin Nutr. 1967;20(11):1212–22.
91. Grunwald GK, Melanson EL, Forster JE, Seagle HM, Sharp TA, Hill JO. Comparison of methods for achieving 24-hour energy balance in a whole-room indirect calorimeter. Obes Res. 2003;11(6):752–9.
92. Hill JO, Melanson EL. Overview of the determinants of overweight and obesity: current evidence and research issues. Med Sci Sports Exerc. 1999;31(11 Suppl):S515–21.
93. Jung CM, Melanson EL, Frydendall EJ, Perreault L, Eckel RH, Wright KP. Energy expenditure during sleep, sleep deprivation and sleep following sleep deprivation in adult humans. J Physiol. 2011;589(Pt 1):235–44.
94. Zhang K, Sun M, Werner P, Kovera AJ, Albu J, Pi-Sunyer FX, et al. Sleeping metabolic rate in relation to body mass index and body composition. Int J Obes. 2002;26:376–83.
95. Goldberg GR, Prentice AM, Davies HL, Murtgatroyed PR. Overnight and basal metabolic rates in men and women. Eur J Clin Nutr. 1988; 42(2):137–44.
96. Everson CA, Smith CB, Sokoloff L. Effects of prolonged sleep deprivation on local rates of cerebral energy metabolism in freely moving rats. J Neurosci. 1994;14(11 Pt 2):6769–78.
97. Bergmann BM, Everson CA, Kushida CA, Fang VS, Leitch CA, Schoeller DA, et al. Sleep deprivation in the rat: V. energy use and mediation. Sleep. 1989;12: 31–41.
98. Spengler CM, Czeisler CA, Shea SA. An endogenous circadian rhythm of respiratory control in humans. J Physiol. 2000;526(Pt 3):683–94.
99. Krauchi K, WirzJustice A. Circadian-rhythm of heat-production, heart-rate, and skin and core temperature under unmasking conditions in men. Am J Physiol. 1994;267(3):R819–29.
100. Baumann CR, Bassetti CL. Hypocretins (orexins): clinical impact of the discovery of a neurotransmitter. Sleep Med Rev. 2005;9(4):253–68.
101. Zheng H, Patterson LM, Berthoud HR. Orexin-A projections to the caudal medulla and orexin-induced c-Fos expression, food intake, and autonomic function. J Comp Neurol. 2005;485(2):127–42.
102. Szekely M, Petervari E, Balasko M, Hernadi I, Uzsoki B. Effects of orexins on energy balance and thermoregulation. Regul Pept. 2002;104(1–3): 47–53.
103. Scammell TE. Wakefulness: an eye-opening perspective on orexin neurons. Curr Biol. 2001;11(19): R769–71.
104. Hakansson ML, Brown H, Ghilardi N, Skoda RC, Meister B. Leptin receptor immunoreactivity in chemically defined target neurons of the hypothalamus. J Neurosci. 1998;18(1):559–72.
105. Bartness TJ, Song CK, Demas GE. SCN efferents to peripheral tissues: implications for biological rhythms. J Biol Rhythms. 2001;16(3):196–204.
106. Chin-Chance C, Polonsky KS, Schoeller DA. Twenty-four-hour leptin levels respond to cumulative short-term energy imbalance and predict

subsequent intake. J Clin Endocrinol Metab. 2000;85(8):2685–91.
107. Havel PJ. Peripheral signals conveying metabolic information to the brain: short-term and long-term regulation of food intake and energy homeostasis. Exp Biol Med (Maywood). 2001;226(11):963–77.
108. Weigle DS, Cummings DE, Newby PD, Breen PA, Frayo RS, Matthys CC, et al. Roles of leptin and ghrelin in the loss of body weight caused by a low fat, high carbohydrate diet. J Clin Endocrinol Metab. 2003;88(4):1577–86.
109. Sinha MK, Opentanova I, Ohannesian JP, Kolaczynski JW, Heiman ML, Hale J, et al. Evidence of free and bound leptin in human circulation. Studies in lean and obese subjects and during short-term fasting. J Clin Invest. 1996;98(6):1277–82.
110. Simon C, Gronfier C, Schlienger JL, Brandenberger G. Circadian and ultradian variations of leptin in normal man under continuous enteral nutrition: relationship to sleep and body temperature. J Clin Endocrinol Metab. 1998;83(6):1893–9.
111. Schoeller DA, Cella LK, Sinha MK, Caro JF. Entrainment of the diurnal rhythm of plasma leptin to meal timing. J Clin Invest. 1997;100(7):1882–7.
112. Kelly TL, Neri DF, Grill JT, Ryman D, Hunt PD, Dijk DJ, et al. Nonentrained circadian rhythms of melatonin in submariners scheduled to an 18-hour day. J Biol Rhythms. 1999;14(3):190–6.
113. Dijk DJ, Neri DF, Wyatt JK, Ronda JM, Riel E, Ritz-De Cecco A, et al. Sleep, performance, circadian rhythms, and light-dark cycles during two space shuttle flights. Am J Physiol Regul Integr Comp Physiol. 2001;281(5):R1647–64.
114. Gavrila A, Peng CK, Chan JL, Mietus JE, Goldberger AL, Mantzoros CS. Diurnal and ultradian dynamics of serum adiponectin in healthy men: comparison with leptin, circulating soluble leptin receptor, and cortisol patterns. J Clin Endocrinol Metab. 2003; 88(6):2838–43.
115. Chan JL, Bluher S, Yiannakouris N, Suchard MA, Kratzsch J, Mantzoros CS. Regulation of circulating soluble leptin receptor levels by gender, adiposity, sex steroids, and leptin: observational and interventional studies in humans. Diabetes. 2002;51(7): 2105–12.
116. Spiegel K, Leproult R, L'Hermite-Baleriaux M, Copinschi G, Penev PD, Van Cauter E. Leptin levels are dependent on sleep duration: relationships with sympathovagal balance, carbohydrate regulation, cortisol, and thyrotropin. J Clin Endocrinol Metab. 2004;89(11):5762–71.
117. Spiegel K, Tasali E, Penev P, Van Cauter E. Brief communication: sleep curtailment in healthy young men is associated with decreased leptin levels, elevated ghrelin levels, and increased hunger and appetite. Ann Intern Med. 2004;141(11): 846–50.
118. Taheri S, Lin L, Austin D, Young T, Mignot E. Short sleep duration is associated with reduced leptin, elevated ghrelin, and increased body mass index. PLoS Med. 2004;1(3):1–8.
119. van der Lely AJ, Tschop M, Heiman ML, Ghigo E. Biological, physiological, pathophysiological, and pharmacological aspects of ghrelin. Endocr Rev. 2004;25(3):426–57.
120. Dzaja A, Dalal MA, Himmerich H, Uhr M, Pollmacher T, Schuld A. Sleep enhances nocturnal plasma ghrelin levels in healthy subjects. Am J Physiol Endocrinol Metab. 2004;286(6):E963–7.
121. Morris CJ, Fullick S, Gregson W, Clarke N, Doran D, MacLaren D, et al. Paradoxical post-exercise responses of acylated ghrelin and leptin during a simulated night shift. Chronobiol Int. 2010;27(3):590–605.
122. Nedeltcheva AV, Kilkus JM, Imperial J, Kasza K, Schoeller DA, Penev PD. Sleep curtailment is accompanied by increased intake of calories from snacks. Am J Clin Nutr. 2009;89(1):126–33.
123. Nedeltcheva AV, Kilkus JM, Imperial J, Schoeller DA, Penev PD. Insufficient sleep undermines dietary efforts to reduce adiposity. Ann Intern Med. 2010;153(7):435–41.
124. Persson M, Martensson J. Situations influencing habits in diet and exercise among nurses working night shift. J Nurs Manag. 2006;14(5):414–23.
125. Faugier J, Lancaster J, Pickles D, Dobson K. Barriers to healthy eating in the nursing profession: Part 1. Nurs Stand. 2001;15(36):33–6.
126. Wong H, Wong MC, Wong SY, Lee A. The association between shift duty and abnormal eating behavior among nurses working in a major hospital: a cross-sectional study. Int J Nurs Stud. 2010; 47(8):1021–7.
127. Van Strien T, Rookus MA, Bergers GP, Frijters JE, Defares PB. Life events, emotional eating and change in body mass index. Int J Obes. 1986;10(1): 29–35.
128. Lennernas M, Hambraeus L, Akerstedt T. Shift related dietary intake in day and shift workers. Appetite. 1995;25(3):253–65.
129. de Assis MA, Nahas MV, Bellisle F, Kupek E. Meals, snacks and food choices in Brazilian shift workers with high energy expenditure. J Hum Nutr Diet. 2003;16(4):283–9.
130. Turek FW, Joshu C, Kohsaka A, Lin E, Ivanova G, McDearmon E, et al. Obesity and metabolic syndrome in circadian Clock mutant mice. Science. 2005;308(5724):1043–5.
131. Arble DM, Bass J, Laposky AD, Vitaterna MH, Turek FW. Circadian timing of food intake contributes to weight gain. Obesity (Silver Spring). 2009;17(11):2100–2.
132. Oishi K. Disrupted light-dark cycle induces obesity with hyperglycemia in genetically intact animals. Neuro Endocrinol Lett. 2009;30(4):458–61.

Sleep Apnea and Obesity

9

Vidya Krishnan and Sanjay R. Patel

Abstract

Obesity and obstructive sleep apnea (OSA) are two extremely common and tightly linked disorders. Obesity, particularly in a pattern that increases fat deposition in the neck and upper body, has strong effects on OSA pathogenesis by narrowing the upper airway and increasing airway collapsibility. The effect of obesity on OSA risk is strongest in younger and male populations. In addition, growing evidence suggests OSA may have effects through intermittent hypoxia, sleep fragmentation or other mechanisms, on adipose tissue or pathways that regulate weight homeostasis. Finally, a number of genetic, perinatal, and environmental factors may independently impact risk of both obesity and OSA. Given their high prevalence and frequent coexistence, as well as the shared complications induced by these two conditions particularly in terms of cardiovascular and metabolic diseases, treatment strategies at both an individual and population level should focus on evaluating and treating both conditions.

Introduction

Obstructive sleep apnea (OSA) is a sleep disorder characterized by repetitive collapse of the upper airway during sleep resulting in complete (apnea) or partial (hypopnea) restriction of airflow. OSA severity is typically quantified using the apnea–hypopnea index (AHI), which measures the number of apneas and hypopneas per hour of sleep. Data from the Wisconsin Sleep Cohort Study suggest OSA is extremely common [1]. Among middle-aged Americans, 9% of women and 24% of men have at least mild OSA (defined as an AHI $\geq$5, while the prevalence of moderate-to-severe OSA (AHI $\geq$ 15) is estimated to be 4% in women and 9% in men. Among the elderly, the prevalence of OSA has been estimated to be as high as 60% or more [2]. (See Chap. 13 for details on clinical diagnosis.)

The recurrent episodes of upper airway collapse in OSA have numerous adverse physiologic

V. Krishnan, MD, MHS
Division of Pulmonary, Critical Care and Sleep Medicine, MetroHealth Medical Center, Case Western Reserve University, 2500 MetroHealth Drive, Room BG3-38, Cleveland, OH 44109, USA

S.R. Patel, MD, MS (✉)
Division of Sleep Medicine, Brigham and Women's Hospital, 221 Longwood Avenue, Room 225-C, Boston, MA 02115, USA
e-mail: spatel@partners.org

P.J. Shiromani et al. (eds.), *Sleep Loss and Obesity: Intersecting Epidemics*,
DOI 10.1007/978-1-4614-3492-4_9, © Springer Science+Business Media, LLC 2012

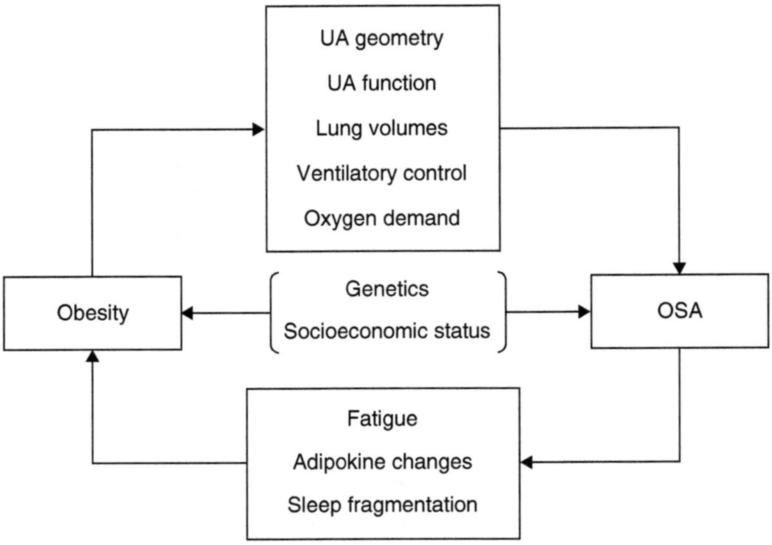

Fig. 9.1 Potential mechanisms linking obesity and OSA. *UA* upper airway, *OSA* obstructive sleep apnea

consequences. These include intermittent episodes of hypoxemia and hypercapnia due to cessations of gas exchange as well as increasing negative intrathoracic pressure swings that result from continued respiratory efforts against a closed pharynx. These in turn contribute to blood pressure swings and sympathetic activation [3, 4]. There are also frequent arousals from sleep necessary to reopen the airway, and these arousals lead to fragmentation of sleep and disruption of sleep architecture. There are often substantial reductions in slow wave and rapid eye movement (REM) sleep with concomitant increases in stages 1 and 2 non-REM sleep as well as an increased amount of wake after sleep onset [5].

OSA has substantial impact on general health. Not only does the poor and fragmented sleep caused by OSA result in daytime sleepiness and consequent increased risk of motor vehicle and occupational accidents [6–8], but neurocognitive function, mood, and quality of life can be impaired [9–12] as well. In addition, OSA has been increasingly recognized as an independent risk factor for hypertension, insulin resistance, heart failure, stroke, and other cardiovascular diseases [13–17]. The independent effect of OSA on many of these conditions has been challenging to measure because of the frequent coexistence of obesity in many patients with OSA and the known effects of obesity on metabolic and cardiovascular disease. In fact, obesity is one of the strongest predictors of OSA and has been utilized in common OSA screening tools [18, 19]. The relationship between OSA and obesity is complex, as illustrated in Fig. 9.1. It is clear that obesity is one of the strongest risk factors for OSA. In addition, there are intriguing data to suggest OSA may impact adiposity. Finally, there are pathways that may simultaneously influence both obesity and OSA status and thus help explain the co-occurrence of these two diseases.

Epidemiology of Obesity and OSA

Numerous cross-sectional studies have identified a relationship between OSA and adiposity [20–24]. In an Australian study of 1,464 consecutive men undergoing sleep studies, 75% of the subjects were overweight or obese, and increasing OSA severity was associated with increasing central obesity [25]. Among 594 patients referred to a sleep clinic in Canada, the mean BMI of patients with OSA was 3 kg/m^2 greater than the mean BMI of those without OSA [26]. In obesity clinics, the prevalence of OSA among those with morbid obesity ($BMI \geq 40\ kg/m^2$) is extremely high ranging from 77% to 98% in studies [27–29].

Similar findings have been obtained from population-based studies which are not susceptible to

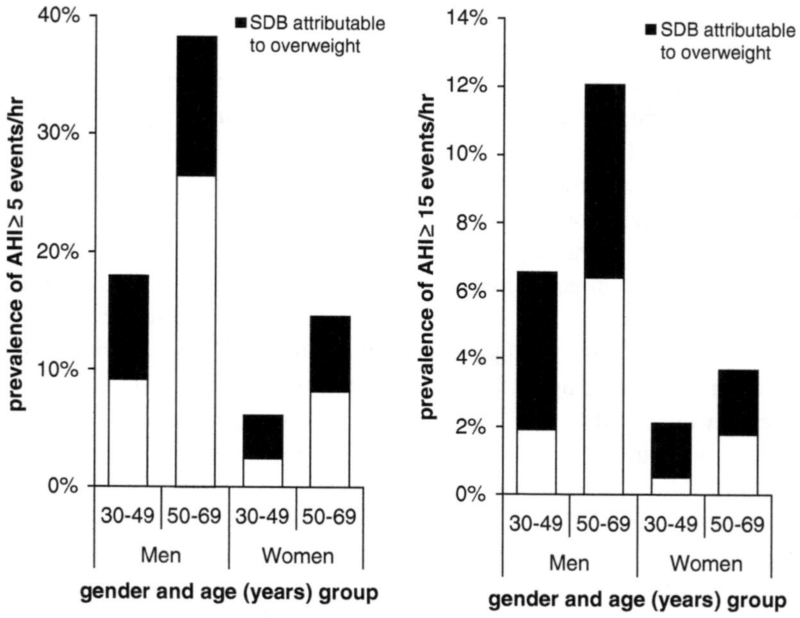

Fig. 9.2 Estimated prevalence of sleep apnea and proportion attributable to overweight by sex and age. This figure displays the total number of cases of sleep apnea and cases of moderate-to-severe sleep apnea as well as the proportion attributable to overweight defined as a body mass index (BMI) ≥ 25 kg/m^2. *AHI* Apnea–hypopnea index, *SDB* Sleep-disordered breathing. Reproduced from [37].

the referral biases of clinic-based studies. In the Wisconsin Sleep Cohort study, a population-based cohort of middle-aged state government workers, each 1 standard deviation increase in BMI (5.7 kg/m^2) was associated with a 4.2-fold increased odds of having OSA [1]. In the Sleep Heart Health Study, though an association between adiposity and OSA was found, this relationship weakened with increasing age. Each 1 standard deviation increase in BMI (5.3 kg/m^2) was associated with a 2.0-fold increased odds of having moderate-to-severe OSA among those in their 40s but only a 1.3-fold increased odds among those in their 80s [30]. Population-based studies from Spain, Italy, Brazil, India, Korea, and Australia confirm obesity is one of the strongest predictors of OSA around the world [31–36].

Assuming a causal relationship, Young et al. have estimated the attributable risk of excess weight on OSA based on data from the Wisconsin Sleep Cohort study [37]. This represents the proportion of OSA cases that would be cured if overweight and/or obesity were eliminated. They estimate that 41% of all OSA cases and 58% of all moderate-to-severe cases of OSA in US adults are attributable to overweight and obesity (i.e., BMI$\geq$25 kg/m^2). The proportions of OSA cases due to excess weight in subgroups stratified by age and gender are shown in Fig. 9.2. In general, as with the Sleep Heart Health Study cohort, these data suggest the association between OSA and excess weight is strongest in younger age groups.

Concerns have been raised by many groups that studies using BMI to measure adiposity may underestimate the co-aggregation of obesity and OSA. Strengths of the BMI metric are its simplicity of measurement allowing for its use in large epidemiologic studies and its well-accepted use by obesity researchers. The index has been used for over 30 years and is recommended for use in guidelines for weight management and weight control [38, 39]. Furthermore, BMI-defined obesity has been shown to predict mortality of all causes, heart disease, and cancer [40]. However, BMI does not account for the distribution of body fat, which may be more relevant to OSA risk than overall level of adiposity. For example, fat deposition in the neck may be more important than deposition in other regions [41, 42]. Central ("apple-shaped") obesity describes a pattern of fat accumulation in the torso including intra-abdominal visceral fat and contrasts with lower body ("pear-shaped") obesity, which describes increased fat accumulation in the hips and thighs. Neck and central obesity are more likely to affect collapsibility of the upper airway and lung mechanics, thereby directly contributing to OSA pathogenesis.

Neck circumference (NC) is a surrogate for the amount of soft tissue, including fat that surrounds the upper airway, and thus may estimate the tissue pressure contributing to upper airway collapsibility [43]. Both waist circumference (WC) and waist-to-hip ratio (WHR) have been used as measures of central obesity [44]. Increased adiposity by any of these measures has been associated with OSA risk [1], which is not surprising since they are all correlated. In the Wisconsin Sleep Cohort, the measure that had the best discriminative ability was neck circumference. Every 1 standard deviation increase in NC was associated with a 5.0-fold increase in OSA risk as compared to odds ratios of 4.2, 4.1, and 3.4 for BMI, WC, and WHR, respectively [1]. In the Sleep Heart Health Study, BMI and NC were found to independently predict OSA status, while WHR contributed no additional information [30]. In contrast, in a clinic-based cohort, WC has been found to predict OSA status independent of BMI and NC [25].

Obesity as a Risk Factor for OSA

A number of mechanisms have been proposed to explain how obesity predisposes to OSA. The most obvious is that with increasing weight, fat is deposited in structures in and around the upper airway leading to both a narrowing of the airway and an increase in collapsibility. Animal studies have demonstrated that mass loading in the neck to simulate increased neck fat leads to narrowing of the upper airway as well as increases in airway resistance and collapsibility [45]. Both obese and nonobese apneics have greater fat deposition in the anterolateral regions of the neck as compared to nonapneics [41]. Fat deposition, particularly within the region enclosed by the mandible, impinges on the upper airway lumen [46]. Fat deposition in the parapharyngeal fat pads as well as thickening of the lateral pharyngeal walls appears to be particularly important in OSA pathogenesis [47, 48].

Patency of the upper airway is also influenced by the caudal traction ("tracheal tug") that results from increased lung volumes [49]. With a reduced functional residual capacity (FRC), as is seen with central obesity [50], there is less caudal traction on the upper airway, making the upper airway more collapsible and worsening OSA severity [51]. In addition, by limiting the body's oxygen stores, a reduced FRC can result in more severe desaturation with each apnea [52]. This in turn can exacerbate the cardiovascular and other sequelae of OSA in obese patients.

Both the narrowing of the upper airway and the reduced traction resulting from lower lung volumes promotes a more collapsible upper airway. Level of adiposity as measured by BMI has been shown to strongly predict the intrinsic collapsibility of the upper airway and weight loss has been shown to make the airway less collapsible [53, 54]. Interestingly, the relationship between adiposity and airway collapsibility is stronger in men than women [53].

It has been hypothesized that another mechanism by which obesity may impact OSA risk is through pathologic effects on upper airway neuromuscular function [55]. Animal studies have demonstrated intramuscular fat deposition into the genioglossus and other upper airway dilator muscles with obesity [56, 57]. However, reduced upper airway muscular function or responsiveness in obesity has not been shown [58].

Increased adiposity may have other potential influences on breathing during sleep. Many adipose-derived proteins (such as lipoprotein lipase, hormone-sensitive lipase, and perilipin) have effects on lipid metabolism and therefore influence CO_2 production, which may secondarily impact ventilatory demand [59]. Another adipose-derived hormone, leptin, has been shown to be a ventilatory stimulant in animal models [60]. In humans, elevated leptin levels as seen in obesity have been

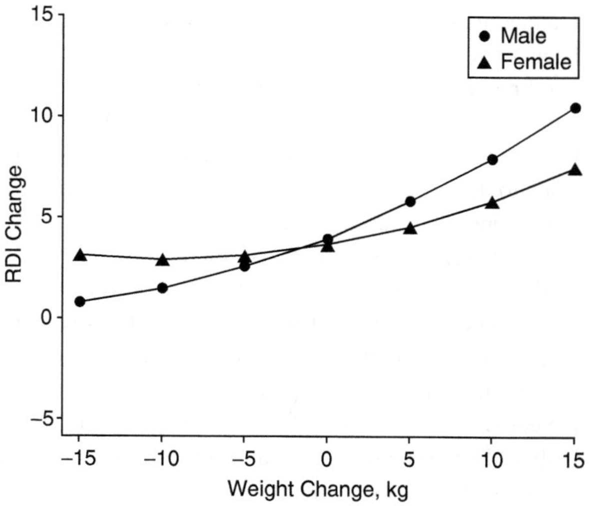

Fig. 9.3 Change in sleep apnea severity as a function of weight change. This graph demonstrates the effect of weight change on sleep apnea severity based on data from the Sleep Heart Health Study comparing two time points 5 years apart. *RDI* Respiratory disturbance index. Reproduced from [66]. Copyright © 2005 American Medical Association. All rights reserved

associated with a blunted ventilatory drive [61, 62], which may reflect a state of leptin resistance. Finally, the systemic inflammation produced by obesity, particularly visceral obesity, may contribute to the sleepiness and fatigue observed in the OSA clinical syndrome [63, 64].

By demonstrating that changes in weight predict a change in OSA severity, longitudinal studies have provided evidence that obesity is a causal risk factor for OSA. In the Wisconsin Sleep Cohort Study, a 10% increase in weight over 4 years predicted a 32% increase in AHI, while a 10% decrease predicted a 26% decrease in AHI [65]. Data from the Sleep Heart Health Study where polysomnography was done 5 years apart confirmed that for the same change in weight, the worsening effect of weight gain on OSA severity is larger than the ameliorating effect of weight loss [66]. This is shown graphically in Fig. 9.3 and suggests the effect of a temporary weight gain on OSA severity may be long lasting. Furthermore, this graph demonstrates that OSA severity is more sensitive to the effects of weight change (both gain and loss) in men as compared to women. This may reflect the differential relationship between adiposity and upper airway collapsibility, whereby the same change in weight leads to a greater change in airway collapsibility in men [53]. Data from the Cleveland Family Study where subjects were studied about 5 years apart support the notion that changes in BMI predict changes in AHI and that the effect of a given change in BMI on AHI is greater in men than women [67]. These investigators also found that OSA severity was more sensitive to changes in BMI among older as compared to younger adults. Several studies have examined the effect of obesity on incident OSA risk. In the Wisconsin study, a 5% increase in weight increased the risk of having moderate-to-severe OSA by 2.5-fold and a 10% increase in weight increased the risk 6-fold [65]. In the Cleveland Family Study, both an increased BMI and an increased WHR independently predicted incident cases of OSA [68].

Interventional studies have further strengthened the evidence that obesity causes OSA. Uncontrolled studies have demonstrated that medical weight loss can lead to substantial improvements in OSA severity. In 14 patients with morbid obesity, a 10% weight loss was associated with a significant reduction in subcutaneous fat in the neck and an improvement in pulmonary function as well as a significant decrease in OSA severity (mean AHI drop from 24.3 to 2.9) [69]. In addition to improvements in OSA severity, weight loss has been shown to increase airway size and reduce airway collapsibility [54, 70, 71].

A large number of trials have evaluated the role of surgically induced weight loss in the management of OSA and positive effects have been seen with both gastric bypass and gastric banding procedures. While early studies reported cure rates above 80% with bariatric procedures [72], these studies relied on self-reported symptoms to diagnose OSA, which is not only insensitive but also susceptible to placebo effects. A subsequent meta-analysis including only studies that objectively assessed OSA severity before and after surgery found that in a morbidly obese population (mean pre-operative BMI was 55.3 kg/m^2), the AHI dropped substantially from a mean of 55 to 16 across all 12 studies. However, it should be noted that residual OSA remained in the majority of these patients [73].

Several controlled randomized clinical trials have been performed assessing the role of medical weight loss by diet and/or exercise on OSA treatment. Although these studies confirm weight loss improves OSA severity, in general these trials have been disappointing in the magnitude of effect obtained with weight loss. An early study compared 15 patients who lost 9.6 kg with eight controls who gained 1.4 kg in follow-up. The weight loss group had improvements in OSA severity in both non-REM and REM sleep and also demonstrated improvements in sleep architecture [74]. In the SleepAHEAD study, 264 patients with OSA, obesity (mean BMI 36.7 kg/m^2) and type 2 diabetes were randomized to an aggressive weight loss treatment arm versus control [75]. After 1 year, weight loss in the two groups was 10.8 kg vs. 0.6 kg. Mean AHI dropped slightly from 22.9 to 18.3 in the active treatment group but increased from 23.5 to 28.3 in the control group. The cure rates for OSA were 14% vs. 3% in the two arms.

A Finnish study evaluating weight loss through a low-calorie diet in patients with mild OSA recruited 62 patients with mean BMI of 32.4 kg/m^2 and mean AHI of 9.6 [76]. At 1 year, the reduction in weight was 10.7 kg vs. 2.4 kg, and this corresponded to a reduction in mean AHI of 4.0 in the active treatment group as compared to an increase in mean AHI of 0.3 in the control group. In terms of cure rates, 63% of the intervention group reduced their AHI below 5 as opposed to 35% in the control group. However, no significant differences were seen between the two groups in terms of improvements in sleepiness or quality of life.

A similar Swedish study assessed the efficacy of a low-calorie diet in moderate-to-severe OSA [77]. A total of 63 patients with mean BMI of 34.6 kg/m^2 and mean AHI of 37 were followed for 9 weeks. The dieting plan was severe resulting in a weight loss of 18.7 kg in the active treatment group as opposed to a 1.1 kg weight gain in the control arm. This translated to a significant reduction in AHI in the active arm from 37 to 12 while no change was seen in the control arm (AHI: 37 to 35). The cure rates were 17% vs. 0% and additionally, a significant improvement in sleepiness compared to the control group was seen.

While weight loss clearly has a therapeutic effect on OSA, the duration of effect has been questioned. Sampol and colleagues studied 24 obese patients who underwent weight loss (mean BMI fall from 31.5 to 25.9 kg/m^2) and managed to achieve complete remission of OSA [78]. Despite the initial success, 50% had recurrence of OSA after a mean of 94 months of follow-up, regardless of whether they had weight gain. An observational report on 14 bariatric patients found that while substantial improvements were seen in OSA severity between 2 and 7 months after surgery, the AHI rose over the next 7 years even without substantial change in weight [79]. While neither study had a control group, these reports suggest excess weight may have a long-term effect on OSA trajectory that cannot be completely altered with weight loss.

OSA as a Risk Factor for Obesity

Although it is clear that obesity causes OSA, there are a number of pathways by which OSA may induce weight gain, potentially creating a vicious cycle of increasing weight and apnea. Proposed pathways by which OSA may promote weight gain include reductions in physical activity, alterations in hunger control, and fragmentation of sleep.

Two of the most common symptoms of OSA are daytime sleepiness and fatigue. Fatigue is commonly associated with reductions in physical activity [80]. In a study comparing 11 patients with OSA and 9 controls, those with OSA had a lower level of fitness as assessed by maximal exercise and oxygen consumption [81]. In addition, a reduced maximal blood lactate concentration and reduced rate of elimination of lactate suggested an impairment in muscle energy metabolism. In a large clinical cohort of 1,149 patients, heart rate and blood pressure responses to exercise were associated with OSA severity [82]. In addition, a trend was found toward reduced maximal exercise capacity with increasing OSA severity. However, because of the substantial energy expended by respiratory muscles in attempting to breathe against an obstructed airway, energy expenditure has been found to be greater in apneics compared to controls and energy expenditure during sleep decreases with OSA treatment using continuous positive airway pressure (CPAP) [83]. Thus, to cause weight gain, any reduction in daytime activity caused by OSA would have to be larger than the increase in energy expenditure caused by ineffective respiratory efforts while asleep. Questionnaire data from 320 adult patients enrolled in a clinical trial found, those with severe OSA (AHI>50) had greater levels of physical activity than less severe apnea but that these differences were due solely to differences in level of obesity [84]. In contrast, a pediatric study found children with OSA were less active as assessed by participation in organized sports [85].

Another possible mechanism by which OSA could promote weight gain is through increases in caloric intake. This may be mediated through pathways similar to those proposed to relate reduced sleep duration with hormonal regulators of appetite. While OSA does not reduce total sleep duration, it does fragment sleep leading to altered sleep architecture including reductions in both slow wave and REM sleep [5]. Reductions in slow wave sleep independent of total sleep time have been associated with impaired glucose homeostasis and obesity [86, 87]. Similarly, reduced amounts of REM sleep have been associated with overweight and obesity [88, 89]. In terms of specific hormonal mediators, OSA has been associated with changes in two hormones, leptin and ghrelin, that have effects on appetite regulation. Leptin, made primarily by subcutaneous adipose tissue, has an appetite suppressing effect at the level of the hypothalamus, while ghrelin, made primarily in the stomach, has a central effect to stimulate appetite and hunger. Multiple studies have demonstrated elevated levels of circulating leptin in OSA and a fall in levels with CPAP therapy [90–93]. Rather than suggesting that OSA is associated with anorexia, these findings have been interpreted to suggest OSA is associated with a state of leptin resistance that improves with treatment. However, no studies have assessed the effect of OSA on leptin signaling in the hypothalamus to verify this interpretation. Circulating levels of ghrelin, and in particular, the biologically active acylated form of the hormone, have also been shown to be elevated in OSA and fall with CPAP treatment [93, 94]. This elevation in ghrelin levels along with worsened leptin resistance in the setting of OSA may serve to increase caloric intake. A cross-sectional study found those with more severe OSA consumed 281 kcal/day more than milder apneics [84]. A trend persisted though was no longer significant after adjusting for differences in BMI. This study also found greater consumption of cholesterol, total fat, and saturated fats in the more severe group. Among children, a pattern of unhealthy eating behaviors has been associated with OSA including reduced consumption of fruits and vegetables and increased consumption of fast food [85].

In terms of assessing the effect of OSA on weight trajectory, a retrospective study suggested those with OSA had a more accelerated rate of weight gain over the year prior to diagnosis [92]. An early study reported that CPAP therapy for OSA for 6 months was associated with significant weight loss [95]. However, a larger study of 228 apneics followed for 1 year found no significant effect of CPAP treatment on weight in the overall cohort and a trend toward weight gain with treatment among women [96]. Furthermore, greater compliance was associated with greater weight gain. Within the format of a randomized controlled trial comparing CPAP to conservative management for OSA, CPAP had no effect on weight over 6 months, while conservative therapy which included weight loss counseling produced a small reduction [97]. Similarly, when added to a behavioral modification program for weight loss, 6 months of CPAP therapy did not confer any additional benefit on weight loss [98]. Thus, overall, there is little evidence that elimination of OSA results in weight loss. However, OSA treatment may produce changes in fat distribution. Two prospective studies have demonstrated reductions in visceral fat volume with CPAP therapy for OSA [90, 99].

Shared Risk Factors for Obesity and OSA

In addition to a causal relationship, the association between obesity and OSA may also be explained by common shared risk factors. For example, a low socioeconomic status (SES) may represent an independent risk factor for both obesity and OSA. The link between SES and obesity has been well studied. While rising SES is associated with greater BMI in many developing nations, in the US and other industrialized nations, the relationship is, in general, the opposite though dependent on gender and race. Among US whites as well as among minority women, low SES is strongly associated with greater obesity risk [100, 101]. However, among African-American and Mexican-American men, low SES predicts a lower BMI [100, 101]. Mechanisms by which low SES might promote obesity have not been completely elucidated. However, neighborhood characteristics have been postulated to be important in creating an obesogenic environment. For example, obesity risk has been found to be greater in neighborhoods with fewer supermarkets and less availability of fresh fruits and vegetables [102, 103]. Conversely, obesity is more common in neighborhoods with a greater density of fast food restaurants [104]. Neighborhoods may also impact physical activity. Poorer neighborhoods are more likely to have problems with safety and upkeep of parks and sidewalks making these areas less "walkable" or conducive to physical activity [105].

Less is known about the relationship between SES and OSA risk. However, in studies adjusting for obesity, neighborhood disadvantage as a proxy for low SES has been found to be an independent risk factor for OSA in children [106]. Among adults, occupation and SES have also been found to predict OSA [107]. Little work has been done to understand the relationship between low SES and OSA risk, but may include increased alcohol and tobacco consumption, as well as increased exposures in the home and at work to chemicals or allergens that might increase upper airway inflammation and thus OSA risk. A recent study, for example, suggests air pollution may exacerbate OSA severity [108].

Adverse intrauterine exposures and premature birth may also represent common risk factors for both obesity and OSA. Maternal smoking is a strong risk factor for childhood obesity [109]. In addition, low birth weight is somewhat paradoxically a risk factor for obesity, in particular central obesity [110]. Similarly, prematurity and low birth weight have been found to predict OSA risk in later life [111–113]. Since the problems of maternal smoking, prematurity, and low birth weight are more prevalent in low SES groups, these factors may also mediate the links between SES, obesity, and apnea.

Another possible common etiology for obesity and OSA is shared genetic risk factors. Studies of extended families selected for OSA suggest that the genetic correlation between OSA severity and obesity measures is about 0.6 [114]. Thus, a substantial portion of the genetic basis for OSA is shared with obesity. This overlap may represent genes that mediate their effects on OSA risk through changes in adiposity (or vice versa). Alternatively, this overlap may represent genetic variants that have pleiotropic effects on both obesity and OSA through independent mechanisms. As of yet, no specific OSA variants have been identified limiting understanding of the relative contributions of each type of genetic risk factor to the overall overlap in risk.

Conclusion

In summary, obesity and OSA are two extremely common and tightly linked disorders. Obesity, particularly in a pattern that increases fat deposition in the neck and upper body, has strong effects on OSA pathogenesis by narrowing the upper airway and increasing airway collapsibility. The effect of obesity on OSA risk is strongest in younger and male populations. In addition, growing evidence suggests OSA may have effects through intermittent hypoxia, sleep fragmentation or other mechanisms, on adipose tissue or pathways that regulate weight homeostasis. Finally, a number of genetic, perinatal, and environmental factors may independently impact risk of both obesity and OSA. Given their high prevalence and frequent coexistence, as well as the shared complications induced by these two conditions particularly in terms of cardiovascular and metabolic diseases, treatment strategies at both an individual and population level should focus on evaluating and treating both conditions.

References

1. Young T, Palta M, Dempsey J, Skatrud J, Weber S, Badr S. The occurrence of sleep-disordered breathing among middle-aged adults. N Engl J Med. 1993; 328(17):1230–5.
2. Ancoli-Israel S, Kripke DF, Klauber MR, Mason WJ, Fell R, Kaplan O. Sleep-disordered breathing in community-dwelling elderly. Sleep. 1991;14(6): 486–95.
3. Somers VK, Dyken ME, Clary MP, Abboud FM. Sympathetic neural mechanisms in obstructive sleep apnea. J Clin Invest. 1995;96(4):1897–904.
4. Tilkian AG, Guilleminault C, Schroeder JS, Lehrman KL, Simmons FB, Dement WC. Hemodynamics in sleep-induced apnea. Studies during wakefulness and sleep. Ann Intern Med. 1976;85(6):714–9.
5. Redline S, Kirchner HL, Quan SF, Gottlieb DJ, Kapur V, Newman A. The effects of age, sex, ethnicity, and sleep-disordered breathing on sleep architecture. Arch Intern Med. 2004;164(4):406–18.
6. Gottlieb DJ, Whitney CW, Bonekat WH, et al. Relation of sleepiness to respiratory disturbance index: the sleep heart health study. Am J Respir Crit Care Med. 1999;159(2):502–7.
7. Teran-Santos J, Jimenez-Gomez A, Cordero-Guevara J. The association between sleep apnea and the risk of traffic accidents. Cooperative group Burgos-Santander. N Engl J Med. 1999;340(11):847–51.
8. Ulfberg J, Carter N, Edling C. Sleep-disordered breathing and occupational accidents. Scand J Work Environ Health. 2000;26(3):237–42.
9. Engleman HM, Martin SE, Deary IJ, Douglas NJ. Effect of CPAP therapy on daytime function in patients with mild sleep apnoea/hypopnoea syndrome. Thorax. 1997;52(2):114–9.
10. Engleman HM, Kingshott RN, Martin SE, Douglas NJ. Cognitive function in the sleep apnea/hypopnea syndrome (SAHS). Sleep. 2000;23 Suppl 4:S102–8.
11. Jenkinson C, Davies RJ, Mullins R, Stradling JR. Comparison of therapeutic and subtherapeutic nasal continuous positive airway pressure for obstructive sleep apnoea: a randomised prospective parallel trial. Lancet. 1999;353(9170):2100–5.
12. Redline S, Adams N, Strauss ME, Roebuck T, Winters M, Rosenberg C. Improvement of mild sleep-disordered breathing with CPAP compared with conservative therapy. Am J Respir Crit Care Med. 1998;157(3 Pt 1):858–65.
13. Peppard PE, Young T, Palta M, Skatrud J. Prospective study of the association between sleep-disordered breathing and hypertension. N Engl J Med. 2000; 342(19):1378–84.
14. Ip MS, Lam B, Ng MM, Lam WK, Tsang KW, Lam KS. Obstructive sleep apnea is independently associated with insulin resistance. Am J Respir Crit Care Med. 2002;165(5):670–6.
15. Gottlieb DJ, Yenokyan G, Newman AB, et al. Prospective study of obstructive sleep apnea and incident coronary heart disease and heart failure: the sleep heart health study. Circulation. 2010;122(4): 352–60.
16. Yaggi HK, Concato J, Kernan WN, Lichtman JH, Brass LM, Mohsenin V. Obstructive sleep apnea as a risk factor for stroke and death. N Engl J Med. 2005;353(19):2034–41.
17. Marin JM, Carrizo SJ, Vicente E, Agusti AG. Long-term cardiovascular outcomes in men with obstructive sleep apnoea-hypopnoea with or without treatment with continuous positive airway pressure: an observational study. Lancet. 2005;365(9464): 1046–53.
18. Netzer NC, Stoohs RA, Netzer CM, Clark K, Strohl KP. Using the Berlin questionnaire to identify patients at risk for the sleep apnea syndrome. Ann Intern Med. 1999;131(7):485–91.
19. Chung F, Yegneswaran B, Liao P, et al. STOP questionnaire: a tool to screen patients for obstructive sleep apnea. Anesthesiology. 2008;108(5):812–21.
20. Davies RJ, Ali NJ, Stradling JR. Neck circumference and other clinical features in the diagnosis of the obstructive sleep apnoea syndrome. Thorax. 1992; 47(2):101–5.
21. Levinson PD, McGarvey ST, Carlisle CC, Eveloff SE, Herbert PN, Millman RP. Adiposity and cardiovascular risk factors in men with obstructive sleep apnea. Chest. 1993;103(5):1336–42.

22. Shelton KE, Woodson H, Gay S, Suratt PM. Pharyngeal fat in obstructive sleep apnea. Am Rev Respir Dis. 1993;148(2):462–6.
23. Millman RP, Carlisle CC, McGarvey ST, Eveloff SE, Levinson PD. Body fat distribution and sleep apnea severity in women. Chest. 1995;107(2): 362–6.
24. Shinohara E, Kihara S, Yamashita S, et al. Visceral fat accumulation as an important risk factor for obstructive sleep apnoea syndrome in obese subjects. J Intern Med. 1997;241(1):11–8.
25. Grunstein R, Wilcox I, Yang TS, Gould Y, Hedner J. Snoring and sleep apnoea in men: association with central obesity and hypertension. Int J Obes Relat Metab Disord. 1993;17(9):533–40.
26. Hoffstein V, Szalai JP. Predictive value of clinical features in diagnosing obstructive sleep apnea. Sleep. 1993;16(2):118–22.
27. Valencia-Flores M, Orea A, Castano VA, et al. Prevalence of sleep apnea and electrocardiographic disturbances in morbidly obese patients. Obes Res. 2000;8(3):262–9.
28. Sareli AE, Cantor CR, Williams NN, et al. Obstructive sleep apnea in patients undergoing bariatric surgery-a tertiary center experience. Obes Surg. 2011;21(3):316–27.
29. Sharkey KM, Machan JT, Tosi C, Roye GD, Harrington D, Millman RP. Predicting obstructive sleep apnea among women candidates for bariatric surgery. J Womens Health (Larchmt). 2010;19(10): 1833–41.
30. Young T, Shahar E, Nieto FJ, et al. Predictors of sleep-disordered breathing in community-dwelling adults: the sleep heart health study. Arch Intern Med. 2002;162(8):893–900.
31. Duran J, Esnaola S, Rubio R, Iztueta A. Obstructive sleep apnea-hypopnea and related clinical features in a population-based sample of subjects aged 30 to 70 yr. Am J Respir Crit Care Med. 2001;163(3 Pt 1):685–9.
32. Tufik S, Santos-Silva R, Taddei JA, Bittencourt LR. Obstructive sleep apnea syndrome in the Sao Paulo epidemiologic sleep study. Sleep Med. 2010;11(5): 441–6.
33. Udwadia ZF, Doshi AV, Lonkar SG, Singh CI. Prevalence of sleep-disordered breathing and sleep apnea in middle-aged urban Indian men. Am J Respir Crit Care Med. 2004;169(2):168–73.
34. Kim J, In K, You S, et al. Prevalence of sleep-disordered breathing in middle-aged Korean men and women. Am J Respir Crit Care Med. 2004;170(10): 1108–13.
35. Bearpark H, Elliott L, Grunstein R, et al. Snoring and sleep apnea. A population study in Australian men. Am J Respir Crit Care Med. 1995;151(5): 1459–65.
36. Ferini-Strambi L, Zucconi M, Palazzi S, et al. Snoring and nocturnal oxygen desaturations in an Italian middle-aged male population. Epidemiologic study with an ambulatory device. Chest. 1994; 105(6):1759–64.
37. Young T, Peppard PE, Taheri S. Excess weight and sleep-disordered breathing. J Appl Physiol. 2005; 99(4):1592–9.
38. Seidell JC, Kahn HS, Williamson DF, Lissner L, Valdez R. Report from a centers for disease control and prevention workshop on use of adult anthropometry for public health and primary health care. Am J Clin Nutr. 2001;73(1):123–6.
39. NHLBI Obesity Education Initiative Expert Panel on the identification E, and treatment of obesity in adults clinical guidelines on the identification, and treatment of overweight and obesity in adults: the evidence report. In: National Institutes of Health. National Heart L, and Blood Institute, ed. Bethesda, MD: NIH Publication no. 98–4083; Sep, 1998.
40. McGee DL. Body mass index and mortality: a meta-analysis based on person-level data from twenty-six observational studies. Ann Epidemiol. 2005;15(2): 87–97.
41. Mortimore IL, Marshall I, Wraith PK, Sellar RJ, Douglas NJ. Neck and total body fat deposition in nonobese and obese patients with sleep apnea compared with that in control subjects. Am J Respir Crit Care Med. 1998;157(1):280–3.
42. Flemons WW, Whitelaw WA, Brant R, Remmers JE. Likelihood ratios for a sleep apnea clinical prediction rule. Am J Respir Crit Care Med. 1994;150(5 Pt 1):1279–85.
43. Katz I, Stradling J, Slutsky AS, Zamel N, Hoffstein V. Do patients with obstructive sleep apnea have thick necks? Am Rev Respir Dis. 1990;141(5 Pt 1): 1228–31.
44. Yusuf S, Hawken S, Ounpuu S, et al. Effect of potentially modifiable risk factors associated with myocardial infarction in 52 countries (the INTERHEART study): case–control study. Lancet. 2004;364(9438): 937–52.
45. Koenig JS, Thach BT. Effects of mass loading on the upper airway. J Appl Physiol. 1988;64(6):2294–9.
46. Shelton KE, Gay SB, Hollowell DE, Woodson H, Suratt PM. Mandible enclosure of upper airway and weight in obstructive sleep apnea. Am Rev Respir Dis. 1993;148(1):195–200.
47. Schwab RJ, Gupta KB, Gefter WB, Metzger LJ, Hoffman EA, Pack AI. Upper airway and soft tissue anatomy in normal subjects and patients with sleep-disordered breathing. Significance of the lateral pharyngeal walls. Am J Respir Crit Care Med. 1995;152(5 Pt 1):1673–89.
48. Schwab RJ, Pasirstein M, Pierson R, et al. Identification of upper airway anatomic risk factors for obstructive sleep apnea with volumetric magnetic resonance imaging. Am J Respir Crit Care Med. 2003;168(5):522–30.
49. Van de Graaff WB. Thoracic traction on the trachea: mechanisms and magnitude. J Appl Physiol. 1991; 70(3):1328–36.

50. Sharp JT, Henry JP, Sweany SK, Meadows WR, Pietras RJ. Effects of mass loading the respiratory system in man. J Appl Physiol. 1964;19:959–66.
51. Heinzer RC, Stanchina ML, Malhotra A, et al. Effect of increased lung volume on sleep disordered breathing in patients with sleep apnoea. Thorax. 2006;61(5):435–9.
52. Peppard PE, Ward NR, Morrell MJ. The impact of obesity on oxygen desaturation during sleep-disordered breathing. Am J Respir Crit Care Med. 2009;180(8):788–93.
53. Kirkness JP, Schwartz AR, Schneider H, et al. Contribution of male sex, age, and obesity to mechanical instability of the upper airway during sleep. J Appl Physiol. 2008;104(6):1618–24.
54. Schwartz AR, Gold AR, Schubert N, et al. Effect of weight loss on upper airway collapsibility in obstructive sleep apnea. Am Rev Respir Dis. 1991;144(3 Pt 1):494–8.
55. Schwartz AR, Patil SP, Squier S, Schneider H, Kirkness JP, Smith PL. Obesity and upper airway control during sleep. J Appl Physiol. 2010;108(2):430–5.
56. Brennick MJ, Pack AI, Ko K, et al. Altered upper airway and soft tissue structures in the New Zealand obese mouse. Am J Respir Crit Care Med. 2009;179(2):158–69.
57. Saito T, Yamane A, Kaneko S, et al. Changes in the lingual muscles of obese rats induced by high-fat diet feeding. Arch Oral Biol. 2010;55(10):803–8.
58. Carrera M, Barbe F, Sauleda J, et al. Effects of obesity upon genioglossus structure and function in obstructive sleep apnoea. Eur Respir J. 2004;23(3):425–9.
59. Ronti T, Lupattelli G, Mannarino E. The endocrine function of adipose tissue: an update. Clin Endocrinol (Oxf). 2006;64(4):355–65.
60. O'Donnell CP, Schaub CD, Haines AS, et al. Leptin prevents respiratory depression in obesity. Am J Respir Crit Care Med. 1999;159(5 Pt 1):1477–84.
61. Phipps PR, Starritt E, Caterson I, Grunstein RR. Association of serum leptin with hypoventilation in human obesity. Thorax. 2002;57(1):75–6.
62. Campo A, Fruhbeck G, Zulueta JJ, et al. Hyperleptinaemia, respiratory drive and hypercapnic response in obese patients. Eur Respir J. 2007;30(2):223–31.
63. Vgontzas AN, Bixler EO, Tan TL, Kantner D, Martin LF, Kales A. Obesity without sleep apnea is associated with daytime sleepiness. Arch Intern Med. 1998;158(12):1333–7.
64. Vgontzas AN, Papanicolaou DA, Bixler EO, et al. Sleep apnea and daytime sleepiness and fatigue: relation to visceral obesity, insulin resistance, and hypercytokinemia. J Clin Endocrinol Metab. 2000;85(3):1151–8.
65. Peppard PE, Young T, Palta M, Dempsey J, Skatrud J. Longitudinal study of moderate weight change and sleep-disordered breathing. JAMA. 2000;284(23):3015–21.
66. Newman AB, Foster G, Givelber R, Nieto FJ, Redline S, Young T. Progression and regression of sleep-disordered breathing with changes in weight: the sleep heart health study. Arch Intern Med. 2005;165(20):2408–13.
67. Redline S, Schluchter MD, Larkin EK, Tishler PV. Predictors of longitudinal change in sleep-disordered breathing in a nonclinic population. Sleep. 2003;26(6):703–9.
68. Tishler PV, Larkin EK, Schluchter MD, Redline S. Incidence of sleep-disordered breathing in an urban adult population: the relative importance of risk factors in the development of sleep-disordered breathing. JAMA. 2003;289(17):2230–7.
69. Hernandez TL, Ballard RD, Weil KM, et al. Effects of maintained weight loss on sleep dynamics and neck morphology in severely obese adults. Obesity (Silver Spring). 2009;17(1):84–91.
70. Suratt PM, McTier RF, Findley LJ, Pohl SL, Wilhoit SC. Changes in breathing and the pharynx after weight loss in obstructive sleep apnea. Chest. 1987;92(4):631–7.
71. Rubinstein I, Colapinto N, Rotstein LE, Brown IG, Hoffstein V. Improvement in upper airway function after weight loss in patients with obstructive sleep apnea. Am Rev Respir Dis. 1988;138(5):1192–5.
72. Buchwald H, Avidor Y, Braunwald E, et al. Bariatric surgery: a systematic review and meta-analysis. JAMA. 2004;292(14):1724–37.
73. Greenburg DL, Lettieri CJ, Eliasson AH. Effects of surgical weight loss on measures of obstructive sleep apnea: a meta-analysis. Am J Med. 2009;122(6):535–42.
74. Smith PL, Gold AR, Meyers DA, Haponik EF, Bleecker ER. Weight loss in mildly to moderately obese patients with obstructive sleep apnea. Ann Intern Med. 1985;103(6 (Pt 1)):850–5.
75. Foster GD, Borradaile KE, Sanders MH, et al. A randomized study on the effect of weight loss on obstructive sleep apnea among obese patients with type 2 diabetes: the sleep ahead study. Arch Intern Med. 2009;169(17):1619–26.
76. Tuomilehto HP, Seppa JM, Partinen MM, et al. Lifestyle intervention with weight reduction: first-line treatment in mild obstructive sleep apnea. Am J Respir Crit Care Med. 2009;179(4):320–7.
77. Johansson K, Neovius M, Lagerros YT, et al. Effect of a very low energy diet on moderate and severe obstructive sleep apnoea in obese men: a randomised controlled trial. BMJ. 2009;339:b4609.
78. Sampol G, Munoz X, Sagales MT, et al. Long-term efficacy of dietary weight loss in sleep apnoea/hypopnoea syndrome. Eur Respir J. 1998;12(5):1156–9.
79. Pillar G, Peled R, Lavie P. Recurrence of sleep apnea without concomitant weight increase 7.5 years after weight reduction surgery. Chest. 1994;106(6):1702–4.
80. Resnick HE, Carter EA, Aloia M, Phillips B. Cross-sectional relationship of reported fatigue to obesity, diet, and physical activity: results from the third

national health and nutrition examination survey. J Clin Sleep Med. 2006;2(2):163–9.
81. Vanuxem D, Badier M, Guillot C, Delpierre S, Jahjah F, Vanuxem P. Impairment of muscle energy metabolism in patients with sleep apnoea syndrome. Respir Med. 1997;91(9):551–7.
82. Grote L, Hedner J, Peter JH. The heart rate response to exercise is blunted in patients with sleep-related breathing disorder. Cardiology. 2004;102(2):93–9.
83. Stenlof K, Grunstein R, Hedner J, Sjostrom L. Energy expenditure in obstructive sleep apnea: effects of treatment with continuous positive airway pressure. Am J Physiol. 1996;271(6 Pt 1):E1036–43.
84. Vasquez MM, Goodwin JL, Drescher AA, Smith TW, Quan SF. Associations of dietary intake and physical activity with sleep disordered breathing in the apnea positive pressure long-term efficacy study (APPLES). J Clin Sleep Med. 2008;4(5):411–8.
85. Spruyt K, Sans Capdevila O, Serpero LD, Kheirandish-Gozal L, Gozal D. Dietary and physical activity patterns in children with obstructive sleep apnea. J Pediatr 2010;156(5):724–730, 730 e721–730 e723.
86. Tasali E, Leproult R, Ehrmann DA, Van Cauter E. Slow-wave sleep and the risk of type 2 diabetes in humans. Proc Natl Acad Sci USA. 2008;105(3):1044–9.
87. Rao MN, Blackwell T, Redline S, Stefanick ML, Ancoli-Israel S, Stone KL. Association between sleep architecture and measures of body composition. Sleep. 2009;32(4):483–90.
88. Theorell-Haglow J, Berne C, Janson C, Sahlin C, Lindberg E. Associations between short sleep duration and central obesity in women. Sleep. 2010;33(5):593–8.
89. Liu X, Forbes EE, Ryan ND, Rofey D, Hannon TS, Dahl RE. Rapid eye movement sleep in relation to overweight in children and adolescents. Arch Gen Psychiatry. 2008;65(8):924–32.
90. Chin K, Shimizu K, Nakamura T, et al. Changes in intra-abdominal visceral fat and serum leptin levels in patients with obstructive sleep apnea syndrome following nasal continuous positive airway pressure therapy. Circulation. 1999;100(7):706–12.
91. Ip MS, Lam KS, Ho C, Tsang KW, Lam W. Serum leptin and vascular risk factors in obstructive sleep apnea. Chest. 2000;118(3):580–6.
92. Phillips BG, Kato M, Narkiewicz K, Choe I, Somers VK. Increases in leptin levels, sympathetic drive, and weight gain in obstructive sleep apnea. Am J Physiol Heart Circ Physiol. 2000;279(1):H234–7.
93. Harsch IA, Konturek PC, Koebnick C, et al. Leptin and ghrelin levels in patients with obstructive sleep apnoea: effect of CPAP treatment. Eur Respir J. 2003;22(2):251–7.
94. Takahashi K, Chin K, Akamizu T, et al. Acylated ghrelin level in patients with OSA before and after nasal CPAP treatment. Respirology. 2008;13(6):810–6.
95. Loube DI, Loube AA, Erman MK. Continuous positive airway pressure treatment results in weight less in obese and overweight patients with obstructive sleep apnea. J Am Diet Assoc. 1997;97(8):896–7.
96. Redenius R, Murphy C, O'Neill E, Al-Hamwi M, Zallek SN. Does CPAP lead to change in BMI? J Clin Sleep Med. 2008;4(3):205–9.
97. Monasterio C, Vidal S, Duran J, et al. Effectiveness of continuous positive airway pressure in mild sleep apnea-hypopnea syndrome. Am J Respir Crit Care Med. 2001;164(6):939–43.
98. Kajaste S, Brander PE, Telakivi T, Partinen M, Mustajoki P. A cognitive-behavioral weight reduction program in the treatment of obstructive sleep apnea syndrome with or without initial nasal CPAP: a randomized study. Sleep Med. 2004;5(2):125–31.
99. Trenell MI, Ward JA, Yee BJ, et al. Influence of constant positive airway pressure therapy on lipid storage, muscle metabolism and insulin action in obese patients with severe obstructive sleep apnoea syndrome. Diabetes Obes Metab. 2007;9(5):679–87.
100. Mujahid MS, Diez Roux AV, Borrell LN, Nieto FJ. Cross-sectional and longitudinal associations of BMI with socioeconomic characteristics. Obes Res. 2005;13(8):1412–21.
101. Zhang Q, Wang Y. Trends in the association between obesity and socioeconomic status in U.S. adults: 1971 to 2000. Obes Res. 2004;12(10):1622–32.
102. Zenk SN, Schulz AJ, Hollis-Neely T, et al. Fruit and vegetable intake in African Americans income and store characteristics. Am J Prev Med. 2005;29(1):1–9.
103. Morland K, Diez Roux AV, Wing S. Supermarkets, other food stores, and obesity: the atherosclerosis risk in communities study. Am J Prev Med. 2006;30(4):333–9.
104. Maddock J. The relationship between obesity and the prevalence of fast food restaurants: state-level analysis. Am J Health Promot. 2004;19(2):137–43.
105. Suminski RR, Poston WS, Petosa RL, Stevens E, Katzenmoyer LM. Features of the neighborhood environment and walking by U.S. adults. Am J Prev Med. 2005;28(2):149–55.
106. Spilsbury JC, Storfer-Isser A, Kirchner HL, et al. Neighborhood disadvantage as a risk factor for pediatric obstructive sleep apnea. J Pediatr. 2006;149(3):342–7.
107. Li X, Sundquist K, Sundquist J. Socioeconomic status and occupation as risk factors for obstructive sleep apnea in Sweden: a population-based study. Sleep Med. 2008;9(2):129–36.
108. Zanobetti A, Redline S, Schwartz J, et al. Associations of PM10 with sleep and sleep-disordered breathing in adults from seven U.S. urban areas. Am J Respir Crit Care Med. 2010;182(6):819–25.
109. Oken E, Levitan EB, Gillman MW. Maternal smoking during pregnancy and child overweight: systematic

review and meta-analysis. Int J Obes (Lond). 2008; 32(2):201–10.
110. Oken E, Gillman MW. Fetal origins of obesity. Obes Res. 2003;11(4):496–506.
111. Calhoun SL, Vgontzas AN, Mayes SD, et al. Prenatal and perinatal complications: is it the link between race and SES and childhood sleep disordered breathing? J Clin Sleep Med. 2010;6(3):264–9.
112. Rosen CL, Larkin EK, Kirchner HL, et al. Prevalence and risk factors for sleep-disordered breathing in 8- to 11-year-old children: association with race and prematurity. J Pediatr. 2003;142(4):383–9.
113. Paavonen EJ, Strang-Karlsson S, Raikkonen K, et al. Very low birth weight increases risk for sleep-disordered breathing in young adulthood: the Helsinki Study of Very Low Birth Weight Adults. Pediatrics. 2007;120(4):778–84.
114. Patel SR. Shared genetic risk factors for obstructive sleep apnea and obesity. J Appl Physiol. 2005;99(4): 1600–6.

The Connection Between Sleep Loss, Obesity, and Type 2 Diabetes

10

Silvana Pannain, Guglielmo Beccuti, and Eve Van Cauter

Abstract

In this review, evidence is presented to support the hypothesis that reduced sleep duration may be part of the behavioral modifications that played a role in the development of the current epidemic of obesity and diabetes. An important consideration when trying to explain the epidemiologic link between sleep loss and metabolic risk is that it is not clear whether the physiological effects of sleep restriction observed under laboratory conditions over a period of a few days can be translated to chronic sleep restriction as it occurs in free-living individuals. Also, when comparing different laboratory studies of sleep restriction, differences in the "dose" of sleep loss relative to the physiological need of the individual are often ignored. While the body of evidence suggestive of an interaction between sleep loss and the epidemics of obesity and diabetes continues to build at a rapid pace, much remains to be discovered as far as mechanisms and the transition from short-term laboratory conditions to chronic partial sleep deprivation in real life. Intervention studies extending sleep in habitual short sleepers and examining the impact on metabolic outcomes are needed to further address the direction of causality of the association between insufficient sleep, obesity, and diabetes and the potential clinical implications.

Secular Trends in Sleep Duration and the Prevalence of Obesity and Diabetes

In the past few decades, the prevalence of obesity and, consequently, of type 2 diabetes mellitus (T2DM) have increased alarmingly worldwide. Such a rapid increase cannot be explained by an alteration in the genetic pool; it is more likely due to environmental, socioeconomic, behavioral, and demographic factors and the interaction between genetics and these factors. Food marketing practices with increased portion size and widespread availability of high caloric fast food are often cited as a major culprit, alongside reduced physical activity. In recent years, there has been an increased interest in nontraditional behavioral and environmental factors that could also contribute to the epidemic of obesity and

S. Pannain, MD • G. Beccuti, MD • E. Van Cauter, PhD (✉)
Department of Medicine, University of Chicago,
5841 S. Maryland Avenue, Chicago, IL 60637, USA
e-mail: evcauter@medicine.bsd.uchicago.edu

P.J. Shiromani et al. (eds.), *Sleep Loss and Obesity: Intersecting Epidemics*,
DOI 10.1007/978-1-4614-3492-4_10,

diabetes [1]. Among these, one novel behavior that seems to have developed during the past few decades and has become highly prevalent is chronic partial sleep curtailment.

Secular trends in sleep duration are poorly documented, but a comparison of surveys conducted in the 1960s and 1970s to those conducted after 2000 suggests a marked decrease in sleep duration. For example, in 1960, a survey study conducted by the American Cancer Society found the modal sleep duration to be 8.0–8.9 h [2] and, in 1975, more than 85% of the participants in the Older Finnish Twin Cohort reported sleeping more than 7 h per night [3]. In contrast, the "Sleep in America" poll conducted by the National Sleep Foundation in 2008 revealed that the average number of hours of sleep on workdays was 6 h 40 min, with an extension to 7 h 25 min on nonworkdays [4]. A report from the National Health Interview Survey indicated that the percentage of adults between the ages of 30 and 65 years who report sleeping 6 h or less increased by approximately 5–6% between 1985 and 2004, such that in 2004, more than 30% of men and women in this age group reported sleeping 6 h or less [5]. According to recent polls from the US Centers for Disease Control and Prevention (CDC), approximately 29% of US adults report sleeping less than 7 h per night, and 50–70 million have chronic sleep and wakefulness disorders [6]. When sleep duration is measured objectively (using wrist actigraphy) rather than self-reported, the findings are not less alarming. For example, the Coronary Artery Risk Development in Young Adults (CARDIA) Sleep Study measured the sleep of adults aged 38–50 years for 3 consecutive days on two occasions spaced approximately 1 year apart. The mean sleep duration was 6.1 (± 1.2) h, and it varied across race–gender groups from 6.7 (± 0.9) h for white women to only 5.1 (± 1.3) h for African-American men [7].

Insufficient sleep may be due to a voluntary restriction of time spent in bed or may be the result of a sleep disorder, such as insomnia or obstructive sleep apnea (OSA). Unfortunately, the vast majority of epidemiologic studies that addressed the relationship between sleep duration and the risk of obesity or diabetes did not distinguish between voluntary sleep curtailment and sleep loss due to a pathological condition. Chronic partial sleep loss in contemporary society is certainly partly self-imposed. Our 24-h society involves demands and opportunities to extend the waking period for evening and nighttime work and leisure activities, and consequently a sacrifice of hours available for sleep. These relatively novel behaviors have had a major impact on bedtime duration and duration of dark exposure, resulting in later bedtimes, reduced total sleep time, and the opportunity to be active and ingest food during the natural night.

The function of sleep is most frequently described as a restorative process for the brain, but there is now abundant evidence that sleep is a healthy behavior that is also important for the rest of the body, consistent with the important modulatory effects of sleep on neuroendocrine function and glucose metabolism [8]. The decrease in sleep duration (and the associated increase in sleep complaints) in modern society [9] may be considered as a sleep disorder because it produces both daytime and nighttime alterations of neurobehavioral and physiological systems and raises concerns for a negative impact on health in general, not only mental health.

The gold standard method for assessing sleep is polysomnography (PSG), which combines an all night recording of the EEG with measures of muscle tone and eye movements and allows for the scoring of sleep in stages I, II, III, IV, REM, and Wake. A single night of PSG does not generally provide a good estimation of habitual sleep duration. Objective estimations of sleep duration and sleep fragmentation may be obtained under ambulatory conditions by wrist actigraphy monitoring (WAM). WAM has been validated against PSG, demonstrating a correlation for sleep duration between 0.82 in insomniacs and 0.97 in healthy subjects [10]. Lastly, a number of validated questionnaires to assess subjective sleep duration and quality have been developed. Subjective sleep duration often overestimates the actual sleep duration [11].

Short Sleep Duration and Obesity: Epidemiologic Evidence

According to recent estimates, the worldwide prevalence of obesity has doubled since 1980 [12]. This obesity epidemic has been mirrored in modern society by a secular trend for reduced sleep duration [9]. Figure 10.1 represents the trends in sleep duration and obesity prevalence from 1960 to the first decade of the twenty-first century. Growing evidence suggests that short sleep duration (SSD) may have played a role in the increased prevalence of obesity [14–17]. This section will summarize the literature examining the link between sleep loss and obesity, focusing on studies in adults.

Of note, a number of studies have reported a U-shaped relationship between sleep duration and obesity, where both SSD (generally $\leq$ 6 h) and long sleep duration (generally > 8 h) were associated with higher body mass index (BMI). There is a general consensus that the mechanisms linking long sleep to obesity are not likely to involve the same pathways linking short sleep and obesity [16, 18–20]. The majority of studies reporting a significant association between long sleep and obesity were based on self-reported sleep duration, and it has been argued that long sleepers may be spending more time in bed without obtaining more sleep. Another putative explanation could be that a subset of obese individuals suffer from fatigue associated with a subclinical condition and spend more time in bed. A further possible explanation is that long sleep may be associated with reduced physical activity and that therefore the association between long sleep and obesity would not persist when adjusting for physical activity. The present chapter focuses on the findings relating short sleep, obesity, and diabetes.

To date, more than 60 epidemiological studies from different geographical regions have examined the association between sleep duration and obesity in adults. In cross-sectional approaches, the vast majority of studies found a significant association between SSD (generally < 6 h per night) and BMI or prevalence of obesity and/or overweight. A systematic review of prospective studies provides similar, albeit not as consistent results, revealing an association with being a short sleeper at baseline and weight gain or the incidence of obesity during the follow-up period.

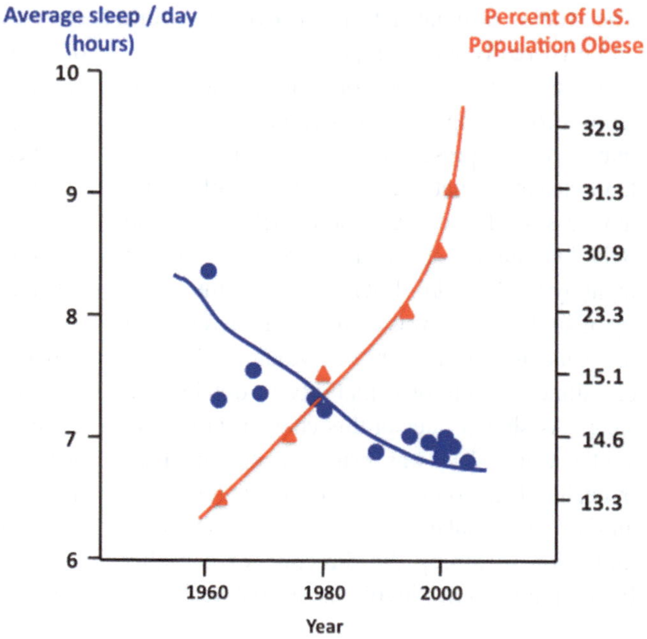

Fig. 10.1 Trends of sleep duration and obesity prevalence in the US population from 1960 to the first decade of the twenty-first century. The sleep duration estimates have been derived from McAllister et al. [13]. US obesity prevalence is from Centers for Disease Control and Prevention (http://www.cdc.gov/obesity/)

A meta-analysis published in 2008 combined data from 18 cross-sectional studies including 604,509 adults from 12 different countries and demonstrated a pooled obesity odds ratio (OR) of 1.55 (CI: 1.43–1.68; $p<0.0001$) for sleep durations <5 h as compared to 7–8 h [21]. A dose–response effect became apparent such that for each additional hour of sleep, the BMI decreased by 0.35 kg/m^2 (95% confidence interval – CI: –0.57 to –0.12), which would translate to a 1.4 kg decrease in weight in an individual of 178 cm of height. This study represents the first systematic review and meta-analysis of the population-based studies published up to 2008 and demonstrates a consistent association between sleep duration and obesity in different populations around the world. Direction of causality cannot be inferred from these cross-sectional studies. Table 10.1 summarizes the prospective and cross-sectional epidemiologic studies published since 2008, which were not included in the meta-analysis by Cappuccio et al. [21]. We were able to identify 31 such studies, of which 10 involved a longitudinal analysis. Only one of the ten prospective studies had negative findings [46]. In the cross-sectional studies, all but one study found a significant association between short sleep and obesity, although significance was not always found for subsets of subjects (e.g., in men vs. women or conversely). The studies originated from all over the world and involve diverse adult populations.

In sum, the number of concordant studies lends strong support to the hypothesis that SSD may indeed represent a risk factor for obesity. One caveat is that the vast majority of studies have assessed sleep duration by self-report with only five studies so far using objective assessment by WAM and/or PSG. Further, in the majority of studies, there was no information regarding the cause for short sleep, i.e., voluntary bedtime curtailment or biologic inability to obtain more sleep. As shown later in this chapter, laboratory studies where sleep was restricted experimentally in healthy lean volunteers and appetite regulation and/or food intake were examined offer some indication regarding the direction of causality. Intervention studies involving sleep extension in short sleepers will be important to further support a causative role for short sleep on the risk of obesity. A National Institute of Health (NIH)-funded randomized control trial [53] has enrolled 150 US short sleeper adults (< 6.5 h per night) to examine the feasibility of increasing sleep duration to a healthy length (approximately 7.5 h) and to determine the effect of sleep extension on body weight. The findings have not yet been published.

In the subsequent subsections, we discuss the findings of the studies that have used objective assessments of sleep duration and then summarize the state of knowledge regarding the possibility of sex differences in the relationship between sleep and obesity. We then briefly discuss studies that have addressed the impact of genetics and race/ethnicity. Lastly, we review the few studies that have examined the contribution of dietary habits to the relationship between SSD and obesity risk.

Studies Using Objective Measurements of Sleep Duration

When examining epidemiologic studies, one concern is that comparative studies have shown that self-reported sleep duration correlates only moderately with more objective estimations of sleep duration such as those derived from WAM or PSG, and that self-report may overestimate the amount of sleep [11, 54]. A discrepancy between self-report and sleep duration derived from WAM has been confirmed in the Rotterdam Study, a population-based cohort of elderly adults [32]. In that cohort, men overestimated sleep duration by self-report by 0.61 h, while the difference between self-reported and measured sleep durations was only 0.14 h for women. The possibility of a systematic bias in the estimation of the relationship between sleep duration and obesity was suggested by Lauderdale et al. who noted that obese persons tend to report shorter sleep duration for the same amount of objectively assessed sleep than non-obese individuals [26].

To date, only five studies have used objective methods to assess sleep duration in population

Table 10.1 Summary of recent epidemiologic studies (published after 2007 and not included in the meta-analysis by Cappuccio et al. [21]) examining the association between short sleep duration and obesity in adults. The association between sleep duration and obesity is expressed as higher probability, prevalence, or adjusted odds ratio (AOR) of obesity (BMI > 30 kg/m^2) or increased waist circumference

Author	Description and data source	Cohort	Sleep assessment	Results	
Prospective studies					
Gunderson et al. [22]	1-year follow-up Project Viva cohort (US)	940 women postpartum Age 33.0 ± 4.7 yrs	Self-report	≤5 h	OR of SPPWR 3.08 (CI: 1.76–5.38; $p<0.001$) at 6 months postpartum; OR of SPPWR 3.38 (CI: 1.66–6.86; $p=0.011$) at 1 year postpartum
				5–≤6 h	NS at 6 months and 1 year postpartum
				7–<8 h	Reference category
Chaput et al. [23]	6-year follow-up Quebec Family Study (Canada)	276 men and women Age 21–64 yrs	Self-report	5–6 h	Increased body weight (+1.98 kg; CI 1.16–2.82) and risk of obesity (+27%)
				7–8 h	Reference category
Lopez-Garcia et al. [24]	2-year follow-up of older adults (age ≥ 60 yrs) (Spain)	1,064 men (age 71.0 ± 8.0 yrs), 1,271 women (age 72.1 ± 7.6 yrs)	Self-report	*In women:*	
				≤5 h	AOR of 5-kg weight gain 3.41 (CI: 1.34–8.69; $p<0.02$)
				6 h	NS
				7 h	Reference category
				In men: no significant association	
Chaput et al. [25]	6-year follow-up Quebec Family Study (Canada)	283 men and women Age 18–64 yrs	Self-report	<6 h	35% higher probability of ≥5 kg weight gain ($p<0.01$)
Lauderdale et al. [26]	5-year follow-up Coronary Artery Risk Development in Young Adults (CARDIA) Study	612 men and women Approximate mean age 45 yrs	WAM	No longitudinal association between sleep measurements and change in BMI	
Watanabe et al. [27]	1-year follow-up of employees of an electric power company (Japan)	31,477 men Age 40 ± 9 yrs 3,770 women Age 38 ± 9 yrs	Self-report	*In men:*	
				<5 h	AOR of obesity 1.91 (CI: 1.36–2.67; $p<0.001$)
				5–5.9 h	AOR of obesity 1.5 (CI: 1.25–1.8; $p<0.001$)
				7–8 h	Reference category
				In women: no significant association	
Nishiura et al. [28]	4-year follow-up of employees of a gas company (Japan)	2,362 men Age 40–59 yrs	Self-report	<6 h	AOR of obesity 2.46 (CI: 1.41–4.31; $p=0.011$)
				7–7.9 h	Reference category

(continued)

Table 10.1 (continued)

Author	Description and data source	Cohort	Sleep assessment	Results	
Hairston et al. [29]	5-year follow-up Insulin Resistance Stherosclerosis Study (IRAS) Family Study (US)	322 African-American men and women and 775 Hispanic-American men and women Age 18–81 yrs	Self-report	*Age < 40 yrs:*	
				≤5 h	Increase in BMI (+1.8 kg/m^2, p<0.001), SAT (+41 cm^2, p<0.0001), and VAT (+13 cm^2, p<0.01)
				6–7 h	Reference category—≤ 8 h
				≥8 h	Increase in BMI (+0.8 kg/m^2, p<0.001), SAT (+20 cm^2, p<0.01), and VAT (+6 cm^2, p<0.05)
				Age > 40 yrs: no significant association	
Bo et al. [30]	6-year follow-up Patients from Local Health Units (Italy)	1,597 men and women Age 45–64 yrs	Self-report	Each hour increase in total sleep time	30% Reduction in incident obesity (AOR 0.7/h; CI: 0.57–0.86; p<0.001)
Chaput et al. [31]	6-year follow-up Quebec Family Study (Canada)	216 men and women Age 18–64 yrs	Self-report	Maintained short sleep (≤6 h)	Increase in BMI (+1.1±0.36 kg/m^2, p<0.05), and fat mass (+2.4±0.64 kg, p<0.05) vs. control group
				Increased sleep to 7–8 h	NS vs. control group
				Control (7–8 h)	Reference category
Cross-sectional studies					
Van den Berg et al. [32]	Rotterdam Study (Netherlands)	471 men, 512 women Age 57–97 yrs	WAM and self-report	<5 h	AOR of obesity 2.76 (CI: 1.38–5.49) [NS after adjusting for fragmentation index]
				5–6 h	AOR of obesity 1.97 (CI: 1.26–3.08) [NS after adjusting for fragmentation index]
				6–7 h	NS
				7 h	Reference category
Patel et al. [55]	Osteoporotic Fractures in Men Study (MrOS) and Study of Osteoporotic Fractures (SOF) (US)	3,055 men Age 67–96 yrs 3,052 women Age 70–99 yrs	WAM in all; PSG in 2,862 men and 455 women	*In men:*	
				<5 h	AOR of obesity 3.70 (CI: 2.72–5.04)
				5–7 h	AOR of obesity 1.51 (CI: 1.18–1.93)
				7–8 h	Reference category
				In women:	
				<5 h	AOR of obesity 2.26 (CI: 1.64–3.13)
				5–7 h	AOR of obesity 1.51 (CI: 1.18–1.93)
				7–8 h	Reference category

Vgontzas et al. [33]	Penn State Cohort (US)	561 men Age 50.8 ± 12.6 yrs 739 women Age 54.9 ± 13.6 yrs	Self report	Compared to the group of subjects who slept > 6 and ≤ 7 h, BMI decreased proportionally to increased sleep for those who slept less ($p<0.05$). BMI remained similar for those who slept more (NS).	
Hall et al. [34]	Adult Health and Behavior Project registry (US)	568 men, 646 women 83.7% Non-Hispanic Caucasian Age 45 ± 7 yrs	Self-report	<6 h	AOR of central adiposity 1.73 (CI: 1.21–2.57)
				6–7 h	AOR of central adiposity 1.64 (CI: 1.22–2.20)
				7–8 h	Reference category
Choi et al. [35]	2001 Korean National Health and Nutrition Survey (KNHNS) (Korea)	1,822 men, 2,400 women Age 44.1 ± 0.4 yrs	Self-report	≤5 h	Prevalence of abdominal obesity 41.4% (CI: 35.9–47.2)
				6 h	Prevalence of abdominal obesity 31.5% (CI: 28.4–34.8)
				7 h	Prevalence of abdominal obesity 29.2% (CI: 26.5–32.2)
Lopez-Garcia et al. [24]	Older adults (age ≥ 60 yrs) from Spain	1,739 men, 2,269 women Age 71.6 ± 7.7 yrs	Self-report	≤5 h	AOR of obesity 1.33 (CI: 1.00–1.77; $p<0.005$) and severe obesity 2.08 (CI: 1.31–3.32; $p<0.004$)
				6 h	NS
				7 h	Reference category
Park et al. [36]	2001 and 2005 KNHNS (Korea)	3,723 men, 4,994 women Age 20–65 yrs	Self-report	≤5 h	AOR of general obesity 1.24 (CI: 1.05–1.47) and abdominal obesity 1.22 (CI: 1.01–1.47)
				6 h	AOR of general obesity 1.16 (CI: 1.02–1.31) and abdominal obesity NS
				7 h	Reference category
Chaput et al. [25]	Quebec Family Study (Canada)	Cross-sectional analysis: 537 men and women Age 18–64 yrs	Self-report	<6 h	6% non-obese vs. 30.1% overweight/obese, OR 4.66 (CI: 2.98–6.48), $p<0.01$
				≥7 h	NS
Lauderdale et al. [26]	CARDIA Study (USA)	612 men and women Approximate mean age 45 yrs	WAM	With increasing category of sleep duration (<4.5 h, 4.5–6 h, 6–7.5 h, ≥ 7.5 h), there was a 0.78 kg/m^2 decrease in BMI	
St Onge et al. [37]	CARDIA study (USA)	3,473 men and women Age 33–45 yrs	Self-report	No associations between sleep measurements and BMI when controlled for physical activity.	
Adamkova et al. [38]	Adults from Czech Republic	2,038 men, 1,932 women Age 18–65 yrs	Self-report	4–6 h	BMI 27.46 ± 4.919
				7 h	BMI 25.40 ± 4.201 $p<0.001$ for trend
				8–11	BMI 25.18 ± 4.868
Di Milia et al. [39]	Employees in coal industry and university (Australia)	292 men, 59 women Age 41 ± 11 yrs	Self-report	AOR of obesity 2.05 (CI: 1.03–3.55, p 0.05) for <6 h sleep	

(continued)

Table 10.1 (continued)

Author	Description and data source	Cohort	Sleep assessment	Results	
Thomas et al. [40]	EADS/Augsburg cohort study (Germany)	1,047 men, 116 women Age 39±11 yrs	Self-report	Significant association between sleep duration and BMI (β_{st}=−0.06, p=0.04) when demographic, health behavior, and work status variables were included	
Watson et al. [41]	University of Washington Twin Registry (US)	1,224 twins: 423 monozygotic, 143 dizygotic, and 46 pairs of unknown zygocity Mean age 36.9 yrs	Self-report	*Dizygotic pairs discordant for sleep duration* (n = 57): no BMI difference. *Monozygotic pairs discordant for sleep duration* (n=167):	
				<7 h	Higher mean BMI (p<0.02)
				7–8.9 h	Reference category
Buxton et al. [42]	National Health Interview Survey (US)	56,507 men and women Age 18–85 yrs	Self-report	<7 h	6% Higher probability of obesity
				7–8 h	Reference category
Magee et al. [43]	"45 and UP Study" (Australia)	40,834 men and women Age 45–65 yrs	Self-report	*In men:*	
				<6 h	AOR of obesity 1.72 (CI: 1.34–2.20; p<0.017)
				6 h	AOR of obesity 1.51 (CI: 1.32–1.73; p<0.017)
				7 h	Reference category
				In women:	
				<6 h	AOR of obesity 1.42 (CI: 1.16–1.75; p<0.017)
				6 h	AOR of obesity 1.35 (CI: 1.19–1.52; p<0.017)
				7 h	Reference category
Magee et al. [44]	"45 and UP Study" (Australia)	45,325 men and women Age 55–95 yrs	Self-report	*Age 55–64 yrs:*	
				<6 h	AOR obesity 1.52 (CI: 1.21–1.89; p<0.001)
				6 h	AOR obesity 1.42 (CI: 1.26–1.61; p<0.001)
				7 h	Reference category
				Age > 65 yrs: no significant association	
Magee et al. [45]	"45 and UP Study" (Australia)	16,951 men and women, full time workers Age 45–65 yrs	Self-report	Inverse association between sleep duration and BMI (p<0.001)	
Anic et al. [46]	Collaborative Breast Cancer Study (US)	5,549 women Age 20–75 yrs	Self-report	<6 h	AOR obesity 1.89 (CI: 1.45–2.47; p<0.0001); AOR extreme obesity 3.12 (CI: 1.70–5.75; p=0.0003)
				6–6.9 h	AOR obesity 1.52 (CI: 1.23–1.89; p=0.0003); AOR extreme obesity 2.22 (CI: 1.27–3.87; p=0.0003)
				7–7.9 h	Reference category

Theorell-Haglöw et al. [47]	"Sleep and Health in woman" study (Sweden)	400 women Age 29–70 yrs	Ambulatory PSG	Inverse association between sleep duration and both waist circumference (Adj. β −1.22 cm/h; $p=0.016$) and sagittal abdominal diameter (Adj. β −0.46 cm/h; $p=0.001$).
Knutson [48]	1982–1984 Hispanic Health and Nutrition Examination Survey (HHANES) (US)	889 Cuban-Americans 3,520 Mexican-Americans 1,316 Puerto Ricans Age 36–44 yrs	Self-report	Association between SSD and BMI only significant in Mexican-Americans (β −0.03; CI: −0.06 to −0.004)
Baron et al. [49]	Adults from US	27 men, 25 women Age 18–71 yrs	WAM	Higher BMI was associated with shorter sleep, later sleep timing, caloric consumption after 8 p.m., and fast food consumption
Kim et al. [50]	Adults from US and Puerto Rico	27,983 women Age 35–74 yrs	Self-report	Decreased eating at conventional times among women sleeping <6 h and >10 h vs. 7–7.9 h
Liu et al. [51]	Twin cohort, (China)	854 men and 640 women Age 20–70 yrs	Self-report	In women, but not in men, sleep duration <7 h was associated with higher insulin resistance (HOMA-IR) than sleep duration >7 h but ≤8 h, even after adjustment for BMI or % trunk fat
Wheaton et al. [52]	2008 Behavioral Risk Factor Surveillance System (BRFSS) (US)	384,541 men and women Age 18 to >65 yrs	Self-reported insufficient sleep	Number of days of insufficient rest or sleep strongly correlated with BMI. The relationship was found in all ethnic groups

Adj. β adjusted beta coefficient, *CI* 95% confidence interval, *NS* not significant, *WAM* wrist activity monitor, *PSG* polysomnography, *SAT* subcutaneous adipose tissue, *SPPWR* substantial postpartum weight retention, *VAT* visceral adipose tissue, *yrs* years

studies. A 2008 study by Van den Berg et al. recorded sleep by WAM in 983 participant of the Rotterdam Study of Aging and found that both short sleepers and long sleepers were more likely to be obese, compared to participants who slept 7 to < 8 h [32]. BMI also increased with sleep fragmentation. After adjustment for sleep fragmentation, the relationship between short sleep and BMI was no longer significant whereas it remained significant for long sleep. Of note, in this cohort of elderly participants, self-reported habitual sleep duration was not associated with BMI or obesity, further suggesting that self-reported sleep duration may not correctly estimate actual sleep duration. The largest study based on objective sleep assessments was published by Patel et al. who analyzed WAM recordings from a cohort of elderly men ($n=3{,}055$; age: 67–96 years) and women ($n=3{,}052$; age: 70–99 years) participating in the Osteoporotic Fracture Study [55]. As summarized in Table 10.1, the study had positive findings in both men and women. The special interest of this cross-sectional study is that a subgroup of 2,862 men and 455 women also underwent a PSG study to assess the presence and severity of sleep apnea. Compared to those sleeping an average of 7–8 h per night, sleep duration (based on WAM) < 5 h was associated with a BMI on average 2.5 kg/m^2 greater in men and 1.8 kg/m^2 greater in women. Additionally the odds of obesity were 3.7-fold greater in men and 2.3-fold greater in women who slept < 5 h. These associations persisted after adjusting for the severity of sleep apnea as assessed by the Apnea Hypopnea Index (AHI), insomnia and daytime sleepiness. This remarkable study was the first to demonstrate that the association between sleep duration and obesity is in part dependent on the presence and severity of sleep apnea but persists after controlling for AHI or when limiting the analysis to participants without significant sleep apnea. While the findings suggest that the association between short sleep and obesity may be stronger in men, a direct comparison is not possible as the women were nearly a decade older than the men. An additional unique contribution of this study is the demonstration that the impact of reduced sleep times (assessed objectively) on obesity risk is also present in older populations. This is in contrast with other reports that assessed sleep duration by self-report and indicated that short sleep may not be relevant to obesity risk in older populations [29, 44, 56]. The CARDIA Sleep Study assessed sleep by WAM for 6 days and involved both a cross-sectional and a longitudinal analysis [26]. The cross-sectional analysis confirmed the association between SSD and higher BMI reported in previous studies (0.78 kg/m^2 decrease in BMI for each increasing sleep duration category). Greater sleep fragmentation was also associated with higher BMI. The presence of snoring (by self-report) significantly affected the cross-sectional association such that the sleep duration-BMI association observed across the entire sample was stronger among the participants who reported snoring. This finding suggests that obesity-related comorbidities such as OSA may affect sleep duration or conversely that OSA has an independent effect on obesity risk. The prospective analysis did not find an association between sleep duration and weight change over the 5-year follow-up [26]. A small actigraphy-based study by Baron et al. [49] was innovative as it looked not only at sleep duration but also sleep timing. Sleep duration was a significant predictor of BMI while sleep timing did not predict BMI after controlling for sleep duration. Calories consumed after 8 p.m. predicted BMI after controlling for sleep timing and duration, suggesting that evening eating may promote obesity risk, consistent with findings from animal models. Lastly, a study by Theorell-Haglöw et al. performed a PSG in 400 women (aged 29–70 years) participants in the Sleep and Health in Women Study [47]. Sleep duration was inversely related to waist circumference, after adjusting for multiple confounders, including AHI. This study is consistent with the findings of Patel et al. [55] who found that the association between short sleep and obesity is not entirely dependent on the presence and severity of OSA.

Sex Differences in the Relationship Between Sleep and Obesity Risk

Sex differences in sleep duration and quality have been well documented. Women have more sleep complaints, particularly insomnia, but are much less likely to have OSA than men. Somewhat paradoxically, objective sleep duration and the amount and intensity of non-REM sleep are higher in women than in men. There are also well-established sex differences in eating behavior. It is therefore logical that the relationship between sleep duration and obesity risk may be sex-dependent. So far, the studies that have addressed this issue have had contradictory or inconclusive results. Of note, all these studies assessed sleep duration by self-report.

Two prospective studies [24, 27] and three cross-sectional studies [45, 51, 55] have explicitly addressed sex differences. A few additional studies have included women only.

A prospective study conducted in Spain reported that women reporting sleeping <5 h per night had increased odds of gaining 5 kg or more over the following 2 years compared to those who reported sleeping 7 h per night. This association was not found in men [24]. Contrasting with these findings, a Japanese prospective study with a 1-year follow-up showed that the increased risk of obesity for self-described short sleepers was present in men but not in women [27]. In this latter study, the lack of a significant finding in women could be due to the small sample size. Vgontzas et al. in a cross-sectional analysis found a negative linear relationship between hours of sleep duration and BMI. When the analysis was stratified by sex, the association was significant only for men [33]. An analysis of the CARDIA Sleep Study indicated the existence of an inverse relationship between reported sleep duration and BMI in both sex groups, but in unadjusted analyses, the findings appeared more robust in women than in men [26]. Liu et al. were the first to examine the gender-specific association of sleep duration with body composition as assessed by Dual-emission X-ray absorptiometry (DXA) [51]. To adjust for the decrease of sleep duration with age, age-specific quartiles of sleep duration were considered. Additionally the analysis also considered sleep quality factors such as sleep disturbance and habitual snoring. Women in the lowest quartile of short-sleep duration had higher overall and central adiposity and lower lean body mass when compared to those with moderate sleep duration (second and third quartiles). The association persisted after excluding subjects who reported either habitual snoring or sleep disturbance, suggesting that the sleep duration itself is a potential determinant of increased adiposity. A similar association was not found in men.

Four studies up to date have included only women [22, 46, 47, 50] and all four had positive findings linking short sleep with obesity risk. Gunderson et al. found that women who reported shorter sleep duration (≤ 5 h within a 24-h period) at 6 months postpartum were 2.3 times more likely to retain at 1 year substantial postpartum weight (≥5 kg above pre-pregnancy weight) independent of potential confounders including maternal sociodemographics, pre-pregnancy BMI, gestational weight gain, parity, and postpartum behaviors [22]. Additionally, women who reported a reduction in hours of sleep at 1 year postpartum were two times more likely to have substantial postpartum weight retention. The study from Theorell-Haglöw et al. performed a PSG in 400 participants (aged 29–70 years) in the Sleep and Health in Women Study [47]. Not only sleep duration but also sleep quality, as determined by sleep efficiency and sleep architecture (specifically minutes of SWS, the "deep restorative sleep"), was inversely related to waist circumference, after adjusting for age, level of physical activity, smoking status, alcohol consumption, and AHI. Associations were stronger in age <50 years. In a cohort of 5,549 US adult women of similar age range, Anic et al. confirmed an association between SSD and obesity [46]. The association was stronger in participants with morbid obesity [46]. The analysis explored a possible causal relationship by examining the association between lifetime sleep duration (possibly preceding the onset of obesity) and obesity and found a weaker association than with sleep duration measured during the study. The authors tentatively concluded that short sleep might have been the consequence of obesity. However, self-reported measures of lifetime sleep duration may be poorly reliable and have never been validated. Lastly, Kim et al. collected information about eating behavior and source of calories and correlated to various sleep categories in a cohort of nearly 28,000 women only [50]. SSD was associated with disrupted eating patterns and poor food choices, and thus potentially to a risk of weight gain and obesity.

Impact of Genetic Factors and Race/Ethnicity

A recent study examined self-reported sleep duration and BMI in 1,224 twins (423 monozygotic, 143 dizygotic, and 46 indeterminate pairs), mean age 36.9 years [41]. In a multivariate adjusted analysis including all twins, the mean BMI was found to be 1.2 kg/m^2 higher in short sleeping twins (< 7 h/night) compared to twins sleeping 7–8.9 h per night. The novelty of this study lies in the within-pair analyses. Even when restricted to monozygotic twins, the short sleeping member of the pair had a significantly elevated BMI by 1.0 kg/m^2 compared to the reference group. The persistence of the association within individuals with an identical genetic background supports the hypothesis that behavioral curtailment of sleep, rather than genetic factors, drives the association. Bivariate analysis revealed little evidence of shared genetics between sleep duration and BMI. Consistent findings were reported in a twin study of a Chinese rural population where heritability of sleep duration appeared to be primarily determined by environmental factors whereas heritability of body composition (assessed by DXA) had a strong genetic component [51].

In a cross-sectional analysis of the CARDIA Sleep Study which by design enrolled similar proportions of middle-aged White and African-American men and women, a significant relationship between objective sleep duration based on WAM and BMI emerged, and this association did not vary by race/sex groups [26]. Another prospective study by Hairston et al. focused on minorities (322 African-Americans and 775 Hispanic Americans men and women) known to be at higher risk of metabolic disorders, and used abdominal computer tomography scans to evaluate visceral and subcutaneous adipose tissue (VAT and SAT, respectively) [29]. After controlling for multiple confounders, short sleep (≤ 5 h) was associated with greater fat accumulation over the 5-year follow-up with increased BMI (+1.8 kg/m^2, $p<0.001$), SAT (+41 cm^2, $p<0.0001$), and VAT (+13 cm^2, $p<0.01$) as compared to > 6–7 h sleepers. There were no significant interactions between sleep duration and race groups, suggesting that the impact of short sleep was similar in African-Americans and Hispanics. The relationship was significant in younger participants only (< 40 years old). Because there is an elevated prevalence of short sleepers in these ethnic minorities, these findings raise the possibility that their increased risk of metabolic disorders may be partly mediated by sleep habits. Knutson et al. explored the impact of ethnicity on the association between sleep and body size measured from BMI, skin folds, arm, and calf circumference using data from the Hispanic Health and Nutrition Examination Survey (HHANES) [48]. In a cross-sectional analysis, SSD was associated with larger body size in Mexican-Americans ($n=3{,}520$), but not in Cubans-Americans ($n=889$) or Puerto Ricans ($n=1{,}316$), indicating that distinct factors (e.g., diet intake vs. physical activity) in different ethnic groups could influence the risk of weight gain. One limitation of this study is that the data analyzed were collected over 25 years ago, which was at the beginning of the obesity epidemic and at the time when the prevalence of short sleepers was very small. In fact only 3–5% of the HHANES ethnic groups reported sleeping less than 6 h per night. It is possible that today the association between sleep duration and body size may be detectable in all Hispanic groups. Most recently, in a large cohort of almost 400,000 US adults, of whom 70% white non-Hispanic, there was a positive-graded relationship between days of perceived insufficient sleep and BMI categories from normal weight through different obesity grades among both men and women and in all ethnic groups [52]. Of note, perceived insufficient sleep does not distinguish between sleep duration and sleep quality.

Role of Dietary Habits in the Relationship Between Short Sleep and Obesity

In 2010 and 2011, four epidemiologic studies examined the contribution of dietary habits to the association between sleep duration and obesity [28, 30, 49, 50]. Nishiura et al. analyzed the dietary patterns of 2,362 non-obese Japanese workers. The increased risk of obesity at 4 years for the short sleepers (AOR 2.46 for < 6 h; CI 1.41–4.31) was slightly attenuated but remained significant after controlling for food preferences and unhealthy behavior such as skipping breakfast, snacking, and eating out [28]. In a prospective study with a 6-year follow-up, Bo et al. showed in an Italian cohort that hours of sleep per night, home temperature, and numbers of restaurant meals were each associated with higher obesity incidence [30]. Kim et al. collected information regarding eating behavior in almost 28,000 women. Short sleep (< 5 h/night) was cross-sectionally associated with an increased tendency for eating at unconventional times and dominance of snacks over meals [50]. These eating patterns were associated with increased caloric intake from sweets and fat and lower intake of fruits and vegetables. The finding suggests that short sleep may promote disrupted eating patterns and unhealthy food choices. Lastly, a small cross-sectional study showed that later sleep time and short sleep were associated with increased BMI, but that the association was mostly due to the increased caloric intake after 8 p.m., suggesting that the relationship between short sleep time and obesity could also be mediated by the opportunity of ingesting food during the natural night [49].

Sleep Duration and Diabetes: Epidemiologic Evidence

As for obesity risk, there is evidence for associations of both short sleep and long sleep with an increased risk of T2DM [57–59]. Very different mechanisms are likely to be involved and the present review will focus on short sleep only and on the recent and best-documented studies (summarized in Table 10.2). Additionally, in the last subsection, we review the relationship between short sleep and gestational diabetes risk.

Prospective Studies

A number of prospective studies have examined the association between SSD and incident diabetes. Ten prospective studies published between 2003 and 2007 are included in a meta-analysis reported in 2010 by Cappuccio et al. [57]. The estimated pooled OR of incident diabetes for short sleep was 1.28 (CI: 1.03–1.6). There was however a significant sex difference. The OR was 2.07 (CI: 1.16–3.72) for men but only 1.07 (CI: 0.90–1.28) for women. Difficulty initiating sleep and difficulty maintaining sleep were also significant predictors of incident diabetes. We will review here the recent prospective epidemiologic studies that were not included in this meta-analysis [57].

In a 2009 article, Chaput et al. examined the predictors of T2DM or impaired glucose tolerance (IGT), as assessed by the oral glucose tolerance test (OGTT), over a 6-year follow-up period in 276 participants of the Quebec Family Study [58]. Sleep was self-reported. After adjusting for multiple confounders, using adults with 7–8 h of sleep as a reference, the adjusted relative risk (RR) for the development of T2DM/IGT was 2.78 (CI: 1.61–4.12) for those with sleep duration ≤ 6 h. The RR was attenuated but remained significant after adjustment for BMI, waist circumference, or percent body fat. The latter finding suggests that obesity could partially mediate the developing of T2DM in short sleepers. Data from a community-based cohort of nondiabetics men and women from the Western New York Health Follow-up Study followed for an average of 6 years were used to examine biomarkers that predicted the incidence of T2DM [61]. Participants who were free of T2DM and cardiovascular disease at baseline (1996–2001) were reexamined in the period 2003–2004. Sleep duration < 6 h was categorized as short sleep and sleep duration of 6–8 h served as the reference. A nested case–control study was used to test the hypothesis that being a short sleeper at baseline is associated

Table 10.2 Summary of the epidemiologic studies (published after April 2009 and not included in the meta-analysis by Cappuccio et al. [57]), examining the association between sleep duration and glucose metabolism in adults. The association is expressed as adjusted odds ratio (AOR), adjusted relative risk (ARR) of diabetes, gestational diabetes mellitus (GDM), impaired fasting glucose (IFG), or prediabetes

Author	Description and data source	Cohort	Sleep assessment	Results	
Prospective studies					
Beihl et al. [60]	5-year follow-up Insulin Resistance Atherosclerosis Study (IRAS) (US)	390 men, 510 women Age 40–69 yrs	Self-report	*Non-Hispanic Whites and Hispanics:*	
				≤ 7 h	AOR of diabetes 2.36 (CI: 1.11–5.00)
				8 h	Reference category
				African-Americans: no significant association	
Chaput et al. [58]	5-year follow-up Quebec Family Study (Canada)	276 men and women Age 21–64 yrs	Self-report	≤ 6 h	ARR for diabetes 2.42 (CI: 1.49–3.33)
				7–8 h	Reference category
Rafalson et al. [61]	6-year follow-up Western New York Health Follow-up Study (US)	1,455 men and women 91 cases developed IFG individually matched to 272 controls. Age 35–79 yrs	Self-report	< 6 h	AOR of IFG 3 (CI: 1.05–8.59, $p=0.022$) in model 1, NS in model 2, which includes insulin resistance
				6–8 h	Reference category
Xu et al. [62]	10-year follow-up NIH-AARP Diet and Health cohort (US)	164,399 men and women Age 50–71 yrs	Self-report	< 5 h	AOR of diabetes 1.34 (CI: 1.20–1.50)
				5–6 h	AOR of diabetes 1.06 (CI: 1.01–1.11)
				7–8 h	Reference category
Bo et al. [30]	6-year follow-up Patients from Local Health Units (Italy)	1,597 men and women Age 45–64 yrs	Self-report	No association between sleep duration and incident fasting hyperglycemia at follow-up	
Cross-sectional studies					
Vgontzas et al. [63]	Penn State Cohort (US)	1,741 men and women Age ≥ 20 yrs	PSG	Chronic insomnia but not poor sleep was associated with a higher risk for diabetes. Compared with normal sleeping with ≥ 6 h sleep duration, the highest risk of diabetes was in individuals with insomnia and ≤ 5 h sleep (OR 2.95; CI: 1.2–7.0) and in insomniacs who slept 5–6 h (OR 2.07; CI: 0.68–6.4)	
Kim et al. [64]	2005 Korean National Health and Nutrition Survey (KNHNS) (Korea)	1,652 men Age 20–60 yrs	Self-report	*No abdominal obesity* ($n=1{,}241$):	
				≤ 5 h	AOR of obesity 2.40 (CI: 1.18–4.91)
				5 h	NS
				6 h	NS
				7 h	Reference category
				Abdominal obesity ($n=411$): no significant association	

Shankar et al. [65]	2008 Behavioral Risk Factor Surveillance System (BRFSS) (US)	372,144 men and women Age > 20 yrs	Self-reported perception of insufficient rest or sleep	0 days of insufficient rest/sleep	Reference category
				14–29 days of insufficient rest/sleep	AOR of diabetes 1.15 (CI: 1.07–1.23)
				30 days of insufficient rest/sleep	AOR of diabetes 1.31 (CI: 1.21–1.41)
Knutson et al. [66]	Coronary Artery Risk Development in Young Adults (CARDIA) Study (US)	200 men; 331 women Age 18–30 yrs	WAM	Absence of diabetes	No association between sleep measures and fasting glucose, insulin, or HOMA
				In diabetics: 10% higher sleep fragmentation	Associated with: 9% higher fasting glucose level 30% higher fasting insulin level 43% higher HOMA level
Chao et al. [59]	2006–2007 health examination in Taiwanese University Hospital (Taiwan)	2,145 men, 1,325 women Age >18 yrs	Self-report	< 6 h	AOR of prediabetes NS AOR of diabetes 1.55 (CI: 1.07–2.24, $p = 0.022$)
				6–8.49 h	Reference category
Pregnancy					
Qiu et al. [67]	Cross-sectional study (US)	1,290 women in the 2nd trimester Mean age 33.3 ± 4.4	Self-report	*Lean (BMI < 25 kg/m²): no significant association* *Overweight (BMI ≥ 25 kg/m²):*	
				≤ 7 h	ARR of GDM 9.83 (CI: 1.12–86.32)
				8 h	NS
				≥ 9 h	Reference category
Facco et al. [68]	Prospective study (US) Longest follow-up 34 weeks	189 healthy nulliparous women Mean age 29.7 ± 5.5 yrs	Self-report	< 7 h	Higher oral glucose tolerance value (116 ± 31 mg/dl vs. 105 ± 23; $p = 0.008$). AOR of 1-h OGT ≥ 130 2.4 (CI: 1.1–5.3) AOR of GDM 11.7 (CI: 1.2–114.5)
				> 7 h	Reference category
Reutrakul et al. [69]	Cross-sectional study (US)	169 women in the 2nd trimester Mean age 28.5 ± 5.5 yrs	Self-report	Each hour of sleep reduction was associated with a 4% increase in 1 h glucose during the screening 50-g OGTT	
				OR of GDM 3.4 (CI: 1.3–8.7; $p = 0.01$) if short sleep was associated with being at increased risk for SDB	

Adj. b adjusted beta coefficient, *CI* 95% confidence interval, *HOMA* homeostatic model assessment, *NS* not significant, *OGT* oral glucose tolerance, *SS* short sleep, *SDB* sleep disordered breathing, *yrs* years

with an increased likelihood of developing impaired fasting glucose (IFG) independently of diabetes risk factors and several confounding variables. From their final cohort of approximately 900 individuals, 91 cases progressed from normal fasting glucose to IFG over the 6-year follow-up. Each case was matched with up to three controls (subjects who had normal fasting glucose at both exams, $n=273$) based on sex, race (white vs. other), and duration of follow-up. The average number of hours of weekday sleep duration was 6.8 vs. 7.1 ($p=0.019$) for cases and controls, respectively. Also the HOMA IR, a measure of insulin resistance, was higher in the cases than in the controls. Short sleep was associated with a threefold increased likelihood of developing IFG at 6 years. When HOMA IR was included in the statistical model, the contribution of short sleep was attenuated and no longer statistically significant, suggesting that insulin resistance explains in part the association.

The National Institutes of Health (NIH)-AARP Diet and Health was a large prospective study established in 1995–1996 to examine the relationship between diet and health behaviors and cancer [62]. Six months into the study (1996–1997), a question on hours of day napping and night sleeping was introduced. In 2004–2005, a questionnaire asking to report major chronic diseases including T2DM was mailed to the participants. The final sample included 164,399 participants without diabetes and 10,143 participants with diabetes diagnosed after 2000. [62]. Both SSD (< 5 h) and daytime napping (≥ 1 h) were independently associated with risk of incident T2DM, after controlling for several variables, including health-related and socioeconomic factors, family history of T2DM, and total energy intake. Duration of daytime napping in 1996–1997 was associated with higher risk of diabetes in 2004–2005 in a dose–response manner and in each subgroup of night sleeping duration, after controlling for variable factors, including physical activity. The adjusted RR was moderately attenuated after adjustment for BMI alone or simultaneously with physical activity. The novelty of this study is the prospective examination of daytime napping as an independent risk factor for T2DM. Daytime napping had been previously linked to diabetes in cross-sectional studies, where the direction of causation could not be inferred and the increased napping was interpreted as a consequence rather than a cause of diabetes [70–72]. Daytime napping could be a marker of poor sleep quality or/and of other conditions such as OSA and depression which have been linked to increased risk of diabetes.

Both SSD and T2DM are more prevalent in ethnic/racial minorities than in whites [7, 73, 74], and therefore there may be an interaction between sleep duration and race/ethnicity as predictors of the incidence of diabetes. Beihl et al. evaluated the association between sleep duration and incident T2DM in the Insulin Resistance Atherosclerosis Study (IRAS), a cohort including African-Americans (AA), Hispanic, and non-Hispanic whites (NHW) [60]. They confirmed that sleep duration differed by race/ethnic group with the longest mean sleep duration of 7.1 h in the NHW, 6.8 h per night for Hispanics, and 6.3 h per night for AA. Furthermore, they observed a strong interaction between short sleep and race/ethnicity as predictors of incident diabetes, with a significant association present in NHW and Hispanics but not in African-Americans, after controlling for multiple variables.

In an Italian study of 1,282 patients recruited from the practice of six independent family physicians and studied at baseline and after 6 years of follow-up [30], predictors of incidence of obesity and fasting hyperglycemia (including IFG and diabetes) were examined. A total of 979 subjects had normal fasting glucose levels at baseline. In a multiple logistic regression analysis, after adjusting for sex, education level, alcohol intake, baseline BMI and glucose, and multiple putative risk factors for diabetes, the incidence of obesity but not of fasting hyperglycemia was related to the hours of sleep. It is noteworthy that laboratory studies have been consistent in indicating that post-challenge glucose levels are more readily increased by sleep restriction than fasting values.

Cross-Sectional Studies

Most cross-sectional population studies of the relationship between sleep and metabolism relied on self-reported sleep duration.

A study reported by Vgontzas et al. in 2009 is unique in using PSG to examine in a cohort of 1,741 adults the joint effects of insomnia and objective SSD on diabetes risk [63], while controlling for sleep apnea, a major confounder for both sleep disturbance and risk of T2DM. Complaining of insomnia for 1 year and having the lowest objectively defined sleep duration (≤ 5 h) increased the odds for prevalent diabetes by nearly threefold (OR 2.95; CI 1.24–7.05), compared with the group who had no insomnia/poor sleep complaint and slept for >6 h. The risk did not change after adjusting for PSG variables such as number of awakenings, number of sleep stage changes, percentage of stage 1 sleep, and periodic limb movements. The second highest OR was found in the group of insomniacs who slept 5–6 h, with a near twofold but nonsignificant increase in the risk of diabetes. Finally, objective SSD in the absence of a sleep complaint was associated with a nonsignificant increase in the odds for diabetes, but this subgroup was of relatively small size. Shankar et al. [65] examined the data on 372,144 participants of the Behavioral Risk Factor Surveillance System (BRFSS), a large multiethnic, nationally representative, cross-sectional survey conducted annually by the Center for Disease Control (CDC) in men and women, of all race-ethnicities from all 50 US states, the District of Columbia, and the three territories. A recent report based on this survey found that an estimated 11.1% of Americans reported experiencing insufficient rest or sleep every day for the preceding 30 days and only 30.7% of respondents reported no days of insufficient rest or sleep [6]. The analysis by Shankar et al. focused on cardiovascular disease, diabetes mellitus, and obesity in relationship to days of insufficient rest or sleep. Increasing categories of self-reported insufficient rest/sleep in the previous month were found to be positively associated with all three outcomes. Specifically 13 days of insufficient sleep in 1 month compared to 0 days increased the risk of diabetes by 30%, with a graded relationship to the number of days, after adjustment for a large number of potential confounders. This relationship persisted when men and women were analyzed separately. The strength of this study is the very large size of the cohort and the equal representation of men and women. One caveat is that perceived insufficient rest or sleep is a subjective measure; additionally, it may be related to other factors such as underlying sleep disordered breathing (SDB), psychosocial stress, depressive symptom, endocrine disorders, or the effect of lifestyle choices which all may predispose to diabetes mellitus and cardiovascular disease. An analysis of data of the 2004–2005 US National Health Interview Survey in 56,507 adults, 49% males, age range 18–85 years, showed that both short sleep (< 7 h) and long sleep (>8 h) were positively associated with the risk of obesity, diabetes, hypertension, and cardiovascular disease [42]. The researchers employed a multilevel logistic regression, simultaneously controlling for individual characteristics (e.g., ethno-racial group, gender, age, education), other health behaviors (e.g., exercise, smoking), family environment (e.g., income, size, education), and geographic context (e.g., census region). The CARDIA Sleep Study collected information on objective sleep duration and fragmentation by WAM in more than 600 middle-aged adults. A recent paper from Knutson et al. [66] examined the association between sleep measures and fasting glucose, fasting insulin, and HOMA-IR in participants with and without diabetes. Sleep fragmentation, but not habitual sleep duration, was related to higher fasting glucose, insulin, and estimated insulin resistance in subjects with diabetes but not in those without diabetes. These findings are consistent with a previous survey study from the same group [75] which showed that self-reported sleep disturbances may adversely affect diabetes control.

Kim et al. examined the data from the Third Korean National Health and Nutrition Examination Survey 2005 in 1,652 male adults [64]. SSD (≤ 5 h, self-reported) was significantly associated with an increased prevalence of diabetes among men without abdominal obesity (AOR of diabetes 2.40; CI: 1.18–4.91) compared to those with sleep duration of 7 h after adjustment of age, smoking, drinking, exercise, education, household income, residential area, hypertension, general obesity, abdominal obesity, high triglyceride, low HDL-C, and high cholesterol. The adjusted OR for diabetes was not significantly elevated in short sleepers with abdominal obesity, suggesting that abdominal obesity may have had a predominant role on diabetes risk. This was the first study that demonstrated an association between sleep duration and diabetes in an Asian population. Chao et al. examined the relationship between sleep duration and prediabetes/newly diagnosed T2DM in a Taiwanese population [59]. After excluding the subjects with a high risk of OSA, those with a positive history of T2DM, and those taking hypnotic drugs, a total of 3,470 adults were recruited. Each subject completed a questionnaire on sleep duration and lifestyle factors. Subjects were classified into short (< 6.0 h), normal (6.0–8.49 h), and long sleepers (≥ 8.5 h). The proportion of subjects with normal glucose tolerance, prediabetes, and newly diagnosed T2DM was 71.9, 22.9, and 5.2%, respectively. There were significant differences in age, sex, weight, education level, BMI, waist to hip ratio, systolic and diastolic blood pressure, alcohol and coffee drinking habits, family history of T2DM, and sleep duration among the three glycemic groups. In a multinomial regression, both short and long sleepers had a higher risk of newly diagnosed T2DM with an OR of 1.55 (CI: 1.07–2.24) and 2.83 (CI: 1.19–6.73), respectively, even after adjustments for age, sex, education level, family history of T2DM, cigarette smoking, alcohol and coffee drinking, and physical exercise. Furthermore, this association remained significant even after controlling for both general and central obesity. Sleep duration was not found to relate to prediabetes.

Studies in Pregnancy

Pregnancy is a condition of increased insulin resistance, and women with risk factors for diabetes such as family history, obesity, and excessive pregnancy-related weight gain may develop hyperglycemia during pregnancy, referred to as gestational diabetes mellitus (GDM). Decreases in both duration and quality of sleep are common in pregnant women [76, 77] as a result of hormonal and physical factors. Based on the epidemiologic data in the general adult population discussed above, it is conceivable that pregnant women with insufficient sleep or poor sleep quality may be at increased risk of GDM. Most epidemiologic studies have not included pregnant women; hence very little is known regarding this question.

We discuss here three studies that have examined the risk of GDM in women with sleep disturbances, including insufficient sleep. In a pilot study, Qiu et al. interviewed 1,290 women during early pregnancy to obtain self-reported measures of habitual sleep duration and snoring behavior [67]. Results from screening and diagnostic testing for GDM were abstracted from medical records. After adjusting for maternal age and race/ethnicity, women who reported sleeping ≤ 4 h per night during early pregnancy had a 5.6-fold increased risk of GDM as compared with those women who reported sleeping 9 h per night (the reference group) (RR = 5.56; CI 1.31–23.69). The positive association remained, although somewhat attenuated, after further adjustment for maternal pre-pregnancy BMI (RR = 4.18; CI 0.94–18.60). Overall, snoring was associated with a nonsignificant 1.86-fold increased risk of GDM (RR = 1.86; CI 0.88–3.94). The risk of GDM was particularly elevated among overweight pregnant women who reported snoring. Compared with lean pregnant women who did not snore, those who were overweight and snored had a 6.9-fold increased risk of GDM (CI 2.87–16.6). Facco et al. conducted a prospective cohort study in a convenience sample of healthy nulliparous women during pregnancy [68]. The women responded to a survey addressing sleep duration and SDB symptoms early and again later in pregnancy. SSD was defined as < 7 h per night. Subjects were asked about snoring and snoring frequency. Frequent snoring, used as a surrogate marker of SDB, was defined as snoring ≥ 3 nights per week. Outcomes in women who reported

SSD or frequent snoring while pregnant (early and/or late pregnancy) were compared to outcomes in women without these sleep complaints. A total of 189 women participated, 48% reported SSD, and 18.5% reported frequent snoring. Impaired glucose tolerance and GDM were more frequent in women with these sleep disturbances. Both SSD (10.2% vs. 1.1%; $p=0.008$) and frequent snoring (14.3% vs. 3.3%; $p=0.009$) were associated with a higher incidence of GDM compared to women without sleep complaints, even after controlling for potential confounders. Reutrakul et al. enrolled pregnant women scheduled to undergo a 50-g OGTT during the second trimester of gestation, according to standard of care [69]. Subjects completed standardized questionnaires assessing daytime sleepiness, SDB risk, sleep quality and duration, and sleep disturbance due to nocturia or other causes. There was an inverse correlation between sleep duration and 1-h glucose values post 50-g OGTT ($r=-0.21$, $p<0.01$) such that each hour of shorter sleep was associated with a 4% glucose increase. They also noted an increased incidence of preterm delivery in short sleepers. In addition, sleep disturbances, including frequent snoring (after adjustment for BMI), increased SDB risk, short sleep, and a combination of increased SDB and short sleep, were associated with a significantly higher risk of developing GDM.

These three studies emphasize the need for more research to characterize the impact of sleep disturbances on the risk of GDM and on pregnancy outcomes as well as the need for intervention studies to examine the possible beneficial effects of optimizing sleep duration and quality during pregnancy.

Impact of Sleep Restriction on Obesity Risk: Laboratory Studies

While a large body of epidemiologic evidence has pointed to an association between sleep loss and the increased risk of obesity, the direction of causality and the underlying mechanisms are still unclear. Theoretically, sleep loss could affect energy balance via a decrease in energy expenditure or an increase in energy intake. To date, laboratory studies examining the impact of experimental sleep deprivation have mainly focused on energy intake and/or the hormonal signals known to regulate hunger and appetite (e.g., leptin, ghrelin, PYY). The potential impact of sleep loss on energy expenditure has been much less explored.

Studies of acute total sleep deprivation (TSD) (as compared to normal nighttime sleep) conducted more than two decades ago demonstrated unequivocally that the presence or absence of sleep has a major impact on pituitary-dependent hormonal regulation and glucose metabolism [78]. In the first part of this section, we review the findings from laboratory studies that have used protocols of TSD to examine the role of sleep and circadian rhythmicity in the 24-h profiles of hormones involved in the neuroendocrine regulation of appetite (first part of Table 10.3). To date, such studies have focused only on leptin and ghrelin. We then summarize the results from a growing number of laboratory studies that have explored the effects of partial sleep deprivation (PSD) on caloric intake, weight gain, hunger, and appetite and the levels of hormones known to affect energy metabolism (second part of Table 10.3).

TSD Studies: Impact on Leptin and Ghrelin

Leptin, a hormone secreted by the adipocytes, provides information about energy status to the neural networks regulating homeostatic feeding in the hypothalamus [99, 100]. In humans acute caloric shortage or surplus leads to decreased or increased circulating leptin levels, respectively [100, 101]. These changes in leptin concentrations have been associated with reciprocal changes in hunger [101]. The 24-h leptin profile is not only dependent on the timing and amount of food intake, but appears to be also modulated by sleep and circadian rhythmicity. In a landmark study, when bedtimes were shifted by 8 h and the impact of meal intake was eliminated by administering continuous enteral nutrition to healthy lean young volunteers, a leptin elevation was

Table 10.3 Summary of the laboratory studies examining possible mechanisms linking sleep duration and obesity in adults

Author	Intervention – time in bed	Subjects	Caloric intake	Changes with sleep deprivation
Total sleep deprivation				
Simon et al. [79]	8 h×1 night (23:00–7:00) 24 h sleep deprivation (8 h shift of sleep) 8 h daytime recovery	7 men Age 21–25 yrs BMI 22.2±0.6 kg/m^2	24 h continuous enteral nutrition (50% carbohydrate, 35% fat, and 15% protein; 378 kilojoules/h)	Weight: n/a Leptin was increased during both the night of total sleep deprivation and daytime recovery sleep. A circadian elevation of leptin independent of sleep was also observed. Ghrelin: n/a Hunger: n/a Food intake: n/a
Mullington et al. [80]	8 h×3 nights (23:30–7.30) 88 h sleep deprivation Either 7 h or 14 h×3 nights of recovery	10 men Age 22–37 yrs BMI 20–34.5 kg/m^2	Three meals/24 h+optional evening snack during baseline and recovery vs. three meals+scheduled late evening snack during sleep deprivation	Weight: unchanged Leptin: decreased during night Ghrelin: n/a Hunger: n/a Food intake: n/a
Dzaja et al. [81]	8 h×1 night (23:00–7:00) 1 night sleep deprivation	10 men Age 28±3.1 yrs BMI 20.5–29.5 kg/m^2	Matched standardized meals (1,800 kcal/24 h; 30% fat, 60% carbohydrates, and 10% protein)	Weight: n/a Leptin: n/a Ghrelin: decreased during night Hunger: n/a Food intake: n/a
Schmid et al. [82]	7 h×1 night (22:00–6:00) 1 night sleep deprivation	10 men Age 20–40 yrs BMI 20.7–25 kg/m^2	No food from 21:00 to hypoglycemic clamp (at 7:30)	Weight: n/a Leptin: n/a Ghrelin: n/a Hunger: increased Food intake: n/a
Pejovic et al. [41]	8 h×4 nights (22:30–6:30) 40 h sleep deprivation 8 h×2 nights of recovery	21 men and women Age 18–30 yrs Divided in 2 groups: BMI 23.2±2.8 and 25.0±2.1 kg/m^2	Uncontrolled food intake	Weight: n/a Leptin (24 h profile): increased Ghrelin: n/a Hunger: unchanged Food intake: unchanged
Partial sleep deprivation				
Guilleminault et al. [83]	8.5 h×2 nights (22:30–7.00) 4 h×7 nights (first group 22:30–02:30; second group 2:15–06:15) 1 night *ad libitum* and 8.5 h×2 nights of recovery	8 men Age 18–25 yrs BMI 22.9±0.7 kg/m^2	Standardized diet (rich in carbohydrates on the blood-drawing days)	Weight: n/a Leptin: reduced Ghrelin: n/a Hunger: n/a Food intake: n/a

Spiegel et al. [84]	8 h × 3 nights (23:00–7:00) 4 h × 6 nights (1:00–5:00) 12 h × 7 nights (21:00–9:00)	11 men Age 22 ± 1 yrs BMI 23.4 ± 0.5 kg/m^2	Weight-maintenance meals on the day preceding and on the day of blood sampling	Weight: unchanged Leptin: reduced Ghrelin: n/a Hunger: n/a Food intake: n/a
Spiegel et al. [85].	10 h × 2 nights (22:00–8:00) 4 h × 2 nights (1:00–5:00)	12 men Age 22 ± 2 yrs BMI 23.6 ± 2 kg/m^2	Matched dinner on the second night Matched glucose infusion during blood sampling	Weight: unchanged Leptin: reduced Ghrelin: increased Hunger: increased Food intake: n/a
Schmid et al. [86]	7 h × 1 night (1st session) 4.5 h × 1 night (2nd session) 1 night of sleep deprivation (3rd session)	9 men Age 20–40 yrs BMI 20.7–25.0 kg/m^2	No food from 21:00 to single morning blood draw	Weight: n/a Leptin: unchanged Ghrelin: increased Hunger: increased Food intake: n/a
Bosy-Westphal et al. [87]	>8 h × 2 nights Four nights of consecutively increasing sleep curtailment (7 h, 6 h, 6 h, 4 h) >8 h × 2 nights of recovery	14 women Age 23–38 yrs BMI 20–36.6 kg/m^2	*Ad libitum* diet prior to OGTT	Weight: increased (+0.4 kg) Leptin: increased Ghrelin: unchanged Hunger: unchanged Food intake: increased (+20% by dietary records) Resting metabolic rate and total energy expenditure: unchanged
Schmid et al. [88]	7 h × 1 night (22.30–6.30) 4.5 h × 1 night (22.30–3.00)	10 men Age 20–40 yrs BMI 20.7–25.0 kg/m^2	No food from 21:00 to hypoglycemic clamp	Weight: n/a Leptin: n/a Ghrelin: n/a Hunger: unchanged Food intake: n/a
Schmid et al. [89]	8 h 15 min × 2 nights (22.45–7.00) 4 h 15 min (2.45–7.00) × 2 nights	15 men Age 20–40 yrs BMI 22.9 ± 0.3 kg/m^2	Uncontrolled food intake until the morning of the 2nd night when blood sampling was initiated and *ad libitum* food offered	Weight: n/a Leptin: unchanged Ghrelin: unchanged Hunger: unchanged Food intake: increased in both sleep conditions (60% excess in energy intake vs. their estimated daily energy demand) Physical activity: decreased during the daytime under free-living and shifted toward lower intensity levels

(continued)

Table 10.3 (continued)

Total sleep deprivation				
Author	Intervention – time in bed	Subjects	Caloric intake	Changes with sleep deprivation
Magee et al. [90]	8 h × 1 night (22.30–6.30) 5 h × 2 nights (1.30–6.30) 8–10 h × 1 night of recovery	10 men Age 19–23 yrs Non-obese	Standardized evening meal (1,546–1,992 KJ)	Weight: n/a Leptin: unchanged Ghrelin: unchanged Hunger: unchanged but significant reduction in satiety Caloric intake: n/a
Tasali et al. [91]	8.5 h × 4 nights 4.5 h × 4 nights	10 men and women Age 18–28 yrs BMI 20–25 kg/m^2	Matched meals. *Ad libitum* buffet at the end	Weight: unchanged Leptin (single AM assessment): increased Ghrelin: n/a Hunger: n/a Caloric intake: increased by >400 Kcal
Nedeltcheva et al. [92]	8.5 h × 14 nights 5.5 h × 14 nights	6 men, 5 women Age 34–49 yrs BMI 24–29 kg/m^2	*Ad libitum* diet. Identical meals on the blood sampling day	Weight: similar increase in both sleep conditions Leptin: unchanged Ghrelin: unchanged Hunger: n/a Caloric intake: increased snacks
Omisade et al. [93]	10 h × 2 nights (22.00–8.00) 3 h × 1 night (5.00–8.00)	15 women Age 18–25 yrs BMI 18.3–51.9 kg/m^2	Matched meals	Weight: n/a Leptin (AM and PM assessment): increased Ghrelin: n/a Hunger: unchanged Caloric intake: n/a
van Leeuwen et al. [94]	8 h × 2 nights (23.00–7.00) 4 h × 5 nights (3.00–7.00) 8 h × 3 nights (23.00–7.00) of recovery	15 men Age 19–29 yrs BMI 23.3 ± 2.7 kg/m^2	Matched meals and snacks + one more snack (fruit; 50 kcal) at 0:30 during sleep restriction	Weight: n/a Leptin: increased Ghrelin: n/a Hunger: unchanged Food intake: n/a
Simpson et al. [95]	10 h × 2 nights (22.00–8.00) 4 h × 5 nights (3.00–8.00)	136 men and women Age 22–45 yrs BMI 17.7–32.6 kg/m^2	*Ad libitum* food access	Weight: n/a Leptin (single AM assessment): increased Ghrelin: n/a Hunger: n/a Caloric intake: n/a

Nedeltcheva et al. [96]	7 h × 2 nights 5.5 h × 14 nights 8.5 h × 14 nights	10 men and women Mean age 41 ± 5 yrs Mean BMI 27.4 ± 2 kg/m^2	Caloric content restricted to 90% of resting metabolic rate	5.5 h (vs. 8.5): Same weight loss, but decreased fat mass loss and increased fat-free mass loss Leptin (24 h profile): unchanged Acylated ghrelin (24 h profile): increased Hunger: increased
Brondel et al. [97]	8 h × 2 nights (00.00–8.00) 4 h × 1 night (2.00–6.00)	12 men Age 18–29 yrs BMI 19–24.6 kg/m^2	*Ad libitum* food intake after sleep restriction.	Weight: n/a Leptin: n/a Ghrelin: n/a Hunger: increased before breakfast and dinner Caloric intake: increases of 560 kcal Physical activity: increased by 48 kcl Positive 24 h energy balance of 510 kcal
St Onge et al. [98]	9 h × 5 nights 4 h × 5 nights	15 men and 15 women Age 30–45 yrs BMI 22–26 kg/m^2	Food intake controlled during first 4 days, then *ad libitum*	Weight: n/a Leptin: n/a Ghrelin: n/a Hunger: increased before breakfast and dinner Caloric intake: increased Resting metabolic rate and total energy expenditure: unchanged

yrs years, *OGTT* oral glucose tolerance test

observed when sleep was allowed during the daytime after the night of TSD [79], suggesting that the sleep state, irrespective of time of day, affects leptin release. The possibility that sleep may promote the release of leptin and thus the control of satiety was further supported by the demonstration that prolonged TSD resulted in a decreased amplitude of the leptin diurnal variation and that sleep recovery restores the normal circadian variation [80]. While the two previous studies [79, 80] included only men, more recent work by Pejovic et al. involving both men and women confirmed a dampening of the 24-h circadian rhythm of leptin following a night of TSD but the flattening of the rhythm was due to higher daytime, rather than lower nighttime, levels [102]. Hunger ratings were unchanged but caloric intake and meal composition were not strictly controlled.

Ghrelin, a peptide produced predominantly by the stomach, is also involved in energy homeostasis, but, in contrast to leptin, ghrelin stimulates appetite [103]. The 24-h profile of ghrelin levels shows a marked nocturnal rise, which reflects at least partly the rebound of ghrelin following suppression by the evening meal. The nocturnal rebound is eventually attenuated as the night progresses, suggesting the inhibitory effects of sleep on ghrelin secretion, and therefore on the hunger-promoting effects of ghrelin [104]. The impact of TSD on the nocturnal ghrelin profile has been examined in only one study in which the nocturnal ghrelin elevation was paradoxically dampened when subjects were sleep deprived [81]. More recently, Schmid et al. reported that a single night of TSD resulted in increased subjective hunger the following morning [82] but neither leptin nor ghrelin levels were assessed.

PSD Studies: Impact on Leptin and Ghrelin

The pioneer "sleep debt study" of Spiegel et al. looked at the impact of recurrent PSD (bedtime restricted to 4 h per night for 6 nights, as compared to a fully rested condition) in healthy young men and demonstrated a robust decrease of leptin levels throughout the 24-h cycle, despite identical amounts of caloric intake, similar sedentary conditions, and stable weight [84]. The magnitude of the decrease was comparable to that observed in a similar subject population after 3 days of underfeeding by approximately 900 cal/day [105]. These observations confirmed preliminary findings by Guilleminault et al. [83] who reported that 7 nights of sleep restriction to 5-h bedtimes led to a reduction in peak nocturnal leptin levels. In the "sleep debt study," the reduction in leptin levels in the debt condition was paralleled by an increase in peripheral sympathetic nervous activity, measured via an analysis of heart rate variability. The findings suggested that repeated PSD could result in a reduced ability of leptin to accurately sense energy balance. The findings suggested that if exposed to *ad libitum* food, the subjects, under sleep restriction, would have increased their food intake and possibly gained weight. This initial demonstration of an adverse impact of sleep loss on appetite regulation was confirmed and extended in a follow-up randomized crossover design study examining the impact of 2 nights of 4 h as compared to 8 h in bed on leptin, ghrelin, and hunger and appetite [85]. Relative to the rested condition, sleep restriction was associated with an 18% decrease in leptin levels, a 28% increase in ghrelin, and more than 70% increase in the ghrelin:leptin ratio [85]. Hunger showed a 23% increase and appetite for nutrients with high carbohydrate content (such as sweets, salty snacks, and starchy foods) was increased by more than 30% [85]. Importantly, there was a remarkable correlation between the increase in subjective hunger ratings and the increase in the ghrelin:leptin ratio.

Based on these initial findings, subsequent PSD studies examined measures of caloric intake. A preliminary study in 10 healthy young adults estimated that after 4 nights of restricted sleep to 4.5 h in bed, participants ingested on average an excess of more than 400 kcal from an *ad libitum* buffet relative to when they were allowed 8.5 h bedtimes. This nearly 14% increase in caloric intake was achieved mainly by excess intake of carbohydrate-rich nutrients [91]. Confirming and extending these findings, a recent randomized crossover study comparing 5 nights of 4 h in bed to 5 nights of 9 h in bed in 15 men and 15 women found that when presented with ad lib food on the fifth day of each condition, the subjects consumed nearly 300 more Kcal when sleep restricted [98]. There was no significant impact of sex on this increase in caloric intake. Of note, in these two studies involving *ad libitum* conditions, caloric intake was controlled and kept identical under both sleep conditions prior to access to *ad libitum* food. In another study, 14 young women were exposed to an *ad libitum* diet during an 8-day at home protocol including 2 days of bedtimes > 8 h, followed by 4 days of bedtime progressively decreased to 7 h, 6 h, 6 h, and 4 h (for a total of 9 h of bedtime loss over a 4-day period relative to 8 h in bed). Sleep was not recorded and caloric intake was self-reported. The women reported on average a 20% increase in food consumption over the 4 days of sleep restriction and 11 of the 14 participants experienced weight gain (mean: +0.4 kg) [87]. A randomized crossover design study of 14 nights of sleep restriction or extension by ±1.5 h per night in overweight middle-aged adults who had *ad libitum* access to palatable food throughout the study demonstrated an increased consumption of carbohydrates and calories mostly from snacks, particularly in the evening and overnight, in the restricted sleep condition [92]. In this study, significant weight gain was observed under both sleep conditions because the participants consumed excessive amounts of calories in the "obesigenic" sedentary environment of the laboratory.

When interpreting the findings regarding the neuroendocrine regulation of appetite from studies that provided *ad libitum* access to food, one must keep in mind that weight gain may obliterate or obscure the impact of sleep loss on leptin and/or ghrelin since the release of both hormones is readily affected by changes in adiposity. A recent study in a large sample ($n = 136$) observed an increase in morning leptin levels after five to 7 nights of bedtime restriction to 4 h per night [95]. This increase in leptin levels was larger in women than in men and also larger in those with higher baseline BMI. The findings are suggestive of increased food intake and consequent weight gain following sleep restriction. However, neither food intake nor changes in body weight across the study period were evaluated.

Infrequent sampling for leptin and/or ghrelin (both hormones are secreted in a pulsatile fashion, and are modulated by circadian rhythmicity) may also complicate data interpretation. For example, in a study involving a single assessment of satiety and leptin levels at 7:30 a.m. after 5 nights of 4 h in bed (bedtimes from 3 a.m. to 7 a.m.) and after five nights of habitual sleep (bedtimes from 11 p.m. to 7 a.m.), no effect of bedtime restriction on satiety could be detected and leptin levels were elevated, rather than decreased [94]. However, bedtime restriction was achieved by delaying the timing of lights off, a condition that usually results in a phase delay of the central circadian pacemaker. As the nocturnal elevation in leptin levels is influenced by circadian rhythmicity [79], the elevated morning leptin level is likely to reflect a phase delay of the nocturnal rise in leptin.

Short-term studies involving one or two nights of bedtime restriction as compared to normal sleep have had variable results. A 2010 study comparing one night of 8 h in bed vs. one night of 4 h in bed in a randomized crossover design conducted in 12 young lean men observed a large increase in caloric intake (+22% or nearly 560 Kcal) and an increase in hunger ratings before breakfast and dinner [97]. In a subsequent study comparing 2 nights of 4 h 15 bedtimes vs. 2 nights of 8 h 15 bedtimes, appetite ratings, daytime levels of leptin and ghrelin, hunger, and calories consumed were similar after both sleep conditions. Of note, the first experimental day was spent under ambulatory conditions and food intake was not controlled [89]. A recent report where bedtimes were restricted to 3 h for only one night and leptin levels were measured in saliva at two isolated time points during the following day did not find significant changes in hunger or craving scores, while morning leptin levels were elevated after short sleep [93].

Lastly, a recent report has examined for the first time the response of PYY levels of sleep restriction. PYY is a peptide secreted by the neuroendocrine

L cells in the ileum and colon in response to a meal. The postprandial release of PYY appears to be involved in meal-related satiety and to contribute to meal termination. Similarly, glucagon-like peptide-1 (GLP-1) is secreted by the same L cells in response to a meal and has multiple actions, mostly related to glucose homeostasis, and decreases food intake by increasing satiety via central nervous system (CNS) mechanisms. Adiponectin, released by adipose tissues, promotes insulin sensitivity. Levels of adiponectin are reduced in obese and diabetic subjects. The study with PYY determinations involved 2 nights of 5 h in bed as compared to a fully rested night (8–10 h in bed) and examined peripheral levels of PYY, ghrelin, adiponectin, and leptin in young healthy men [90]. Satiety was reduced and levels of PYY were lower in the sleep loss condition. The other hormones were not affected. Although hormonal levels were assessed at a single time point upon awakening, this is the first report of decreased PYY levels after sleep restriction in humans, which could represent another mechanism underlying the reduced feeling of satiety consistently reported by sleep-deprived individuals. Lastly, young healthy men and women studied in a forced desynchrony protocol (involving over 1 month of laboratory conditions on a 28-h sleep–wake and dark–light cycle with four isocaloric meals per 28-h cycle) exhibited lower leptin levels when they ate and slept 12 h out of phase from their usual schedule [106]. Sleep efficiency was 67% when circadian disruption was maximal, compared to 84% in conditions of circadian alignment. The findings are consistent with an inhibition of leptin levels by sleep disruption but the relative contributions of circadian misalignment and sleep loss in such a protocol cannot be unambiguously dissected.

Cross-sectional population studies that examined leptin levels in relation to sleep duration have had conflicting findings. In the Wisconsin Sleep Cohort study, 5 h of habitual sleep time as compared to 8 h of sleep and both self-reported habitual sleep duration and PSG were obtained. Short habitual sleep was associated with a 15.5% decrease in morning leptin levels, while short PSG-based sleep duration was associated with 14.9% increase in morning ghrelin levels, after controlling for BMI [19]. In contrast, in a more recent study by Hayes et al. on data from the Cleveland Family Study, for each hour of decreased sleep there was a 6% increase in leptin levels after controlling for obesity and associated comorbidities [107]. A recent paper by Knutson et al. suggests that the data obtained in lean subjects may not be easily extrapolated to individuals with obesity [108]. The authors performed a cross-sectional analyses of data from participants in an ongoing sleep extension study of obese men and women, aged 18–50 years, who report sleeping less than 6.5 h per night on average [53]. Habitual nocturnal sleep duration and quality were also estimated using WAM. SDB was assessed over one night using a portable screening device. Using the baseline data available on 80 participants at the time of the analysis, no significant associations between leptin levels adjusted for the degree of adiposity and any of the sleep measures, including sleep duration, sleep efficiency, and SDB, were found.

In summary, the bulk of the current evidence from laboratory studies of sleep restriction points to a dysregulation of appetite. Inconsistent findings regarding the neurohormonal control of appetite during partial sleep restriction may be attributed to differences in the study design such as the duration of sleep restriction (1–2 vs. multiple days), the circadian timing of the restricted bedtimes, caloric intake and weight changes during the study, and finally the timing and frequency of hormonal measurements. The original finding of a decrease in leptin levels after PSD was obtained under conditions of strictly controlled caloric intake, fixed circadian timing, and BMI was unchanged [84, 85]. When feeding is *ad libitum*, an increase in weight generally occurs and has therefore the opposite effect on leptin levels, which may be more responsive to changes in adiposity than to changes in sleep duration. Ghrelin levels were measured in only one of the six studies. It is possible that the impact of sleep restriction on the neuroendocrine regulation of appetite is more clearly apparent in conditions of weight maintenance caloric intake or in conditions where caloric intake is lower than energy requirements.

If this was the case, sleep restriction could undermine the success of a reduced calorie diet by decreasing the compliance to the dietary regimen and its efficacy. The findings of a recent study [96] support this hypothesis.

Sleep Loss and Energy Expenditure

Beside the changes in neurohormones involved in the regulation of food intake, reduced energy expenditure (EE) is to date a poorly explored pathway that could also link short sleep and the risk of obesity. The amount of total daily energy expenditure (TEE) comprises three components: (1) *Resting metabolic rate* (RMR, 60% of TEE) defined as the energy expenditure of an individual under basal conditions (at rest, after an overnight fast); (2) *Thermic effects of meal* (TEM 10% of TEE), which includes the energy expenditure involved in digestion, absorption metabolism, and storage of food; (3) *Activity-related energy expenditure* (AEE, 30% of TEE), which involves all volitional and non-volitional activities. For most individuals, AEE is not accounted for by physical exercise, but rather by low-moderate intensity activities of daily living such as sitting, standing, walking, and other occupational, volitional, and spontaneous activities, all together referred to as nonexercise activity thermogenesis (NEAT) [109]. AEE is the most variable component of TEE, has a major weight in the energy balance equation, and is critical for long-term weight maintenance.

Subjects with sleep problems and/or excessive daytime sleepiness have reported significant reductions in energy ratings and in levels of physical activity [110, 111], which could reflect both reduced amounts of exercise and reductions in NEAT, and thus an overall decrease in AEE. Subjective sleepiness and fatigue increase immediately and significantly with sleep deprivation [112], however, is not clear if these would affect volitional or non-volitional daily activities or other components of TEE. Prospective data from the Nurses' Health Study showed differences in risk of weight gain in short sleepers but no difference in self-reported levels of voluntary activity in the women sleeping ≤ 6 h per day vs. those sleeping 7 h per day [113]. In the cross-sectional analysis of the CARDIA sleep study, BMI was independently associated with sleep duration and sleep fragmentation in over 600 early-middle-aged adults, and this association was not modified by accounting for self-reported levels of physical activity [26]. In participants in the Third National Health and Nutrition Examination Survey, self-reported fatigue was associated with a higher BMI, higher waist circumference, and a reduced likelihood of getting recommended levels of physical activity [114].

The findings from the five studies that examined the impact of short-term sleep restriction on physical activity have not been entirely consistent. In comparison with a rested night (7 to > 8 h in bed), Schmid et al. demonstrated that sleep restriction to 4 h for 2 nights led to a reduction in physical activity measured by accelerometry under free-living conditions, but there were no significant changes in food intake, hunger and appetite, and levels of leptin and ghrelin [89]. In contrast, Brondel et al., also using accelerometry, observed increased physical activity in the afternoon and evening after one night of PSD [97], but caloric intake increased by 560 kcal with sleep deprivation, likely resulting in a overall positive energy balance. Bosy-Westphal et al. studied 14 healthy lean and obese women after 4 nights of 5.5 h in bed, by indirect calorimetry; compared to the rested condition (9 h sleep for 2 nights), there was no change in resting EE, even when adjusted for fat-free mass or total EE. In a protocol involving a more prolonged sleep restriction (14 nights of 5.5 h vs. 14 nights of 8.5 h in bed) in healthy overweight subjects who remained in the laboratory under sedentary conditions, total EE assessed by the gold standard doubly labeled water method, RMR assessed by indirect calorimetry, and the TEM were not affected by the bedtime condition [87]. The most recent study [98] examined EE by the doubly labeled water method in subjects who participated in a randomized crossover design comparison of 5 nights of 4 h in bed vs. 5 nights of 9 h in bed and did not detect a difference in TEE.

In sum, the bulk of the evidence points at reduced or unchanged energy expenditure in subjects submitted to repeated partial sleep loss. Of note, a recent study that compared total energy expenditure during a night of sleep and during a night of TSD in subjects who remained in a whole room indirect calorimeter for 3 days found that the energy cost of sustained wakefulness across the night under sedentary conditions was only 134±2 Kcal [115]. The energy cost of sleep restriction by 2–4 h per night is likely to be less, may be as low as 50–70 Kcal, in sharp contrast with 300–600 Kcal increases in energy intake which were observed in several laboratory studies of partial sleep restriction [87, 98].

Impact of Sleep Restriction on Diabetes Risk: Laboratory Studies

Studies in healthy volunteers who underwent experimental sleep restriction have unequivocally demonstrated that insufficient sleep may cause alterations in glucose metabolism and have suggested mechanisms by which sleep loss might increase the risk of diabetes.

Total Sleep Deprivation

Kuhn et al. published in 1969 the very first laboratory study of the effect of prolonged TSD (for 72–126 h) on oral glucose tolerance and showed that TSD leads to a marked increase in glucose levels [116]. These findings were ignored for a long time, most probably because such extended periods of TSD are uncommon in real life. In 1981, another study involving 120 h of TSD demonstrated alterations of glucose metabolism at the level of the muscle consistent with a prediabetic state and increased fasting glucose levels at the end of the sleep deprivation period [117]. In 1993, a study involving 60 h of TSD observed increases in fasting insulin levels, as well as in the insulin response to OGTT, without change in glucose levels, suggesting decreased insulin sensitivity [118]. These important studies may not have had the scientific impact they deserved because TSD is a condition invariably followed by sleep recovery and a correction of metabolic abnormalities. Chronic PSD is much more common and may involve irreversible alterations.

Partial Sleep Deprivation

The first laboratory study of PSD in healthy lean adults [18] found that restricting sleep to 4 h per night for 6 nights resulted in a 40% decrease in glucose tolerance, impaired beta-cell function, reduced noninsulin-dependent glucose utilization, and a trend for decreased insulin sensitivity (SI) as assessed by minimal model analysis of a frequently sampled intravenous glucose tolerance test (ivGTT). The ivGTT is a validated tool that provides assessments of SI, pancreatic beta-cell responsiveness (referred to as "acute insulin response to glucose", AIRg), and glucose effectiveness (SG), a measure of noninsulin-dependent glucose disposal [119]. The SG was 30% lower in the state of sleep debt. AIRg was reduced by more than 30% after sleep restriction despite a trend for decreased SI. The disposition index (DI), i.e., the product of SI and AIRg, is a validated marker of diabetes risk [120]. In the state of sleep debt, the DI was decreased by an average of about 40% as compared to the fully rested state. The glucose tolerance values observed after 5 nights of 4-h bedtimes in the young lean participants were similar to those reported in older adults with impaired glucose tolerance [121]. The metabolic findings in the sleep debt condition were paralleled by an increase in the activity of the sympathetic nervous system. At the end of the recovery phase, glucose tolerance normalized to levels expected for healthy young adults [122]. A criticism of this initial "sleep debt study" is that sleep restriction (6 nights of 4-h bedtimes) was more severe than commonly occurring in real life. Also, the study did not follow a randomized crossover design and therefore the possibility of an order effect (sleep restriction preceded sleep extension) could not be excluded. These issues were addressed in a follow-up study of 2 nights with 10 h in bed vs. 2 nights with 4 h in a randomized crossover design [85]. After the second night of each bedtime condition, caloric intake was replaced by an intravenous glucose infusion at a constant rate to avoid fluctuations of hunger and appetite related to meal ingestion. Even though sleep duration was restricted for only two nights, glucose tolerance was decreased as observed in the initial study, partly as a result of inadequate insulin secretion [123]. In a recent study in non-obese healthy men, sleep restriction to 5 h per night for 1 week resulted in a significant reduction in SI as assessed by hyperinsulinemic euglycemic clamp, considered the gold standard method for SI determination [124]. The volunteers also underwent an ivGTT on a separate day and again, SI was decreased following sleep restriction, without adequate compensation by insulin release and therefore diabetes risk, as assessed by the DI, was elevated. In a 2008 study involving women only and performed under ambulatory conditions without objective sleep assessment and without control of caloric intake, progressive sleep curtailment over 4 nights (for an average bedtime restriction of 2.5 h per night, relative to 8-h bedtimes) had no impact on oral glucose tolerance [87]. More recently, Nedeltcheva et al. [125] examined the effects of moderate but prolonged sleep curtailment (5.5 h per night for 14 nights) in sedentary middle-aged men and women, and observed a decrease in glucose tolerance due to decreased SI in the absence of adequate beta-cell compensation. In addition, SG was increased. Such recurrent bedtime restriction is closer to the sleep curtailment experienced by many people in everyday life, and in people at risk it may facilitate the development of insulin resistance, reduced glucose tolerance, and ultimately diabetes. Indeed, epidemiologic studies suggest that people who sleep less than 6 h per night are at higher risk of T2DM. Consistent findings were reported by Van Leeuwen et al. who simulated in healthy young men the cumulative sleep debt as it can occur during a regular five working days schedule [94] with bed times restricted to 4 h per night. After the fifth day of sleep restriction, morning fasting glucose levels were unchanged, but fasting insulin concentrations were increased, suggesting reduced insulin sensitivity. After two nights of recovery sleep, fasting glucose was lower than at baseline, while insulin returned to baseline levels. The authors suggested that the effects of one workweek of sleep restriction could be reversed by recovery sleep on weekends. Donga et al. evaluated SI in middle-aged men and women after one single night of partial sleep restriction with the hyperinsulinemic euglycemic clamp and observed a reduction in glucose infusion and disposal rates, indicating a deterioration of glucose tolerance and peripheral insulin sensitivity [126]. They also assessed endogenous hepatic glucose production rate, by continuous infusion of [6,6-$^2H_2$]-glucose, and found an increase by approximately 22% after sleep restriction. Free fatty acid levels were also increased. These findings point to increased insulin resistance at the level of the liver and adipose tissue, respectively.

Putative Mechanisms and Implication

Multiple pathways are likely to mediate the adverse effects of sleep loss on the risk of obesity and diabetes, and much work is needed to elucidate their respective roles and interactions. Figure 10.2 presents a simplified schematic representation. Among the effects of insufficient sleep that have been documented are alterations of the central neurohormonal control of energy homeostasis and glucose metabolism, a decrease in brain glucose utilization, an increase in sympathetic activity and a decrease in vagal tone, increases in the levels of circulating hormones counter-regulatory to insulin action (cortisol, growth hormone, and catecholamines), a putative decrease in EE, an increase in inflammation, and finally more time to eat.

An upregulation of the activity of orexin neurons, concentrated in the lateral hypothalamus, may be one of the primary mechanisms linking sleep deprivation and some of its adverse metabolic effects. Indeed, the orexin system plays a key role in the interaction between sleeping and feeding. Orexin producing neurons have an extensive and divergent projection system innervating numerous structures in the CNS including all the components of the ascending arousal system and the entire cortex [127]. This system is involved in the regulation of many functions such

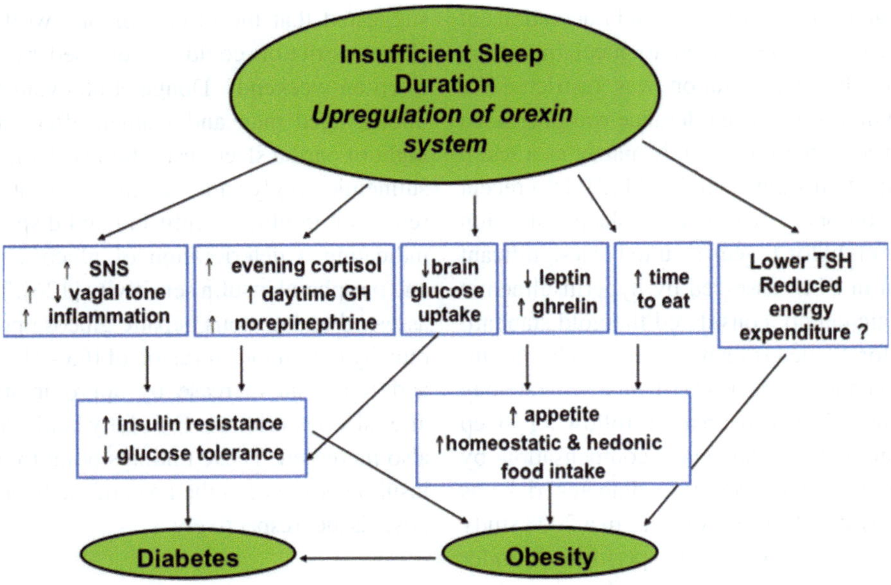

Fig. 10.2 Schematic representation of the multiple pathways likely to mediate the adverse effects of sleep loss on the risk of obesity and diabetes

as sleep–wakefulness, locomotor activity, feeding, thermoregulation, sympatho-vagal balance, and neuroendocrine and cardiovascular control [128]. Orexigenic neurons are firing during the wake period and are inactive during deep non-REM sleep due to a direct inhibition by GABA-ergic hypothalamic neurons [129]. Orexin-containing neurons play a central role in the maintenance of arousal. Deficiencies in the orexin system are associated with sleep disorders that involve chronic excessive daytime sleepiness, including narcolepsy and OSA [130, 131]. In contrast, when sleep deprivation is enforced behaviorally, the orexin system is overactive, most likely to maintain wakefulness against the increased sleep pressure [132–134]. There is evidence that orexins may stimulate food intake, particularly the early part of the usual sleep period, which is when voluntary sleep deprivation most often occurs in humans [130, 135]. Orexigenic neurons regulate the homeostatic feeding center in the hypothalamic arcuate nucleus (ARC), and concurrently affect hedonic feeding mediated by the "reward centers" (ventro-tegmental area and nucleus accumbens) [136, 137]. During starvation, the orexin neurons may be disinhibited by low levels of the anorexigenic hormone leptin and low glucose levels [129], and are excited by the hunger-promoting hormone ghrelin [138]. The peripheral metabolic cues, including insulin, leptin, and ghrelin, directly interact with ARC in the hypothalamus and ultimately indirectly modulate the activity of the orexigenic neurons to regulate food intake [135]. Additionally, the peripheral hormonal signals may influence the activity of orexigenic neurons via vagal afferents to the nucleus of the solitary tract (NST) [139].

As reviewed in this chapter, multiple studies of experimental sleep restriction have shown alterations in the metabolic hormones that are involved in the regulation of energy balance, including elevated evening cortisol levels, extended duration of daytime elevated growth hormone (GH) levels, and reductions in thyroid-stimulating hormone (TSH), lower leptin levels, and higher ghrelin levels [18, 84]. Insufficient sleep also results in elevations of markers of sympathetic nervous activity and in decreases in vagal tone. At the level of the pancreatic beta cell, this altered sympatho-vagal balance is likely to impair the expected compensatory hyperinsulinemia needed to compensate the reduced insulin sensitivity associated with sleep loss. Furthermore, elevated cortisol levels have been shown to promote increased food intake and the accumulation of visceral fat in humans [140, 141]. Similarly, since TSH normally functions to stimulate basal metabolic rate, the reductions in TSH resulting from sleep restriction [84] may lead to a reduction in EE.

Another important mechanism that may promote hyperglycemia, considering that brain is the major user of glucose, is reduced brain glucose utilization after sleep deprivation, as shown by PET studies [142]. Finally, sleep loss and sleep disturbances have been associated with increased concentrations of C-reactive protein (CRP) [143]. Both partial and total sleep loss, in young, healthy individuals results in elevation of the levels of the inflammatory cytokine Il-6, which will in turn increase CPR production [144, 145]. TSD also increases the plasma levels of TNF-alpha soluble receptor 1 [146]. Low-grade inflammation predispose to both insulin and leptin resistance [147]. Thus, sleep disturbances appear to promote systemic inflammation that could, over time, further contribute to metabolic disturbances and increase the risk of obesity and diabetes. CRP has been recently proposed as a leptin-binding protein, and thus the increase in CRP resulting from sleep restriction may further limit the amount of free leptin that is able to penetrate the blood–brain barrier and inhibit central orexigenic activity.

Conclusion

In sum, the evidence reviewed in this chapter support the hypothesis that reduced sleep duration may be part of the behavioral modifications that played a role in the development of the current epidemics of obesity and diabetes. An important consideration when trying to explain the epidemiologic link between sleep loss and metabolic risk is that it is not clear whether the physiological effects of sleep restriction observed under laboratory conditions over a period of a few days can be translated to chronic sleep restriction as it occurs in free-living individuals. Also, when comparing different laboratory studies of sleep restriction, differences in the "dose" of sleep loss relative to the physiological need of the individual are often ignored. While the body of evidence suggestive of an interaction between sleep loss and the epidemics of obesity and diabetes continues to build at a rapid pace, much remains to discover as far as mechanisms and the transition from short-term laboratory conditions to chronic PSD in real life. Intervention studies extending sleep in habitual short sleepers and examining the impact on metabolic outcomes are needed to further address the direction of causality of the association between insufficient sleep, obesity, and diabetes and the potential clinical implications.

Acknowledgments The preparation of this chapter was partly supported by NIH grants PO1 AG-11412, RO1 DK-071696, RO1 OH-009482, P50-HD057796, and P60-DK20595.

References

1. Keith SW, Redden DT, Katzmarzyk PT, et al. Putative contributors to the secular increase in obesity: exploring the roads less traveled. Int J Obes (Lond). 2006;30(11):1585–94.
2. Kripke D, Simons R, Garfinkel L, Hammond E. Short and long sleep and sleeping pills. Is increased mortality associated? Arch Gen Psychiatr. 1979;36(1):103–16.
3. Hublin C, Partinen M, Koskenvuo M, Kaprio J. Sleep and mortality: a population-based 22-year follow-up study. Sleep. 2007;30(10):1245–53.
4. National Sleep Foundation. 2008 Sleep in America poll. www.sleepfoundation.org. 2008.
5. National Center Health Statistics. QuickStats: percentage of adults who reported an average of ≤ 6 hours of sleep per 24-hour period, by sex and age group - United States, 1985 and 2004. MMWR Morb Mortal Wkly Rep. 2005;54:933.
6. Center for Disease Control. Perceived insufficient rest or sleep among adults - United States, 2008. MMWR Morb Mortal Wkly Rep. 2009;58(42): 1175–9.
7. Lauderdale DS, Knutson KL, Yan LL, et al. Objectively measured sleep characteristics among early-middle-aged adults: the CARDIA study. Am J Epidemiol. 2006;164(1):5–16.
8. Pannain S, Van Cauter E. Modulation of endocrine function by sleep-wake homeostasis and circadian rhythmicity. Sleep Med Clin. 2007;2(2):147–59.

9. Center for Disease Control. Unhealthy sleep-related behaviors–12 States, 2009. MMWR Morb Mortal Wkly Rep. 2011;60(8):233–8.
10. Jean-Louis G, von Gizycki H, Zizi F, Spielman A, Hauri P, Taub H. The actigraph data analysis software: I. A novel approach to scoring and interpreting sleep-wake activity. Percept Mot Skills. 1997;85: 207–16.
11. Lauderdale DS, Knutson KL, Yan LL, Liu K, Rathouz PJ. Self-reported and measured sleep duration: how similar are they? Epidemiology. 2008;19(6):838–45.
12. Finucane MM, Stevens GA, Cowan MJ, et al. National, regional, and global trends in body-mass index since 1980: systematic analysis of health examination surveys and epidemiological studies with 960 country-years and 9.1 million participants. Lancet. 2011;377(9765):557–67.
13. McAllister EJ, Dhurandhar NV, Keith SW, et al. Ten putative contributors to the obesity epidemic. Crit Rev Food Sci Nutr. 2009;49(10):868–913.
14. Morselli L, Leproult R, Balbo M, Spiegel K. Role of sleep duration in the regulation of glucose metabolism and appetite. Best Pract Res Clin Endocrinol Metab. 2010;24(5):687–702.
15. Knutson KL. Sleep duration and cardiometabolic risk: a review of the epidemiologic evidence. Best Pract Res Clin Endocrinol Metab. 2010;24(5):731–43.
16. Pannain S, Van Cauter E. Sleep loss, obesity and diabetes: prevalence, association and emerging evidence for causation. Obesity Metabol. 2008;4(1): 28–41.
17. Patel SR, Hu FB. Short sleep duration and weight gain: a systematic review. Obesity (Silver Spring). 2008;16(3):643–53.
18. Spiegel K, Leproult R, Van Cauter E. Impact of sleep debt on metabolic and endocrine function. Lancet. 1999;354(9188):1435–9.
19. Taheri S, Lin L, Austin D, Young T, Mignot E. Short sleep duration is associated with reduced leptin, elevated ghrelin, and increased body mass index. PLoS Med. 2004;1(3):e62.
20. Beccuti G, Pannain S. Sleep and obesity. Curr Opin Clin Nutr Metab Care. 2011;14(4):402–12.
21. Cappuccio FP, Taggart FM, Kandala NB, et al. Meta-analysis of short sleep duration and obesity in children and adults. Sleep. 2008;31(5):619–26.
22. Gunderson EP, Rifas-Shiman SL, Oken E, et al. Association of fewer hours of sleep at 6 months postpartum with substantial weight retention at 1 year postpartum. Am J Epidemiol. 2008;167(2): 178–87.
23. Chaput JP, Despres JP, Bouchard C, Tremblay A. The association between sleep duration and weight gain in adults: a 6-year prospective study from the Quebec Family Study. Sleep. 2008;31(4):517–23.
24. Lopez-Garcia E, Faubel R, Leon-Munoz L, Zuluaga MC, Banegas JR, Rodriguez-Artalejo F. Sleep duration, general and abdominal obesity, and weight change among the older adult population of Spain. Am J Clin Nutr. 2008;87(2):310–6.
25. Chaput JP, Leblanc C, Perusse L, Despres JP, Bouchard C, Tremblay A. Risk factors for adult overweight and obesity in the Quebec Family Study: have we been barking up the wrong tree? Obesity (Silver Spring). 2009;17(10):1964–70.
26. Lauderdale DS, Knutson KL, Rathouz PJ, Yan LL, Hulley SB, Liu K. Cross-sectional and longitudinal associations between objectively measured sleep duration and body mass index: the CARDIA Sleep Study. Am J Epidemiol. 2009;170(7):805–13.
27. Watanabe M, Kikuchi H, Tanaka K, Takahashi M. Association of short sleep duration with weight gain and obesity at 1-year follow-up: a large-scale prospective study. Sleep. 2010;33(2):161–7.
28. Nishiura C, Hashimoto H. A 4-year study of the association between short sleep duration and change in body mass index in Japanese male workers. J Epidemiol. 2010;20(5):385–90.
29. Hairston KG, Bryer-Ash M, Norris JM, Haffner S, Bowden DW, Wagenknecht LE. Sleep duration and five-year abdominal fat accumulation in a minority cohort: the IRAS family study. Sleep. 2010;33(3): 289–95.
30. Bo S, Ciccone G, Durazzo M, et al. Contributors to the obesity and hyperglycemia epidemics. A prospective study in a population-based cohort. Int J Obes (Lond). 2011;35(11):1442–9.
31. Chaput JP, Despres JP, Bouchard C, Tremblay A. Longer sleep duration associates with lower adiposity gain in adult short sleepers. International Journal of Obesity advance online publication, 7 June 2011; doi:10.1038/ijo.2011.110.
32. van den Berg JF, Knvistingh Neven A, Tulen JH, et al. Actigraphic sleep duration and fragmentation are related to obesity in the elderly: the Rotterdam study. Int J Obes (Lond). 2008;32(7): 1083–90.
33. Vgontzas AN, Lin HM, Papaliaga M, et al. Short sleep duration and obesity: the role of emotional stress and sleep disturbances. Int J Obes (Lond). 2008;32(5):801–9.
34. Hall MH, Muldoon MF, Jennings JR, Buysse DJ, Flory JD, Manuck SB. Self-reported sleep duration is associated with the metabolic syndrome in midlife adults. Sleep. 2008;31(5):635–43.
35. Choi KM, Lee JS, Park HS, Baik SH, Choi DS, Kim SM. Relationship between sleep duration and the metabolic syndrome: Korean National Health and Nutrition Survey 2001. Int J Obes (Lond). 2008;32(7):1091–7.
36. Park SE, Kim HM, Kim DH, Kim J, Cha BS, Kim DJ. The association between sleep duration and general and abdominal obesity in Koreans: data from the Korean National Health and Nutrition Examination Survey, 2001 and 2005. Obesity (Silver Spring). 2009;17(4):767–71.
37. St-Onge MP, Perumean-Chaney S, Desmond R, et al. Gender differences in the association between sleep duration and body composition: the Cardia study. Int J Endocrinol. 2010:726071.

38. Adamkova V, Hubacek JA, Lanska V, et al. Association between duration of the sleep and body weight. Physiol Res. 2009;58 Suppl 1:S27–31.
39. Di Milia L, Mummery K. The association between job related factors, short sleep and obesity. Ind Health. 2009;47(4):363–8.
40. Thomas A, Schussler MN, Fischer JE, Terris DD. Employees' sleep duration and body mass index: potential confounders. Prev Med. 2009;48(5):467–70.
41. Watson NF, Buchwald D, Vitiello MV, Noonan C, Goldberg J. A twin study of sleep duration and body mass index. J Clin Sleep Med. 2010;6(1):11–7.
42. Buxton OM, Marcelli E. Short and long sleep are positively associated with obesity, diabetes, hypertension, and cardiovascular disease among adults in the United States. Soc Sci Med. 2010;71(5):1027–36.
43. Magee CA, Iverson DC, Caputi P. Sleep duration and obesity in middle-aged Australian adults. Obesity (Silver Spring). 2010;18(2):420–1.
44. Magee CA, Caputi P, Iverson DC. Is sleep duration associated with obesity in older Australian adults? J Aging Health. 2010;22(8):1235–55.
45. Magee CA, Caputi P, Iverson DC. Short sleep mediates the association between long work hours and increased body mass index. J Behav Med. 2011;34:83–91.
46. Anic GM, Titus-Ernstoff L, Newcomb PA, Trentham-Dietz A, Egan KM. Sleep duration and obesity in a population-based study. Sleep Med. 2010;11(5):447–51.
47. Theorell-Haglow J, Berne C, Janson C, Sahlin C, Lindberg E. Associations between short sleep duration and central obesity in women. Sleep. 2010;33(5):593–8.
48. Knutson KL. Association between sleep duration and body size differs among three Hispanic groups. Am J Hum Biol. 2011;23(1):138–41.
49. Baron KG, Reid KJ, Kern AS, Zee PC. Role of sleep timing in caloric intake and BMI. Obesity (Silver Spring). 2011;19(7):1374–81.
50 Kim S, DeRoo LA, Sandler DP. Eating patterns and nutritional characteristics associated with sleep duration. Public Health Nutr. 2011;14(5):889–95.
51. Liu R, Liu X, Arguelles LM, et al. A population-based twin study on sleep duration and body composition. Obesity (Silver Spring). 2011;20:192–9.
52. Wheaton AG, Perry GS, Chapman DP, McKnight-Eily LR, Presley-Cantrell LR, Croft JB. Relationship between body mass index and perceived insufficient sleep among U.S. adults: an analysis of 2008 BRFSS data. BMC Public Health. 2011;11:295.
53. Cizza G, Marincola P, Mattingly M, et al. Treatment of obesity with extension of sleep duration: a randomized, prospective, controlled trial. Clin Trials. 2010;7(3):274–85.
54. Silva GE, Goodwin JL, Sherrill DL, et al. Relationship between reported and measured sleep times: the Sleep Heart Health Study (SHHS). J Clin Sleep Med. 2007;3(6):622–30.
55. Patel SR, Blackwell T, Redline S, et al. The association between sleep duration and obesity in older adults. Int J Obes (Lond). 2008;32(12):1825–34.
56. Gangwisch JE, Malaspina D, Boden-Albala B, Heymsfield SB. Inadequate sleep as a risk factor for obesity: analyses of the NHANES I. Sleep. 2005;28(10):1289–96.
57. Cappuccio FP, D'Elia L, Strazzlillo P, Miller MA. Quantity and quality of sleep and incidence of type 2 diabetes - a systematic review and meta-analysis. Diabetes Care. 2010;33(2):414–20.
58. Chaput JP, Despres JP, Bouchard C, Astrup A, Tremblay A. Sleep duration as a risk factor for the development of type 2 diabetes or impaired glucose tolerance: analyses of the Quebec Family Study. Sleep Med. 2009;10(8):919–24.
59. Chao CY, Wu JS, Yang YC, et al. Sleep duration is a potential risk factor for newly diagnosed type 2 diabetes mellitus. Metabolism. 2011;60(6):799–804.
60. Beihl DA, Liese AD, Haffner SM. Sleep duration as a risk factor for incident type 2 diabetes in a multiethnic cohort. Ann Epidemiol. 2009;19(5):351–7.
61. Rafalson L, Donahue RP, Stranges S, et al. Short sleep duration is associated with the development of impaired fasting glucose: the Western New York Health Study. Ann Epidemiol. 2010;20(12):883–9.
62. Xu Q, Song Y, Hollenbeck A, Blair A, Schatzkin A, Chen H. Day napping and short night sleeping are associated with higher risk of diabetes in older adults. Diabetes Care. 2010;33(1):78–83.
63. Vgontzas AN, Liao D, Pejovic S, Calhoun S, Karataraki M, Bixler EO. Insomnia with objective short sleep duration is associated with type 2 diabetes: a population-based study. Diabetes Care. 2009;32(11):1980–5.
64. Kim J, Kim HM, Kim KM, Kim DJ. The association of sleep duration and type 2 diabetes in Korean male adults with abdominal obesity: the Korean National Health and Nutrition Examination Survey 2005. Diabetes Res Clin Pract. 2009;86(2):e34–6.
65. Shankar A, Syamala S, Kalidindi S. Insufficient rest or sleep and its relation to cardiovascular disease, diabetes and obesity in a national, multiethnic sample. PLoS One. 2010;5(11):e14189.
66. Knutson KL, Van Cauter E, Zee P, Liu K, Lauderdale DS. Cross-sectional associations between measures of sleep and markers of glucose metabolism among subjects with and without diabetes: the Coronary Artery Risk Development in Young Adults (CARDIA) Sleep Study. Diabetes Care. 2011;34(5):1171–6.
67. Qiu C, Enquobahrie D, Frederick IO, Abetew D, Williams MA. Glucose intolerance and gestational diabetes risk in relation to sleep duration and snoring during pregnancy: a pilot study. BMC Womens Health. 2010;10:17.
68. Facco FL, Grobman WA, Kramer J, Ho KH, Zee PC. Self-reported short sleep duration and frequent snoring in pregnancy: impact on glucose metabolism. Am J Obstet Gynecol. 2010;203(2):142 e141–5.

69. Reutrakul S, Zaidi N, Wroblewski K, et al. Sleep disturbances and their relationship to glucose tolerance in pregnancy. Diabetes Care. 2011;34(11):2454–7.
70. Picarsic JL, Glynn NW, Taylor CA, et al. Self-reported napping and duration and quality of sleep in the lifestyle interventions and independence for elders pilot study. J Am Geriatr Soc. 2008;56(9):1674–80.
71. Goldman SE, Hall M, Boudreau R, et al. Association between nighttime sleep and napping in older adults. Sleep. 2008;31(5):733–40.
72. Asplund R. Daytime sleepiness and napping amongst the elderly in relation to somatic health and medical treatment. J Intern Med. 1996;239(3):261–7.
73. Hicks RA, Lucero-Gorman K, Bautista J, Hicks GJ. Ethnicity, sleep duration, and sleep satisfaction. Percept Mot Skills. 1999;88(1):234–5.
74. Cowie CC, Rust KF, Ford ES, et al. Full accounting of diabetes and pre-diabetes in the U.S. population in 1988-1994 and 2005-2006. Diabetes Care. 2009;32(2):287–94.
75. Knutson KL, Ryden AM, Mander BA, Van Cauter E. Role of sleep duration and quality in the risk and severity of type 2 diabetes mellitus. Arch Intern Med. 2006;166(16):1768–74.
76. Lee KA, Zaffke ME, McEnany G. Parity and sleep patterns during and after pregnancy. Obstet Gynecol. 2000;95(1):14–8.
77. Sahota PK, Jain SS, Dhand R. Sleep disorders in pregnancy. Curr Opin Pulm Med. 2003;9(6):477–83.
78. Van Cauter E, Polonsky KS, Scheen AJ. Roles of circadian rhythmicity and sleep in human glucose regulation. Endocr Rev. 1997;18(5):716–38.
79. Simon C, Gronfier C, Schlienger JL, Brandenberger G. Circadian and ultradian variations of leptin in normal man under continuous enteral nutrition: relationship to sleep and body temperature. J Clin Endocrinol Metab. 1998;83:1893–9.
80. Mullington JM, Chan JL, Van Dongen HP, et al. Sleep loss reduces diurnal rhythm amplitude of leptin in healthy men. J Neuroendocrinol. 2003;15(9):851–4.
81. Dzaja A, Dalal MA, Himmerich H, Uhr M, Pollmacher T, Schuld A. Sleep enhances nocturnal plasma ghrelin levels in healthy subjects. Am J Physiol Endocrinol Metab. 2004;286(6):E963–7.
82. Schmid SM, Hallschmid M, Jauch-Chara K, Bandorf N, Born J, Schultes B. Sleep loss alters basal metabolic hormone secretion and modulates the dynamic counterregulatory response to hypoglycemia. J Clin Endocrinol Metab. 2007;92(8):3044–51.
83. Guilleminault C, Powell NB, Martinez S, et al. Preliminary observations on the effects of sleep time in a sleep restriction paradigm. Sleep Med. 2003;4(3):177–84.
84. Spiegel K, Leproult R, L'Hermite-Baleriaux M, Copinschi G, Penev PD, Van Cauter E. Leptin levels are dependent on sleep duration: relationships with sympathovagal balance, carbohydrate regulation, cortisol, and thyrotropin. J Clin Endocrinol Metab. 2004;89(11):5762–71.
85. Spiegel K, Tasali E, Penev P, Van Cauter E. Brief communication: sleep curtailment in healthy young men is associated with decreased leptin levels, elevated ghrelin levels, and increased hunger and appetite. Ann Intern Med. 2004;141(11):846–50.
86. Schmid SM, Hallschmid M, Jauch-Chara K, Born J, Schultes B. A single night of sleep deprivation increases ghrelin levels and feelings of hunger in normal-weight healthy men. J Sleep Res. 2008;17(3):331–4.
87. Bosy-Westphal A, Hinrichs S, Jauch-Chara K, et al. Influence of partial sleep deprivation on energy balance and insulin sensitivity in healthy women. Obes Facts. 2008;1(5):266–73.
88. Schmid SM, Jauch-Chara K, Hallschmid M, Schultes B. Mild sleep restriction acutely reduces plasma glucagon levels in healthy men. J Clin Endocrinol Metab. 2009;94(12):5169–73.
89. Schmid SM, Hallschmid M, Jauch-Chara K, et al. Short-term sleep loss decreases physical activity under free-living conditions but does not increase food intake under time-deprived laboratory conditions in healthy men. Am J Clin Nutr. 2009;90(6):1476–82.
90. Magee C, Huang X, Iverson D, Caputi P. Acute sleep restriction alters neuroendocrine hormones and appetite in healthy male adults. Sleep Biol Rhythms. 2009;7:125–7.
91. Tasali E, Broussard J, Day A, Kilkus J, Van Cauter E. Sleep curtailment in healthy young adults is associated with increased ad lib food intake. Sleep. 2009;32(Suppl):A163.
92. Nedeltcheva AV, Kilkus JM, Imperial J, Kasza K, Schoeller DA, Penev PD. Sleep curtailment is accompanied by increased intake of calories from snacks. Am J Clin Nutr. 2009;89(1):126–33.
93. Omisade A, Buxton OM, Rusak B. Impact of acute sleep restriction on cortisol and leptin levels in young women. Physiol Behav. 2010;99(5):651–6.
94. van Leeuwen WM, Hublin C, Sallinen M, Harma M, Hirvonen A, Porkka-Heiskanen T. Prolonged sleep restriction affects glucose metabolism in healthy young men. Int J Endocrinol. 2010;2010:108641.
95. Simpson NS, Banks S, Dinges DF. Sleep restriction is associated with increased morning plasma leptin concentrations, especially in women. Biol Res Nurs. 2010;12(1):47–53.
96. Nedeltcheva AV, Kilkus JM, Imperial J, Schoeller DA, Penev PD. Insufficient sleep undermines dietary efforts to reduce adiposity. Ann Intern Med. 2010;153(7):435–41.
97. Brondel L, Romer MA, Nougues PM, Touyarou P, Davenne D. Acute partial sleep deprivation increases food intake in healthy men. Am J Clin Nutr. 2010;91(6):1550–9.
98. St-Onge MP, Roberts AL, Chen J, et al. Short sleep duration increases energy intakes but does not change energy expenditure in normal-weight individuals. Am J Clin Nutr. 2011;94(2):410–6.

99. Ahima RS, Saper CB, Flier JS, Elmquist JK. Leptin regulation of neuroendocrine systems. Front Neuroendocrinol. 2000;21(3):263–307.
100. Kolaczynski JW, Considine RV, Ohannesian J, et al. Responses of leptin to short-term fasting and refeeding in humans: a link with ketogenesis but not ketones themselves. Diabetes. 1996;45(11):1511–5.
101. Chin-Chance C, Polonsky KS, Schoeller D. Twenty-four hour leptin levels respond to cumulative short-term energy imbalance and predict subsequent intake. J Clin Endocrinol Metab. 2000;85:2685–91.
102. Pejovic S, Vgontzas AN, Basta M, et al. Leptin and hunger levels in young healthy adults after one night of sleep loss. J Sleep Res. 2010;19(4):552–8.
103. Havel PJ. Peripheral signals conveying metabolic information to the brain: short-term and long-term regulation of food intake and energy homeostasis. Exp Biol Med. 2001;226(11):963–77.
104. Spiegel K, Tasali E, Leproult R, Scherberg N, Van Cauter E. Twenty-four-hour profiles of acylated and total ghrelin: relationship with glucose levels and impact of time of day and sleep. J Clin Endocrinol Metab. 2010;96(2):486–93.
105. Chin-Chance C, Polonsky KS, Schoeller DA. Twenty-four-hour leptin levels respond to cumulative short-term energy imbalance and predict subsequent intake. J Clin Endocrinol Metabol. 2000;85(8):2685–91.
106. Scheer FA, Hilton MF, Mantzoros CS, Shea SA. Adverse metabolic and cardiovascular consequences of circadian misalignment. Proc Natl Acad Sci USA. 2009;106(11):4453–8.
107. Hayes AL, Xu F, Babineau D, Patel SR. Sleep duration and circulating adipokine levels. Sleep. 2011;34(2):147–52.
108. Knutson KL, Galli G, Zhao X, Mattingly M, Cizza G. No association between leptin levels and sleep duration or quality in obese adults. Obesity (Silver Spring). 2011;19(12):2433–5.
109. Rising R. Total daily energy expenditure. J Am Coll Nutr. 1994;13(4):309–10.
110. Weaver T, Laizner A, Evans L, et al. An instrument to measure functional status outcomes for disorders of excessive sleepiness. Sleep. 1997;20(10):835–43.
111. Briones B, Adams N, Strauss M, et al. Sleepiness and health: relationship between sleepiness and general health status. Sleep. 1996;19(7):583–8.
112. Dinges D, Pack F, Williams K, et al. Cumulative sleepiness, mood disturbance, and psychomotor vigilance performance decrements during a week of sleep restricted to 4-5 hours per night. Sleep. 1997;20:267–77.
113. Patel SR, Malhotra A, White DP, Gottlieb DJ, Hu FB. Association between reduced sleep and weight gain in women. Am J Epidemiol. 2006;164(10):947–54.
114. Resnick HE, Carter EA, Aloia M, Phillips B. Cross-sectional relationship of reported fatigue to obesity, diet, and physical activity: results from the third national health and nutrition examination survey. J Clin Sleep Med. 2006;2(2):163–9.
115. Jung CM, Melanson EL, Frydendall EJ, Perreault L, Eckel RH, Wright KP. Energy expenditure during sleep, sleep deprivation and sleep following sleep deprivation in adult humans. J Physiol. 2011;589(Pt 1):235–44.
116. Kuhn E, Brodan V, Brodanova M, Rysanek K. Metabolic effects of sleep deprivation. Activitas Nervosa Superior. 1969(11-3-1969):165–74.
117. Vondra K, Brodan V, Bass A, et al. Effects of sleep deprivation on the activity of selected metabolic enzymes in skeletal muscle. Eur J Appl Physiol Occup Physiol. 1981;47(1):41–6.
118. VanHelder T, Symons JD, Radomski MW. Effects of sleep deprivation and exercise on glucose tolerance. Aviat Space Environ Med. 1993;64:487–92.
119. Bergman RN. Minimal model: perspective from 2005. Horm Res. 2005;64 Suppl 3:8–15.
120. Lyssenko V, Almgren P, Anevski D, et al. Predictors of and longitudinal changes in insulin sensitivity and secretion preceding onset of type 2 diabetes. Diabetes. 2005;54(1):166–74.
121. Garcia G, Freeman R, Supiano M, Smith M, Galecki A, Halter J. Glucose metabolism in older adults: a study including subjects more than 80 years of age. J Am Geriatr Soc. 1997;45:813–7.
122. Prigeon RL, Kahn SE, Porte Jr D. Changes in insulin sensitivity, glucose effectiveness, and B-cell function in regularly exercising subjects. Metabolism. 1995;44(10):1259–63.
123. Spiegel K, Knutson K, Leproult R, Tasali E, Van Cauter E. Sleep loss: a novel risk factor for insulin resistance and type 2 diabetes. J Appl Physiol. 2005;99(5):2008–19.
124. Buxton OM, Pavlova M, Reid EW, Wang W, Simonson DC, Adler GK. Sleep restriction for 1 week reduces insulin sensitivity in healthy men. Diabetes. 2010;59(9):2126–33.
125. Nedeltcheva AV, Kessler L, Imperial J, Penev PD. Exposure to recurrent sleep restriction in the setting of high caloric intake and physical inactivity results in increased insulin resistance and reduced glucose tolerance. J Clin Endocrinol Metab. 2009;94(9):3242–50.
126. Donga E, van Dijk M, van Dijk JG, et al. A single night of partial sleep deprivation induces insulin resistance in multiple metabolic pathways in healthy subjects. J Clin Endocrinol Metab. 2010;95(6):2963–8.
127. Saper CB, Scammell TE, Lu J. Hypothalamic regulation of sleep and circadian rhythms. Nature. 2005;437(7063):1257–63.
128. Nunez A, Rodrigo-Angulo ML, Andres ID, Garzon M. Hypocretin/orexin neuropeptides: participation in the control of sleep-wakefulness cycle and energy homeostasis. Curr Neuropharmacol. 2009;7(1):50–9.
129. Saper CB, Chou TC, Elmquist JK. The need to feed: homeostatic and hedonic control of eating. Neuron. 2002;36(2):199–211.
130. Sakurai T. Roles of orexin/hypocretin in regulation of sleep/wakefulness and energy homeostasis. Sleep Med Rev. 2005;9(4):231–41.

131. de Lecea L, Sutcliffe JG. The hypocretins and sleep. FEBS J. 2005;272(22):5675–88.
132. Wu MF, John J, Maidment N, Lam HA, Siegel JM. Hypocretin release in normal and narcoleptic dogs after food and sleep deprivation, eating, and movement. Am J Physiol Regul Integr Comp Physiol. 2002;283(5):R1079–86.
133. Estabrooke IV, McCarthy MT, Ko E, et al. Fos expression in orexin neurons varies with behavioral state. J Neurosci. 2001;21(5):1656–62.
134. Zeitzer JM, Buckmaster CL, Lyons DM, Mignot E. Increasing length of wakefulness and modulation of hypocretin-1 in the wake-consolidated squirrel monkey. Am J Physiol Regul Integr Comp Physiol. 2007;293(4):R1736–42.
135. Adamantidis A, de Lecea L. The hypocretins as sensors for metabolism and arousal. J Physiol. 2009;587(Pt 1):33–40.
136. Harris GC, Aston-Jones G. Arousal and reward: a dichotomy in orexin function. Trends Neurosci. 2006;29(10):571–7.
137. Harris GC, Wimmer M, Aston-Jones G. A role for lateral hypothalamic orexin neurons in reward seeking. Nature. 2005;437(7058):556–9.
138. Yamanaka A, Beuckmann CT, Willie JT, et al. Hypothalamic orexin neurons regulate arousal according to energy balance in mice. Neuron. 2003;38(5):701–13.
139. Badman MK, Flier JS. The gut and energy balance: visceral allies in the obesity wars. Science. 2005;307(5717):1909–14.
140. Rosmond R, Dallman MF, Bjorntorp P. Stress-related cortisol secretion in men: relationships with abdominal obesity and endocrine, metabolic and hemodynamic abnormalities. J Clin Endocrinol Metab. 1998;83(6):1853–9.
141. Bjorntorp P, Rosmond R. Obesity and cortisol. Nutrition. 2000;16(10):924–36.
142. Thomas M, Sing H, Belenky G, et al. Neural basis of alertness and cognitive performance impairments during sleepiness. I. Effects of 24 h of sleep deprivation on waking human regional brain activity. J Sleep Res. 2000;9(4):335–52.
143. Meier-Ewert HK, Ridker PM, Rifai N, et al. Effect of sleep loss on C-reactive protein, an inflammatory marker of cardiovascular risk. J Am Coll Cardiol. 2004;43(4):678–83.
144. Vgontzas AN, Papanicolaou DA, Bixler EO, et al. Circadian interleukin-6 secretion and quantity and depth of sleep. J Clin Endocrinol Metab. 1999;84(8):2603–7.
145. Vgontzas AN, Zoumakis E, Bixler EO, et al. Adverse effects of modest sleep restriction on sleepiness, performance, and inflammatory cytokines. J Clin Endocrinol Metab. 2004;89(5):2119–26.
146. Shearer WT, Reuben JM, Mullington JM, et al. Soluble TNF-alpha receptor 1 and IL-6 plasma levels in humans subjected to the sleep deprivation model of spaceflight. [see comments]. J Allergy Clin Immunol. 2001;107(1):165–70.
147. Chen K, Li F, Li J, et al. Induction of leptin resistance through direct interaction of C-reactive protein with leptin. Nat Med. 2006;12(4):425–32.

Sleep Disturbances: Impact on Risk and Severity of Diabetes

11

Kelly A. Holes-Lewis

Abstract

There has been a significant increase in people diagnosed with diabetes. It is estimated that by 2025, 300 million people worldwide will be diagnosed as diabetic. In the U.S. alone, 19.9 million people are expected to be diagnosed with diabetes by the year 2025, and 29 million cases by the year 2050. There are two types of diabetes, type 1 and type 2. Both types of diabetes are diagnosed when blood glucose levels are above normal. Both types of diabetes have serious medical sequelae including neurologic problems with neuropathies, retinopathies, kidney failure, gastroparesis, impaired wound healing, bladder control problems, erectile dysfunction, and cardiovascular complications. Given this broad spectrum of complications, research regarding modifiable risk factors is of paramount importance. Recent research has led to the important discovery of an association between sleep loss and type 2 diabetes. There are many categories of sleep disorders, though the types of sleep disorders that have been associated with type 2 diabetes include: short sleep duration, obstructive sleep apnea, insomnia, and restless legs syndrome. While requiring further exploration, this association is of great clinical significance as it may have numerous implications in both the treatment and diagnosis of both disorders. This review outlines the most cutting-edge research involving the relationship between the above sleep disorders and the risk of diabetes. What is clear from the research is that sleep of sufficient quantity and quality is essential for health and well-being, especially in those with diabetes. Sleep disorders are associated with type 2 diabetes, impairments in glucose metabolism, elevations of glucose and insulin, and insulin resistance, and may serve as a modifiable risk factor. Efforts made to improve sleep may prevent the development of these medical sequelae, though more research is needed.

K.A. Holes-Lewis, MD (✉)
Department of Psychiatry and Behavioral Sciences,
Medical University of South Carolina,
Charleston, SC, USA
e-mail: holeslew@musc.edu

P.J. Shiromani et al. (eds.), *Sleep Loss and Obesity: Intersecting Epidemics*,
DOI 10.1007/978-1-4614-3492-4_11, © Springer Science+Business Media, LLC 2012

Introduction

Diabetes is a prevalent medical condition with serious consequences. There are two types of diabetes: type 1 and type 2. Both types of diabetes are diagnosed when blood glucose levels are above normal. Type 1 diabetes occurs when there is an immune-mediated destruction of insulin-producing cells that leads to a near-total loss of endogenous insulin production. In this condition, exogenous insulin is necessary to achieve glycemic control to prevent diabetic ketoacidosis and to sustain life. This type of diabetes most commonly occurs in childhood or adolescence, but can occur at any age. Type 2 diabetes, formerly called adult-onset diabetes, is the most common form of diabetes and also can develop at any age. Type 2 diabetes typically begins with insulin resistance, a condition in which muscle, liver, and fat cells do not use insulin properly. As a result, the body needs more insulin to help glucose enter cells to be used for energy. At first, the pancreas keeps up with the added demand by producing more insulin. In time, however, the pancreas loses its ability to secrete enough insulin in response to meals. Both types of diabetes have serious medical sequelae including neurologic problems with neuropathies, retinopathies, kidney failure, gastroparesis, impaired wound healing, bladder control problems, erectile dysfunction, and cardiovascular complications. Given this broad spectrum of complications, research regarding modifiable risk factors is of paramount importance. Recent research has led to the important discovery of an association between sleep loss and type 2 diabetes [1]. While requiring further exploration, this association may have numerous implications in both the treatment and diagnosis of both disorders.

Prevalence of Diabetes

There has been a significant increase in people diagnosed with diabetes. It is estimated that by 2025, 300 million people worldwide will be diagnosed as diabetic. In the U.S. alone, 19.9 million people are expected to be diagnosed with diabetes by the year 2025, and 29 million cases by the year 2050 [2].

Disorders of Sleep Associated with Type 2 Diabetes

There are many categories of sleep disorders, though the types of sleep disorders that have been associated with type 2 diabetes include: short sleep duration, obstructive sleep apnea, insomnia, and restless legs syndrome. Each of these types will be described in detail below.

Short Sleep Duration and Type 2 Diabetes

Adequate duration of sleep is important to the normal function of numerous biologic processes. Furthermore, sleep of deficient quantity has been associated with illnesses such as coronary heart disease [3], high blood pressure [4], obesity [5], and other chronic conditions. Additionally, short sleep duration has been linked with the potential increased risk of type 2 diabetes mellitus. Though short sleep has been defined in the literature in several different ways, it is usually defined as a sleep duration of less than 6 h per day. This designation is inclusive of naps, resting, and sleeping [6–10].

The association between short sleep as a risk factor for the development of type 2 diabetes has been investigated in a number of research studies. Mallon et al. investigated the relationship among sleep complaints, sleep duration, and the development of diabetes prospectively over a 12-year period in a middle-aged Swedish population. The study was composed of a random sample of 2,663 subjects living in mid-Sweden who were between the ages of 45 and 65 years. All subjects were sent a questionnaire by mail that included questions about sleep complaints, duration, sociodemographic characteristics, behavioral risk factors, medical conditions, and depression. Twelve years later, the same group was sent a near identical questionnaire in follow-up. This questionnaire was completed by 77.6% of those who completed the first survey. In this study that defined short sleep as a duration of ≤5 h, the relative risk for the development of diabetes was higher in men with short sleep duration or difficulty maintaining sleep after adjustment for age, hypertension, snoring, body mass index (BMI), depression, and other relevant risk factors. Interestingly, in this study, short or long sleep duration or sleep complaints did not influence the risk of new diabetes in women. They hypothesize that the increased risk of diabetes associated with short sleep may be secondary to sympathetic overactivity and activation of the hypothalamic–pituitary–adrenal axis [11].

Ayas et al. performed a study which aimed to test whether habitually short sleep duration increases the risk of developing diabetes. In this study, a cohort of 70,026 women, who were enrolled in the Nurse's Health Study, were investigated. All participants were without the diagnosis of diabetes at baseline. All participants completed a question detailing their total amount of actual daily sleep in a 24-h period and were followed for 10 years for the development of diabetes. Over the 10-year period, 1,969 incident cases of diabetes were diagnosed. After adjusting for age, both long and short sleep duration were associated with a significantly increased risk of incident diabetes diagnosis. The relative risk was higher for short sleepers than long sleepers. When adjustment was made for BMI, the multivariate relative risk of diabetes was reduced and no longer significant for short sleepers, but remained significant for long sleepers. The authors postulate that the increased risk for diabetes in short sleepers may be secondary to weight gain and elevated BMI predisposing to pre-diabetes, rather than the short duration of sleep functioning as an independent risk factor for the development of diabetes. They also speculate that the sleep restriction may lead to the development of diabetes through its effect on weight via a reduction of the hormone leptin [3].

Research has shown that sleep restriction results in a reduction of the hormone leptin which can result in increased appetite and weight gain. Leptin is an anorexigenic hormone produced primarily by adipose tissue which functions to decrease appetite [12]. Through this function, it provides regulation of food intake and therefore body weight [13]. Therefore, one mechanism as mentioned above through which short sleep and sleep restriction may indirectly lead to diabetes is through a reduction of leptin which in turn leads to a cascade of increases in appetite, food intake, weight gain, and BMI which can predispose to diabetes and metabolic syndrome. This association between short sleep and alteration of adipokine levels which can predispose to abnormal metabolic regulation was investigated by Mullington et al. Although this study was only composed of 10 healthy men, it was found that the diurnal amplitude of leptin was reduced during total sleep deprivation and returned toward normal during the period of recovery sleep. They conclude that this finding is evidence that sleep influences the nocturnal leptin profile, and may have implications for the understanding of the role of sleep in metabolic regulation [14]. In two separate studies by Spiegel et al., the hypothesis that sleep curtailment can cause alteration in leptin levels and appetite was tested. In the first study, 11 subjects were studied following 6 days of short sleep (defined as 4 h) and after 6 days of 12 h sleep. In addition to other parameters, this study found that mean, maximum and rhythm amplitude of leptin were decreased (−19%, −26%, and −20%, respectively) during sleep restriction compared with sleep extension [15]. In the second study, 12 healthy men with a mean age of 22 and a mean BMI of 23.6 underwent 2 days of sleep restriction and 2 days of sleep extension under controlled conditions of caloric intake and physical activity. The objective was to determine whether sleep curtailment alters appetite regulation. The results of the study demonstrated that sleep restriction was associated with average reductions in leptin, elevations in ghrelin (a hormone that functions to increase appetite), and increased hunger and appetite, specifically for calorie-dense foods with high carbohydrate content [16]. These studies serve to underscore the relationship between short sleep and the metabolic alterations that can predispose to weight gain and increased risk for the development of diabetes.

Obstructive Sleep Apnea and Type 2 Diabetes

Obstructive sleep apnea is a chronic condition that affects people of any age, but primarily middle-aged obese men. It is a common disorder that affects 2–4% of the U.S. population. The prevalence appears to increase as age increases. This sleep disorder is characterized by apneic and hypopneic episodes (see Chap. 13 for details on clinical diagnosis of OSA). An apneic episode is defined by airflow cessation for greater than 10 s. Hypopneic episodes demonstrate a nasal pressure that is reduced by at least 30% of baseline for a duration of at least 10 s. This is accompanied by an oxygen desaturation of at least 4%. The Apnea–Hypopnea Index (AHI) is used as an indicator for severity of fragmented sleep and chronic sleep deprivation, which lead to daytime sleepiness. Of great clinical significance, OSA has been shown to confer an increased risk of serious medical illness including cerebrovascular accidents and cardiovascular disease. Some of the proposed mechanisms are an intermittent hypoxia as a result of the repetitive upper airway obstruction during sleep in OSA that results in sympathetic nervous system hyperactivity and alterations in function of the autonomic nervous system [17]. In addition to the increased risk of stroke and cardiovascular disease, it has also been associated with diabetes and insulin resistance, and metabolic syndrome as outlined below.

Much research has been undertaken to investigate the relationship between obstructive sleep apnea and its relationship to glucose homeostasis (see also Chap. 9). One such study by Schober et al. investigated 498 patients with DM type 2 and 58 patients with DM type 1 and found that 37.4% had an AHI ≥ 15/h suggestive of OSA [18]. Aronsohn et al. designed a study that aimed to determine whether or not the severity of OSA serves as a predictor of glycemic control in diabetic patients. Hemoglobin A1c was utilized as the main clinical indicator of glycemic control. Sixty diabetic patients recruited from outpatient clinics underwent an initial interview and completed the University of Chicago Diabetes/Quality of Life Survey, the Berlin Questionnaire, and the Center for Epidemiologic Studies Depression Scale. Height, weight, and waist circumference were also measured in subjects. Ambulatory wrist actigraphy was performed for 5 consecutive days in subjects to determine individual habitual sleep habits. The presence and severity of OSA was determined by an overnight laboratory polysomnogram. Hemoglobin A1c levels were either obtained from the patient's chart if performed within the previous 3 months, or else drawn and analyzed following the polysomnogram. The results of this study revealed that OSA is highly prevalent in type 2 diabetics. Seventy-seven percent of their subjects were definitively diagnosed with OSA. This study was also the first to demonstrate a clear, graded, inverse relationship between the severity of obstructive sleep apnea and glycemic control in type 2 diabetics. The demonstration of this relationship through this important study elucidates the clinical significance of effectively evaluating and treating OSA as it may improve glycemic control [19].

In an observational cross-sectional study of 52 consecutive subjects that had risk factors for sleep apnea and who were attending a diabetes and obesity clinic, Pillai et al., aimed to investigate the association between the severity of OSA and HbA1c (glycosylated hemoglobin) in patients with DM type 2. This study demonstrated a prevalence of OSA of 58% in the cohort. An increased severity of OSA was associated with increased HbA1c levels after adjusting for age, gender, body mass index, duration of diabetes, and dose of insulin. They conclude that increased severity of OSA is independently associated with worsening glycemic control. As discussed below, the investigators postulate that improvements in glycemic control may occur with the accurate identification and treatment of obstructive sleep apnea [20].

Given the association between diabetes and OSA, one might postulate whether treatment of the OSA would attenuate the negative effects of diabetes and improve glycemic control. Numerous studies have aimed to investigate this clinical question. Continuous positive airway pressure or CPAP is a common and effective treatment of obstructive sleep apnea. CPAP uses mild air pressure to keep the respiratory airways open and to prevent the hypoxemic episodes that can occur as a part of obstructive sleep apnea. One such study that has been undertaken to investigate this was performed by Gracia et al. in a pre- and post-treatment, within-subject study design. In this study, 20 obese subjects with diagnosed OSA were evaluated before and following 6 months of CPAP. Insulin, glucose, and insulin resistance levels were all assessed as primary outcome measures. Additional parameters evaluated included weight change, and levels of Ghrelin, leptin, adiponectin, and resistin levels. This study demonstrated that following 6 months of CPAP, numerous metabolic parameters were affected. Hypoxia was reversed in all subjects. Fasting insulin levels had a significant increase after CPAP treatment. Of interest, the subjects as a group gained weight following CPAP. Of the parameters studied, the following remained unchanged after 6 months of CPAP: systolic and diastolic blood pressure, heart rate, fasting and post-prandial glucose levels, circulating leptin, adiponectin, and resistin levels. Fasting active ghrelin levels decreased significantly after CPAP use. This study concludes that the alterations of insulin resistance in obese patients with OSA are modulated by changes in weight rather than the elimination of hypoxia [21]. A systematic review and meta-analysis of the effects of CPAP-respiration on markers of glucose metabolism in patients with obstructive sleep apnea syndrome was undertaken by Hecht et al. Following a review of studies that included randomized and nonrandomized trials that compared CPAP with inactive control or placebo CPAP in adults with OSAS (Obstructive Sleep Apnoea Syndrome), they found that the studies did not support the hypothesis that OSAS independently influences glucose metabolism. Given the limited amount of data in this area, more controlled trials are needed to investigate this association more fully [22].

Insomnia and Type 2 Diabetes

Insomnia is a frequent cause of sleep disruption in diabetics. Though not clearly defined, insomnia refers to an individual's perception of their difficulty with sleep, such as by querying if they have difficulty falling or staying asleep or more generally, if they have difficulty sleeping. Polysomnographic evidence of insomnia is often characterized by the presence of frequent nocturnal awakenings, prolonged periods of wakefulness during the sleep period, a long sleep latency and frequent arousals of a transient nature [23]. An increasing body of literature has demonstrated an association between insomnia and type 2 diabetes. This association is of great clinical importance as treatment of one may provide a reciprocal improvement of the other. As discussed above, sleep duration and quality are associated with fasting glucose, fasting insulin, and estimated insulin resistance. In a community-based sample of early middle-aged adults, poor sleep quality with sleep fragmentation resulted in a 9% higher fasting glucose level, 30% higher fasting insulin level and a 43% higher HOMA level (measure of estimating insulin resistance) in those with type 2 diabetes. Insomnia was associated with a 23% higher fasting glucose level, a 48% higher fasting insulin level, and an 82% higher HOMA level [24].

In a population-based study designed to examine the joint effects of insomnia and objective short sleep duration on diabetes risk, Vgontzas et al., studied 1,741 randomly selected men and women in the sleep laboratory. The parameters studied were defined as follows: Insomnia was defined as ≥1 year of complaints of insomnia. Poor sleep was defined as a complaint of difficulty falling asleep, staying asleep, or early final awakening. This study demonstrated that chronic insomnia was associated with a higher risk for diabetes. Also when the subjects within the insomnia and <5 h sleep duration group were compared with the normal sleeping group and the ≥6 h sleep duration group, it was found that the individuals with insomnia and ≤5 h sleep duration group had the highest risk of diabetes. This same level of risk of diabetes was also identified in individuals with insomnia who slept 5–6 h per night. This highly powered study demonstrates the association between insomnia with short sleep duration and the increased risk of diabetes [25].

The association between insomnia and diabetes was also investigated by Budhiraja et al. A community-based sample of 3,282 men and women between the ages of 18 and 65 were studied in this prospective cross-sectional study. Polysomnographic sleep variables and self-reported measures of sleep habits and current health were measured in all subjects. This study revealed that the prevalence of insomnia in subjects with diabetes (OR = 1.4 [10.5–20] $p = 0.04$) is greater than in those without medical disorders [26].

As an ancillary study to the Coronary Artery Risk Development in Young Adults (CARDIA) Study, Knutson et al. examined the association between sleep duration and quality with fasting glucose, fasting insulin, or estimated insulin resistance in a community-based sample of early middle-aged adults. Six days of wrist actigraphy was used to assess habitual sleep duration and fragmentation. They defined insomnia as self-reported difficulty falling asleep or waking up in the middle of the night three or more times per week in addition to an average sleep efficiency of <80% based on the data from the wrist actigraphy. Insulin and glucose levels were assessed via fasting blood samples and the homeostatic model assessment method (HOMA) was used to estimate insulin resistance. The results of this study revealed that in diabetics, sleep fragmentation and insomnia are both associated with higher fasting glucose, insulin, and estimated insulin resistance [24].

Given the clinical importance and what appears to be a bi-directional association between insomnia and diabetes, further studies are warranted to further examine this important association.

Restless Legs Syndrome and Type 2 Diabetes

Restless legs syndrome (RLS) is another sleep disorder that has shown association with diabetes. By definition, RLS is characterized by an urge to move the legs and is usually accompanied by sensations that are unpleasant in the lower extremities. It typically begins or worsens during times of rest and is found to be worse in the evening and nighttime. Movement of the legs may partially or completely relieve the sensation [27, 28]. The accurate identification and management of the presence of RLS in patients with diabetes and vice versa is of clinical significance as it allows for optimization of the course and prognosis of the disorders [29].

In a study by Cuellar and Ratcliff that investigated 121 patients with type 2 diabetes, they found that 45% of the patients met criteria for RLS. From this study, they impress the importance of recognizing RLS as a sleep disorder that has the potential to affect the management of type 2 diabetes [30].

In a case–controlled study by Merlino et al., three objectives were identified: to assess the association between RLS and type 2 diabetes, to analyze the characteristics of RLS in diabetics and to identify risk factors that may be associated with the development of RLS in patients with type 2 diabetes. The study was composed of 124 consecutive patients with diabetes and 87 consecutive controls with a previous diagnosis of other endocrine disease. The results of this study revealed that 17.7% of the diabetics met diagnostic criteria for RLS and only 5.5% in the control group. They found that RLS was independently associated with type 2 diabetes ($p < 0.04$). It was also demonstrated that polyneuropathy was present in 27% of the diabetics with RLS, although this was found to only partially explain the increased prevalence of RLS in the type 2 diabetics [31].

Given the known association between RLS and type 2 diabetes, Lopes et al. designed a study to assess the clinical significance and quality of sleep in type 2 diabetics. One hundred consecutive patients at a diabetes clinic were assessed for presence of RLS and quality of sleep. RLS was diagnosed using the criteria defined by the International Restless Legs Syndrome Study Group. Quality of sleep was assessed utilizing the Pittsburgh Sleep Quality Index and the Epworth Sleepiness Scale measured excessive daytime sleepiness. This study demonstrated a prevalence of RLS in the type 2 diabetics.

Forty-five percent of the patients were identified as having poor sleep quality. Excessive daytime sleepiness was present in 26% of the diabetics. This study elucidates the importance of recognizing sleep disturbance in diabetics and that RLS is a major cause of sleep disruption in patients with type 2 diabetes [32].

Laboratory Studies of Reduced Sleep Quality

Numerous factors can impair sleep quality, including altered sleep schedules due to shift work or lifestyle changes, and various social, behavioral, and psychological problems. Additionally, medications, illnesses, and pain can all serve to reduce the quality of sleep. Recent research has been undertaken to understand the physiologic effects of impaired sleep quality, especially the impact on endocrinologic functions such as glycemic control (for additional information, see Chap. 10). One such study conducted by Cappuccio et al. was a systematic review and meta-analysis that was designed to assess the relationship between habitual sleep disturbances and the incidence of type 2 diabetes and to also obtain an estimate of risk. Ten studies encompassing 107,756 subjects were included in this review of the Medline, EMBASE, Cochrane Library, and manual searches using the terms "sleep" and "diabetes" and "prospective" or "cohort" or "longitudinal." This review used the following definitions to define sleep disturbances that were assessed by self-reported habitual sleep duration with the use of questionnaires. Short sleep was defined by either <5, <6, or <7 h of sleep per night. Sleep of a duration greater than 8 or 9 h per night defined long sleep. The questionnaires also evaluated difficulty in initiating or maintaining sleep. The results of this review revealed a clear and consistent pattern of increased risk of the development of type 2 diabetes with both short and long patterns of sleep impairment and with difficulty in initiating and maintaining sleep. They conclude from the demonstrated association that disrupted quality and quantity of sleep should be regarded as a behavioral risk factor for the development of type 2 diabetes. Furthermore, modifications that would provide for sufficient sleep may decrease the risk of diabetes [7].

Tasali et al., performed a study aimed to test the hypothesis that slow wave sleep (often referred to as "restorative sleep") plays a role in the regulation of glucose homeostasis. They hypothesize that suppression of the slow wave sleep may adversely affect glucose homeostasis and increase the risk of type 2 diabetes. To test this hypothesis, they studied nine healthy volunteers (five men and four women) who had normal results on the Pittsburgh Sleep Quality Index, Berlin Questionnaire, Epworth Sleepiness Scale, Center for Epidemiologic Studies and Beck Depression scales, and functional outcome of sleep questionnaire. Normal glucose tolerance was identified at baseline via an oral glucose tolerance test. The absence of sleep disorders was identified by an overnight screening sleep study at baseline. Each subject underwent testing with sleep recordings using a digital EEG acquisition system in two conditions separated by 4 weeks, after two consecutive nights of undisturbed baseline sleep and after three consecutive nights of experimental suppression of slow wave sleep. All subjects wore a wrist activity monitor to assess adherence to schedule and bedtimes and naps were not permitted. Slow wave sleep was suppressed by delivering acoustic tones of varying frequency and intensity to speakers that were placed on each side of the bed. Glucose metabolism was also assessed, as was plasma cortisol. The results of this study indicate marked decreases in insulin sensitivity without adequate compensatory increase in insulin release and therefore reduced glucose tolerance and increased risk of diabetes in young healthy adults with all-night selective suppression of slow wave sleep. These findings are of significant clinical importance as it may prove a modifiable risk factor for the development of diabetes. They also suggest that the reduced quality of sleep with low levels of slow wave sleep as is seen clinically in obese and aging populations may contribute to an increased risk of diabetes [33].

Fragmented sleep has been shown to have numerous physical and psychological consequences, though the endocrinologic effects of sleep fragmentation have been less investigated. The effects of sleep fragmentation on glucose metabolism in normal subjects were recently investigated by Stamatakis et al. In this study, 11 healthy, normal volunteers underwent two nights of experimentally fragmented sleep through all stages using auditory and mechanical stimuli. Insulin sensitivity, glucose effectiveness, and insulin secretion were the primary outcome measures to assess glucose metabolism. Secondary measures included serum levels of inflammatory markers, adipokines and cortisol, and measures of sympathovagal balance. The results of this study revealed that sleep fragmentation across all stages of sleep was associated with a decrease in insulin sensitivity and glucose effectiveness. Increases in sympathetic nervous system activity as well as morning cortisol release were also demonstrated and are postulated to mediate the adverse metabolic effects of the impaired sleep quality [34]. As fragmented sleep is a common disorder of sleep for many individuals, these findings hold great clinical significance. Based on these results, attention and modification of sleep patterns to reduce fragmentation may prove to reduce the risk of diabetes and abnormal glucose metabolism.

Conclusion

Sleep of sufficient quantity and quality is essential for health and well-being, including in those with diabetes. Sleep disorders are associated with type 2 diabetes and with impairments in glucose metabolism, elevations of glucose, and insulin and insulin resistance and may serve as a modifiable risk factor. Efforts made to improve sleep may reduce the incidence of these medical sequelae, though more research is needed.

References

1. Van Cauter E, et al. Metabolic consequences of sleep and sleep loss. Sleep Med. 2008;9 Suppl 1:S23–8.
2. Boyle JP, et al. Projection of diabetes burden through 2050: impact of changing demography and disease prevalence in the U.S. Diabetes Care. 2001;24(11): 1936–40.
3. Ayas NT, et al. A prospective study of sleep duration and coronary heart disease in women. Arch Intern Med. 2003;163(2):205–9.
4. Gangwisch JE, et al. Short sleep duration as a risk factor for hypertension: analyses of the first National Health and Nutrition Examination Survey. Hypertension. 2006;47(5):833–9.
5. Knutson KL, Van Cauter E. Associations between sleep loss and increased risk of obesity and diabetes. Ann N Y Acad Sci. 2008;1129:287–304.
6. Leger D, et al. Short sleep in young adults: insomnia or sleep debt? Prevalence and clinical description of short sleep in a representative sample of 1004 young adults from France. Sleep Med. 2011;12(5):454–62.
7. Cappuccio FP, et al. Quantity and quality of sleep and incidence of type 2 diabetes: a systematic review and meta-analysis. Diabetes Care. 2010;33(2): 414–20.
8. Gangwisch JE, et al. Sleep duration associated with mortality in elderly, but not middle-aged, adults in a large US sample. Sleep. 2008;31(8):1087–96.
9. Singh M, et al. The association between obesity and short sleep duration: a population-based study. J Clin Sleep Med. 2005;1(4):357–63.
10. Luckhaupt SE, Tak S, Calvert GM. The prevalence of short sleep duration by industry and occupation in the National Health Interview Survey. Sleep. 2010;33(2):149–59.
11. Mallon L, Broman JE, Hetta J. High incidence of diabetes in men with sleep complaints or short sleep duration: a 12-year follow-up study of a middle-aged population. Diabetes Care. 2005;28(11):2762–7.
12. Ahima RS, Flier JS. Leptin. Annu Rev Physiol. 2000;62:413–37.
13. Friedman JM, Halaas JL. Leptin and the regulation of body weight in mammals. Nature. 1998;395(6704): 763–70.
14. Mullington JM, et al. Sleep loss reduces diurnal rhythm amplitude of leptin in healthy men. J Neuroendocrinol. 2003;15(9):851–4.
15. Spiegel K, et al. Leptin levels are dependent on sleep duration: relationships with sympathovagal balance, carbohydrate regulation, cortisol, and thyrotropin. J Clin Endocrinol Metab. 2004;89(11):5762–71.
16. Spiegel K, et al. Brief communication: sleep curtailment in healthy young men is associated with decreased leptin levels, elevated ghrelin levels, and increased hunger and appetite. Ann Intern Med. 2004;141(11):846–50.
17. Peltier AC, et al. Effect of mild hyperglycemia on autonomic function in obstructive sleep apnea. Clin Auton Res. 2011;22:1–8.
18. Schober AK, Neurath MF, Harsch IA. Prevalence of sleep apnoea in diabetic patients. Clin Respir J. 2011;5(3):165–72.
19. Aronsohn RS, et al. Impact of untreated obstructive sleep apnea on glucose control in type 2 diabetes. Am J Respir Crit Care Med. 2010;181(5):507–13.

20. Pillai A, et al. Effects of sleep apnea severity on glycemic control in patients with type 2 diabetes prior to continuous positive airway pressure treatment. Diabetes Technol Ther. 2011;13(9):945–9.
21. Garcia JM, et al. Weight and metabolic effects of CPAP in obstructive sleep apnea patients with obesity. Respir Res. 2011;12(1):80.
22. Hecht L, Mohler R, Meyer G. Effects of CPAP-respiration on markers of glucose metabolism in patients with obstructive sleep apnoea syndrome: a systematic review and meta-analysis. Ger Med Sci. 2011;9:Doc20.
23. Roth T. Insomnia: definition, prevalence, etiology, and consequences. J Clin Sleep Med. 2007;3(5 Suppl): S7–10.
24. Knutson KL, et al. Cross-sectional associations between measures of sleep and markers of glucose metabolism among subjects with and without diabetes: the Coronary Artery Risk Development in Young Adults (CARDIA) Sleep Study. Diabetes Care. 2011;34(5):1171–6.
25 Vgontzas AN, et al. Insomnia with objective short sleep duration is associated with type 2 diabetes: a population-based study. Diabetes Care. 2009;32(11): 1980–5.
26. Budhiraja R, et al. Prevalence and polysomnographic correlates of insomnia comorbid with medical disorders. Sleep. 2011;34(7):859–67.
27. Innes KE, Selfe TK, Agarwal P. Restless legs syndrome and conditions associated with metabolic dysregulation, sympathoadrenal dysfunction, and cardiovascular disease risk: a systematic review. Sleep Med Rev. 2011. Jul 4 [Epub ahead of print].
28. Allen RP, et al. Restless legs syndrome: diagnostic criteria, special considerations, and epidemiology. A report from the restless legs syndrome diagnosis and epidemiology workshop at the National Institutes of Health. Sleep Med. 2003;4(2):101–19.
29. Dikeos D, Georgantopoulos G. Medical comorbidity of sleep disorders. Curr Opin Psychiatry. 2011;24(4): 346–54.
30. Cuellar NG, Ratcliffe SJ. Restless legs syndrome in type 2 diabetes: implications to diabetes educators. Diabetes Educ. 2008;34(2):218–34.
31. Merlino G, et al. Association of restless legs syndrome in type 2 diabetes: a case–control study. Sleep. 2007;30(7):866–71.
32. Lopes LA, et al. Restless legs syndrome and quality of sleep in type 2 diabetes. Diabetes Care. 2005;28(11): 2633–6.
33. Tasali E, et al. Slow-wave sleep and the risk of type 2 diabetes in humans. Proc Natl Acad Sci USA. 2008;105(3):1044–9.
34. Stamatakis KA, Punjabi NM. Effects of sleep fragmentation on glucose metabolism in normal subjects. Chest. 2010;137(1):95–101.

Insufficient Sleep and Cardiovascular Disease

12

Susan Redline

Abstract

In aggregate, studies conducted across multiple settings in diverse populations indicate that extremes of sleep duration are associated with numerous cardiovascular risk factors, including hypertension. However, even after considering these risk factors, the predominance of data indicates that extreme sleep duration is also associated with clinical cardiovascular disease and cardiovascular mortality. The overall "dose–response" for these associations is not clear. While some studies suggest monotonic associations, others suggest a J- or U-shaped distribution. Also, while the most extreme durations (e.g., less than 5 or more than 10 h of sleep per night) have been fairly consistently associated with adverse cardiovascular outcomes, the magnitude of effect of less extreme deviations from 7 to 8 h per night is less apparent. The sources of population heterogeneity in susceptibility to the effects of sleep deprivation also are not well understood. While some data indicate stronger effects in women compared to men, it is plausible that such differences could relate to measurement or greater misclassification of exposures in men. A further understanding of the inter-relationship of sleep patterns with other important health behaviors (activity patterns and diet) and more comprehensive assessments of sleep in terms of duration, quality, timing, and symptoms may help further clarify which aspects of sleep are most important in the pathogenesis of cardiovascular disease.

Introduction

The associations between insufficient sleep and numerous adverse health outcomes, including associations with established cardiovascular risk factors such as diabetes and obesity, are reviewed in other chapters in this book. In addition to effects associated with increased risk of

S. Redline, MD, MPH (✉)
Department of Medicine,
Harvard Medical School, Brigham and Women's Hospital and Beth Israel Deaconess Medical Center,
Boston, MA, USA
e-mail: sredline1@partners.org

P.J. Shiromani et al. (eds.), *Sleep Loss and Obesity: Intersecting Epidemics*,
DOI 10.1007/978-1-4614-3492-4_12, © Springer Science+Business Media, LLC 2012

diabetes and obesity, insufficient sleep may increase the risk of hypertension and cardiovascular disease through effects on sympathetic nervous system activation; alterations of the hypothalamic–pituitary–adrenal axis influencing secretion of cortisol and the renin–angiotensin system; and augmented systemic levels of inflammation such as elevations in CRP levels [1, 2]. These physiological perturbations may in turn contribute to renal dysfunction, endothelial dysfunction, and atherosclerosis. This chapter reviews the evidence linking insufficient sleep to hypertension and to direct measures of clinical cardiovascular disease and cardiovascular mortality. It also provides an overview of mechanisms underlying the biological plausibility for such associations. Challenges are discussed in understanding how to best conceptualize the pertinent stresses associated with "insufficient sleep" – specifically, whether risk is associated with absolute sleep duration, disruptions in sleep continuity, or subjectively experienced poor sleep, or to combinations of each. The impact of potential confounders is reviewed.

Acute Sleep Deprivation and Blood Pressure Elevation

Acute sleep deprivation has been shown to result in modest to high elevations in blood pressure measured within 24 h of the sleep-restricted period. In one experimental study comparing the effects on blood pressure of a single night of total sleep deprivation to a night of 7 h of sleep in eight middle-aged men, systolic and diastolic blood pressure readings were recorded to increase on average by 4 and 3 mm Hg, respectively [3]. Two Italian studies reported that sleep restriction to 4 h per night resulted in average increases in systolic and diastolic blood pressure of 7 and 4 mm Hg, respectively [4, 5]. Furthermore, these changes were accompanied by an increased secretion of urinary norepinephrine secretion, suggesting a mediating role for enhanced sympathetic activation. Additionally, in the latter study, 24-h blood pressure profiles displayed a nondipping pattern [5], suggesting that increased wakefulness attenuates the normal decrease in blood pressure during the night, potentially increasing the overall 24 h blood pressure "burden" and causing a nocturnal blood pattern profile that has been associated with increased cardiovascular disease.

In a natural experiment of sleep deprivation occurring in Japanese workers intermittently exposed to long shifts, a single night in which sleep duration was curtailed to 3.6 h compared to nights when sleep was reported to be 8 h, was associated with an average increase in systolic and diastolic blood pressure of 6 and 3 mm Hg, respectively [6]. However, it is possible that some of this effect was due to the stresses of long work hours or misaligned sleep-work schedule.

Few studies have examined the impact of experimentally restricted sleep over multiple nights. In one study of only four volunteers, 10 days of sleep restriction to 4 h per night was reported to result in marked increases in blood pressure increases of 22 mm Hg for systolic blood pressure and 17 mm Hg for diastolic pressure [2]. Control subjects also experienced some increase in blood pressure, suggesting that other factors may have also contributed to changes in blood pressure in this protocol.

Although in aggregate these studies provide evidence for a causal role of sleep deprivation in influencing blood pressure, these studies do not estimate the public health burden of curtailed sleep which in natural settings is often milder in degree, and experienced as a chronic exposure, which could induce adaptive or nonadaptive responses, and variably associated with other stresses.

Epidemiological Studies Examining the Association Between Insufficient Sleep and Hypertension

The influence of chronic reductions in sleep duration has been addressed in several large epidemiological studies conducted in the U.S., Europe, and Asia. As shown below, the overall consistency of findings from these studies provides evidence for generalizability of results across geographic areas and sociodemographic environments and cultures.

In the Sleep Heart Health Study (SHHS), a large community-based study of predominantly older adults, the group with the lowest prevalence of hypertension was comprised of individuals who reported sleeping 7–8 h per night [7]. Individuals with both shorter and longer sleep had an increased odds of hypertension. After adjusting for numerous potential confounders, the odds of hypertension increased by approximately 70% and 20% for those reporting sleep durations of less than 6 h and between 6 and 7 h per night, respectively. Those reporting more than 9 h of sleep per night had an approximately 30% increased odds of hypertension. A decreased odds of hypertension also was observed in association with improved sleep efficiency measured by polysomnography. However, considering sleep efficiency in the statistical models did not alter the observed associations between sleep duration and hypertension, suggesting that sleep duration and poor sleep quality may independently influence pathophysiological processes associated with blood pressure control.

The Whitehall II Cohort study reported both cross-sectional and longitudinal data on a large U.K. longitudinal cohort of civil servants followed for up to 10 years [8]. Cross-sectional analyses adjusted for multiple potential confounders showed that compared to women reporting sleep durations of 7 h, women reporting a sleeping duration of less than 5 h per night had an approximately twofold increased likelihood of being hypertensive. In contrast, an insignificant association between sleep duration and hypertension was observed in male cohort members. In longitudinal analyses of this cohort, an increased incidence of hypertension in women sleeping less than 5 h per night was observed in analyses adjusted for age and employment status, but this association was attenuated and not significant after adjusting for multiple additional factors, including BMI, health status, and depression. It is possible that some of these factors were not confounders but were mediators, representing intermediate mechanism on the causal pathway, and thus the effects reported may have been under-estimated. A basis for gender differences was not evident.

Stranges and colleagues attempted to further address potential gender differences in the association between sleep duration and hypertension in a cohort of over 3,000 men and women in Western New York [9]. Similar to the Whitehall cohort, a significant cross-sectional association was observed: while the odds of hypertension was approximately 70% elevated in women reporting <6 h per night as compared to those reporting 6–8 h per night, a significant association between short sleep and hypertension was not observed in men. Furthermore, associations were almost twofold stronger in premenopausal compared to postmenopausal women, suggesting that potential vulnerability to the effects of sleep deprivation on blood pressure control may be mediated by sex hormones.

In addition to gender, there may be population heterogeneity in the effects of sleep curtailment on blood pressure associated with age. Potential modification of the effects of sleep deprivation on blood pressure by age was demonstrated in the First National Health and Nutrition Examination Survey (NHANES-I) [10]. In this analysis of nearly 5,000 adults followed for 8–10 years, a significant increased incidence of hypertension was observed in individuals 32–59 years of age reporting 5 or fewer hours of sleep per night compared to those reporting 7–8 h of sleep per night. Insignificant associations were observed among those more than 60 years of age, possibly due to survival biases or to differences in the effects or responses to sleep curtailment in older as compared to younger individuals. These data also underscore the importance in quantifying thresholds for optimal sleep duration across the age span.

Most of the research in adults has been limited by the use of questionnaire-based measurements of sleep duration. One important exception is the Coronary Artery Risk Development in Young Adults (CARDIA) study, a bi-racial cohort of younger adults. Over a 5-year period, 14% of the cohort developed new hypertension. A CARDIA sleep substudy was conducted in 578 cohort members that entailed collection of wrist actigraphy on study participants over several days to objectively estimate sleep duration [11]. Strikingly, with this assessment, 43% of subjects were estimated to obtain an average of less than 6 h of sleep per night; i.e., -underscoring a potentially large segment of the population with moderate to severe levels of insufficient sleep. From age- and gender-adjusted statistical models, each decrease in average sleep duration by 1 h was estimated to increase the odds of hypertension incidence by 37%. These associations were somewhat attenuated after adjusting for physical activity, smoking, BMI, daytime sleepiness, and other factors, although significant associations with sleep duration persisted for longitudinal change of diastolic blood pressure. Low sleep maintenance also was associated with incident hypertension. In contrast to the Whitehall and New York State studies, no evidence for gender differences was demonstrated. Since these studies differed in how sleep duration was ascertained (by questionnaire vs. by actigraphy), it is possible that the stronger associations reported for women in studies that assessed exposure by questionnaire relate to less misclassification of sleep patterns using self-report data in women compared to men. The CARDIA study did not identify an association between long sleep and hypertension, perhaps reflecting the relatively young population studied compared to prior studies that included larger numbers of older individuals.

Since poor health and low socioeconomic status are correlates of extreme sleep durations in adults, all studies of adult populations could be biased by residual confounding. For example, in epidemiological studies that have addressed sleep duration as a risk factor, physical activity has been most often assessed by self-report rather than with objective measurements, and general health and psychiatric history have usually been ascertained by only a few questions. To partly address concerns regarding residual confounding as well as exposure misclassification, Javaheri and colleagues examined the association between sleep duration and sleep efficiency measured objectively using wrist actigraphy in a sample of adolescents recruited from the community who were largely free of chronic illness (Cleveland Children's Sleep and Health Study) [12]. Approximately 10% of the adolescents in this study had markedly short sleep, as defined by an average sleep duration of less than 6.5 h per night. These adolescents had an approximately 2.5-fold increased odds of pre-hypertension or hypertension compared to adolescents sleeping longer. Even stronger associations were seen for low sleep efficiency (<85%) and elevated blood pressure.

The CARDIA and Cleveland Children's Sleep and Health Study both identified a potential role of sleep fragmentation as a hypertension risk factor. In the large GAZEL study, a cohort of French workers, an increased incidence of hypertension was observed in both men and women, who reported hypnotic medication use [13], which may be a marker of poor quality sleep or, itself, may adversely affect blood pressure control. In this study, in women, three or more self-reported sleep disturbances were associations with an approximately 50% increased incidence of hypertension, suggested that poor or fragmented sleep may adversely affect blood pressure control.

Epidemiological Associations Between Insufficient Sleep and Cardiovascular Disease

A meta-analysis examining the association between sleep duration and cardiovascular disease has been conducted of 15 studies, which together include 474,684 individuals [14]. Based on this analysis, short sleep duration was estimated to increase risk of dying from or developing coronary heart disease and stroke by almost 50% and 15%, respectively. Long sleep duration also was associated with increased risk of these outcomes as well as was associated with an increase in total cardiovascular disease mortality. A summary of the effects for each study that reported on coronary artery disease and stroke are shown in Figs. 12.1 and 12.2, respectively. Although the reference group in these studies was defined as those reporting sleeping 7–8 h per

a

First author	Year	Country	Sample size	Total CHD events	Follow-up (years)	Relative risk (95% CI)
Qureshi	1997	USA	7844	413	10.0	1.30 (0.97–1.74)
Mallon (men)	2002	Sweden	906	71	12.0	0.70 (0.29–1.65)
Mallon (women)	2002	Sweden	964	20	12.0	1.20 (0.37–3.88)
Ayas	2003	USA	71617	934	10.0	1.39 (1.06–1.83)
Amagai (men)	2004	Japan	4419	26	8.2	6.17 (1.29–29.6)
Meisinger (men)	2007	Germany	3508	295	10.1	1.13 (0.66–1.91)
Meisinger (women)	2007	Germany	3388	85	10.1	2.97 (1.47–6.02)
Shankar (men)	2008	Singapore	25552	846	7 to 13	1.70 (1.34–2.15)
Shankar (women)	2008	Singapore	32492	570	7 to 13	1.43 (1.09–1.89)
Ikehara (men)	2009	Japan	41489	508	14.3	0.29 (0.04–2.05)
Ikehara (women)	2009	Japan	57145	373	14.3	2.32 (1.19–4.51)
Combined effect: $P<0.0001$			249324	4141		1.48 (1.22–1.80)

Heterogeneity: $I^2=44\%$; $P=0.059$
Egger's test: $P=0.95$

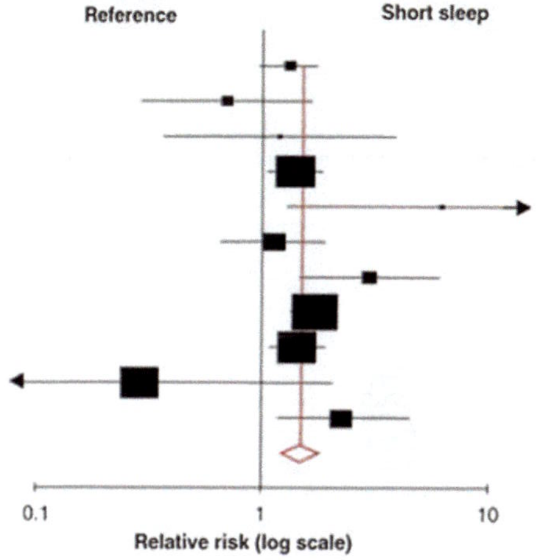

b

First author	Year	Country	Sample size	Total CHD events	Follow-up (years)	Relative risk (95% CI)
Qureshi	1997	USA	7844	413	10.0	1.11 (0.81–1.51)
Mallon (men)	2002	Sweden	906	71	25.0	2.20 (1.05 to 4.64)
Mallon (women)	2002	Sweden	964	20	25.0	0.70 (0.101–5.05)
Ayas	2003	USA	71617	934	10.0	1.36 (1.021–.83)
Amagai (men)	2004	Japan	4419	26	8.2	1.30 (0.381–4.37)
Amagai (women)	2004	Japan	6906	28	8.2	15.96 (2.00 1–127.4)
Meisinger (men)	2007	Germany	3508	295	10.1	1.07 (0.75 1–1.53)
Meisinger (women)	2007	Germany	3388	85	10.1	1.40 (0.751–2.63)
Shankar (men)	2008	Singapore	25552	846	7 to 13	1.88 (1.481–2.38)
Shankar (women)	2008	Singapore	32492	570	7 to 13	1.67 (1.241–2.23)
Ikehara (men)	2009	Japan	41489	508	14.3	1.12 (0.771–1.62)
Ikehara (women)	2009	Japan	57145	373	14.3	1.04 (0.631–1.73)
Combined effect: $P=0.0005$			256230	4169		1.38 (1.15–1.66)

Heterogeneity: $I^2=49\%$; $P=0.028$
Egger's test: $P=0.92$

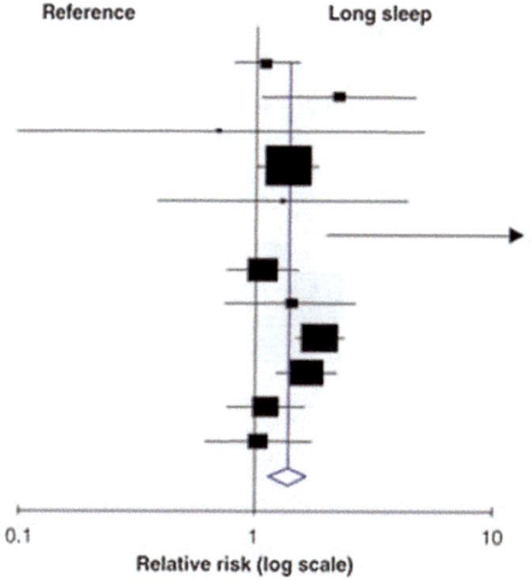

Fig. 12.1 Forest plot of risk showing relative risk (and 95% confidence intervals) for developing or dying from coronary artery disease for short sleep (**a**) and long sleep (**b**), relative to the reference group. Reprinted with permission from [15]

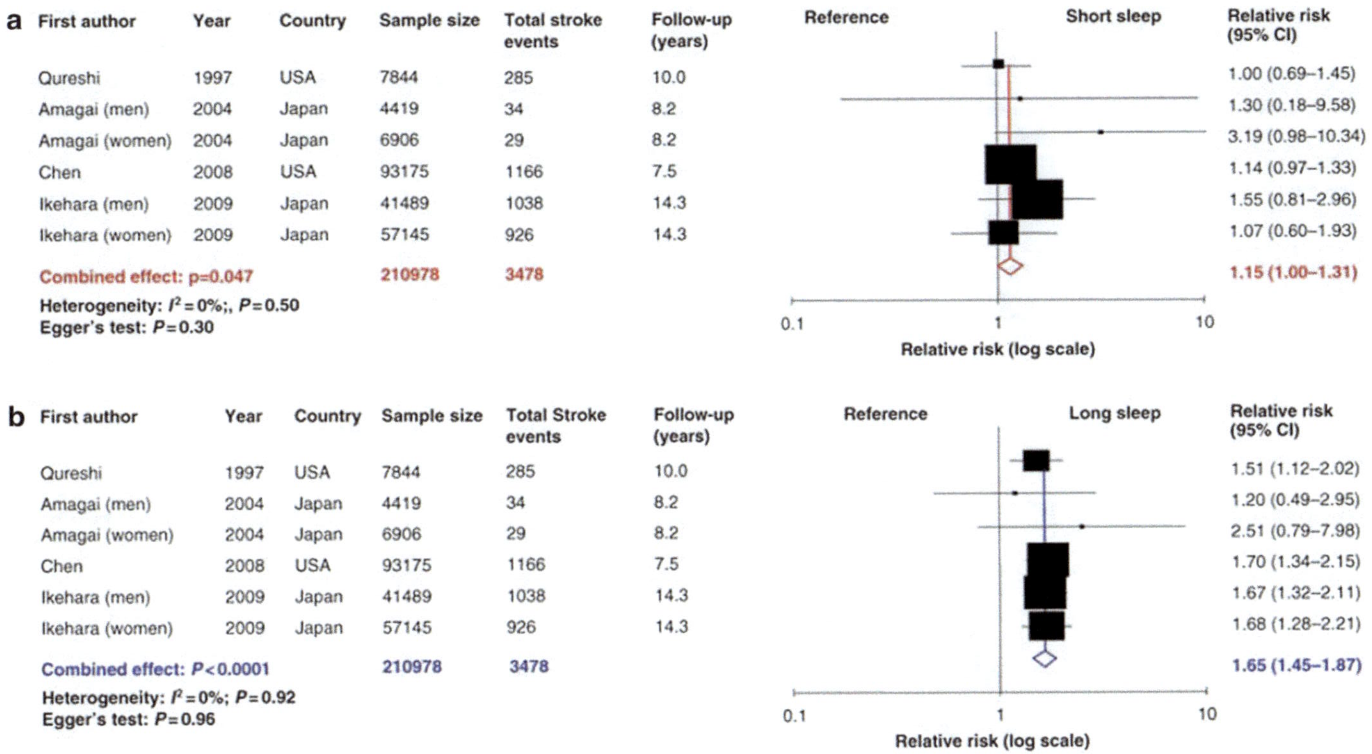

Fig. 12.2 Forest plot of risk showing relative risk (and 95% confidence intervals) for developing or dying from stroke for short sleep (**a**) and long sleep (**b**), relative to the reference group. Reprinted with permission from [15]

night, the thresholds for defining short and long sleep varied across studies. Some heterogeneity across studies was noted. Overall, there was no significant evidence for differences by gender. Some of the key studies are further discussed below.

The U.S. National Health Interview Survey assessed risk factors for cardiovascular disease (myocardial infarction, angina, and stroke) in over 31,000 individuals more than18 years of age [16]. Statistical models included adjustment for multiple covariates, including self-reported depression, diabetes, hypertension, weight, and physical activity. Potential differences across age, gender, and obesity subgroups were examined. Both short and long sleep durations were associated with increases in cardiovascular disease. Consistent associations were observed for men and women and obese and nonobese individuals. In contrast, stronger associations for short sleep and cardiovascular disease were observed for those less than 60 years of age compared to older individuals. Compared to individuals less than age 60 years reporting 7 h of sleep per 24-h period, those reporting less than 5 h of sleep had an almost threefold increased odds of cardiovascular disease. The magnitude of association was approximately 50% less for older individuals. Sleeping longer than 9 h per night was associated with an approximately 80% increased odds of cardiovascular disease, with effects similar in younger and older individuals.

The Nurses Health Study provided longitudinal data on over 71,000 women followed for more than 10 years. After adjusting for multiple covariates, both self-reported short sleep (<5 h) and long sleep (>9 h) were each associated with an almost 40% increased risk of developing coronary artery events [17].

Incident stroke and insufficient sleep were examined in 93,175 women participating in the Women's Health Initiative [18]. At their baseline exam when the women were aged 50–79 years, 8.3% reported sleeping less than 5 h per night and 4.6% reported sleeping more than 9 h per night. Short sleep was more common among nonwhite women and associated with low socioeconomic status, physical inactivity, obesity, and increased health problems. Over an average follow-up period of 7.5 years, incidence of ischemic stroke was observed to be lowest among those getting 7 h of sleep per night. In adjusted analyses, only a modest association was observed for short sleep (<6 h per night; relative risk increased by approximately 10–20%), while long sleep (≥9 h) was associated with an approximately 70% increased rate of stroke. An increased risk of stroke in association with long sleep (defined here as greater than 8 h per night) but not short sleep was demonstrated in the NHANES-1 dataset [19]. Sleepiness occurring with long sleep was associated with an even higher (90%) increased risk of stroke, suggesting that long sleep may be a marker for poor quality sleep. In contrast in a study of hypertensive patients undergoing brain MRI studies, those individuals reporting less than 7.5 h of sleep per night were more than twofold and more likely to have MRI evidence for ischemic vascular disease. Thus, it is not clear whether stroke risk is more strongly related to long or short sleep.

Associations Between Insufficient Sleep and Cardiovascular Mortality

The Japanese Collaborative Cohort Study (JAAC) reported mortality patterns in relationship to self-reported sleep duration for almost 100,000 adults, aged 40–79 years followed for a median of 14.3 years) [20]. Although short and long sleep durations were both associated with increased age and depression, shorter sleep was associated with increased mental stress while longer sleep was associated with lower education level. The lowest mortality rates were observed for those individuals reporting 7 h of sleep per night. Compared to women in this reference group, women reporting less than 4 h of sleep per night had an approximately 2.3-fold increased coronary heart disease mortality rate, and women getting less than 5 h of sleep had an approximate 60% increased coronary heart disease rate. A similar relationship between short sleep and mortality due to coronary heart disease was not observed in men. However, noncardiovascular mortality and mortality due to stroke was increased in both men and women who reported more than 10 h of sleep per night.

Increased mortality rates including all-cause, cardiovascular, and noncardiovascular mortality have been demonstrated in the Whitehall II study for individuals with sleep durations at either end of the continuum, a pattern referred to as a "J" or "U" shape [8, 21]. In this study, the association between mortality and change in sleep duration over time also was evaluated. Compared to the reference group where no change in sleep duration was observed, a decrease in sleep duration from 7 or 8 h over time was associated with an approximately twofold increase in all-cause and noncardiovascular mortality rate while an increase in sleep duration over time was associated with an increase in all-cause and noncardiovascular disease mortality. A U-shaped total mortality distribution in association with sleep duration was also demonstrated in two large Finnish cohort studies in both men and women [22]. However, increased cardiovascular mortality was only observed in women, and was strongest for the most extreme ends of the sleep duration distributions (<5 h and >10 h).

Intermediate Markers of Cardiovascular Disease

In addition to "hard" cardiovascular endpoints, several studies have examined the association of subclinical measures of cardiovascular disease and insufficient sleep. C-Reactive protein, interleukin-6, and TNF-alpha, and proteins associated with increased cardiovascular morbidity have been shown to increase after experimental sleep restriction [2, 23, 24]. The extent to which chronically insufficient sleep influences the levels of inflammatory markers is less clear, but there are some data suggesting that inflammatory and prothrombotic markers may be elevated in conditions of chronically short or long sleep duration [23, 25]. Given that the pathogenesis of atherosclerosis includes inflammation, it is possible that a portion of the aforementioned associations with cardiovascular disease may be mediated through an effect of sleep duration on systemic inflammation.

Coronary artery calcium (CAC) visualized on imaging studies is considered a subclinical marker associated with increased risk for myocardial infarction. In the CARDIA study, the 5-year incidence rate of CAC was shown to be inversely linearly associated with sleep duration estimated by wrist actigraphy [26]. In this study, each additional hour of sleep was associated with an approximately 70% decreased incidence of CAC. Subanalyses suggested that this association was stronger in women compared to men.

Intima-media thickness assessed by carotid ultrasound, another measure of subclinical atherosclerosis, has been assessed in association with self-reported sleep in a German sample, aged 20–79 years [27]. Increased intima thickness was demonstrated in groups reporting both less than 5 h of sleep per night and more than 11 h of sleep per night [27]. Those with extreme sleep durations also have higher levels of hemoglobin A1C, blood pressure, and diabetes.

Thus, associations of sleep duration with biomarkers associated with cardiovascular disease as well as with subclinical markers of atherosclerosis further support the epidemiological data showing an increased in clinical cardiovascular disease and cardiovascular mortality in association with short and/or long sleep duration.

Association of Symptoms of Poor Sleep/Insomnia with Hypertension and Cardiovascular Disease

Insomnia symptoms, particularly reported difficulties in initiating sleep (DIS) and difficulties in maintaining sleep (DMS), have been associated with both short and long sleep durations [28]. Since insomnia is associated with heightened states of arousal and is often accompanied by anxiety or depression, which also are associated with numerous adverse health outcomes, it is plausible that a portion of the relationships observed between sleep duration and cardiovascular disease may reflect insomnia or residual confounding due to mood disorders.

One systemic review estimated that symptoms suggestive of insomnia were associated with relative risks for cardiovascular disease varying from 1.5 to 3.9 [29]. Symptoms of problems initiating sleep and maintaining sleep have also been associated with increased cardiovascular mortality [30, 31]. One study showed that this relationship was only significant for women [30]. It is unclear whether hypnotic use may have confounded the latter associations. It has noted that most published studies of insomnia and cardiovascular disease have not accounted for depression, which may increase risk for both conditions.

Only a few studies have examined both insomnia and sleep duration in the same cohort. The MONICA Ausberg study of nearly 7,000 adults followed for a mean of 10 years demonstrated in analyses adjusted for multiple confounders an almost threefold increased incidence of myocardial infarction in women who reported sleeping ≤5 h compared to those sleeping 8 h per night [28]. In women, the symptom of DMS was associated with incident myocardial infarction; however, this association was attenuated in models that adjusted for other health attributes and behaviors. DIS was unassociated with incident myocardial infarction in men and women. Analyses stratified by insomnia symptoms, although underpowered, showed similar associations, suggesting an independent effect of sleep duration.

In a Pennsylvania community cohort, Vgontzas and colleagues performed a rigorous set of analyses attempting to further dissect the influences of sleep duration from insomnia on hypertension [32]. In statistical models adjusted for multiple potential confounders, a 1-year history of "insomnia" but not symptoms of DMS and DIS, was associated with hypertension. Short sleep duration also was associated with a higher prevalence of hypertension. The highest risk of hypertension, however, was observed in a group who had polysomnography-determined sleep duration of less than 5 h and reported a history of insomnia for one or more years in duration. In subjects sleeping more than 6 h, symptoms of poor sleep/insomnia were not associated with hypertension. Similarly, short sleep in the absence of sleep complaints was not associated with hypertension. Since these analyses used polysomnographically defined sleep from a single night of monitoring, the results cannot be directly compared to the other epidemiological studies reviewed earlier. However, this study suggests that a combination of short sleep and subjective sleep complaints may be associated with the highest risk of hypertension.

Potential Impact of Confounding and Effect Moderation

Numerous epidemiological studies have identified that extremes of sleep duration are associated with other indices of poor health. Invariably, these include low socioeconomic class, obesity, minority race, poorer mental health (including depressive symptoms) and poorer overall general health. Tobacco and alcohol use have also been associated with extreme sleep duration. Although a number of the large epidemiological studies have adjusted for many of these potential confounders, often the measurements used have been based on self-report and could have resulted in some degree of exposure misclassification. Risk factors that have not been rigorously assessed in most studies that have examined sleep duration are physical activity, sedentary activity, and dietary composition—all of which strongly influence cardiovascular disease incidence. It is quite likely that the associations among these factors and sleep duration and quality are complex, and potentially bi-directional. Further, defining these inter-relationships may be critical for creating public policies that provide comprehensive health recommendations. Additionally, better measurements and modeling of psychological and environmental stresses may help to further understand whether some of the differences in study results are due to confounding or mediation by these factors. Finally, it is possible that sleep deficiency preferentially affects certain subgroups. For example, some studies indicate stronger effects in women than men, and in younger compared to older individuals. Whether background hormonal levels, genetic factors or other background risk factors protect or increase susceptibility to insufficient is not understood.

Conclusion

In aggregate, studies conducted across multiple settings in diverse populations indicate that extremes of sleep duration are associated with numerous cardiovascular risk factors, including hypertension. However, even after considering these risk factors, the predominance of data indicates that extreme sleep duration is also associated with clinical cardiovascular disease and cardiovascular mortality. The overall "dose–response" for these associations is not clear. While some studies suggest monotonic associations, others suggest a J- or U-shaped distribution. Also, while the most extreme durations (e.g., less than 5 or more than 10 h of sleep per night) have been fairly consistently associated with adverse cardiovascular outcomes, the magnitude of effect of less extreme deviations from 7 to 8 h per night is less apparent. The sources of population heterogeneity in susceptibility to the effects of sleep deprivation also are not well understood. While some data indicate stronger effects in women compared to men, it is plausible that such differences could relate to measurement or greater misclassification of exposures in men. A further understanding of the inter-relationship of sleep patterns with other important health behaviors (activity patterns and diet) and more comprehensive assessments of sleep in terms of duration, quality, timing, and symptoms may help further clarify which aspects of sleep are most important in the pathogenesis of cardiovascular disease.

References

1. Mullington JM, Haack M, Toth M, Serrador JM, Meier-Ewert HK. Cardiovascular, inflammatory, and metabolic consequences of sleep deprivation. Prog Cardiovasc Dis. 2009;51:294–302.
2. Meier-Ewert HK, Ridker PM, Rifai N, et al. Effect of sleep loss on C-reactive protein, an inflammatory marker of cardiovascular risk. J Am Coll Cardiol. 2004;43:678–83.
3. Kato M, Phillips BG, Sigurdsson G, Narkiewicz K, Pesek CA, Somers VK. Effects of sleep deprivation on neural circulatory control. Hypertension. 2000;35: 1173–5.
4. Lusardi P, Mugellini A, Preti P, Zoppi A, Derosa G, Fogari R. Effects of a restricted sleep regimen on ambulatory blood pressure monitoring in normotensive subjects. Am J Hypertens. 1996;9:503–5.
5. Lusardi P, Zoppi A, Preti P, Pesce RM, Piazza E, Fogari R. Effects of insufficient sleep on blood pressure in hypertensive patients: a 24-h study. Am J Hypertens. 1999;12:63–8.
6. Tochikubo O, Ikeda A, Miyajima E, Ishii M. Effects of insufficient sleep on blood pressure monitored by a new multibiomedical recorder. Hypertension. 1996;27:1318–24.
7. Gottlieb DJ, Redline S, Nieto FJ, et al. Association of usual sleep duration with hypertension: the sleep heart health study. Sleep. 2006;29:1009–14.
8. Cappuccio FP, Stranges S, Kandala NB, et al. Gender-specific associations of short sleep duration with prevalent and incident hypertension: the Whitehall II Study. Hypertension. 2007;50:693–700.
9. Stranges S, Dorn JM, Cappuccio FP, et al. A population-based study of reduced sleep duration and hypertension: the strongest association may be in premenopausal women. J Hypertens. 2010;28: 896–902.
10. Gangwisch JE, Heymsfield SB, Boden-Albala B, et al. Short sleep duration as a risk factor for hypertension. Analyses of the first national health and nutrition examination survey. Hypertension. 2006;47:1–7.
11. Knutson KL, Van Cauter E, Rathouz PJ, et al. Association between sleep and blood pressure in midlife: the CARDIA sleep study. Arch Intern Med. 2009;169:1055–61.
12. Javaheri S, Storfer-Isser A, Rosen CL, Redline S. Sleep quality and elevated blood pressure in adolescents. Circulation. 2008;118:1034–40.
13. Rod NH, Vahtera J, Westerlund H, et al. Sleep disturbances and cause-specific mortality: results from the GAZEL cohort study. Am J Epidemiol. 2011; 173:300–9.
14. Cappuccio FP, Taggart FM, Kandala NB, et al. Meta-analysis of short sleep duration and obesity in children and adults. Sleep. 2008;31:619–26.
15. Cappuccio FP, Cooper D, D'Elia L, Strazzullo P, Miller MA. Sleep duration predicts cardiovascular outcomes: a systematic review and meta-analysis of prospective studies. Eur Heart J. 2011;32(12): 1484–92.
16. Sabanayagam C, Shankar A. Sleep duration and cardiovascular disease: results from the National Health Interview Survey. Sleep. 2010;33:1037–42.
17. Ayas NT, White DP, Manson JE, et al. A prospective study of sleep duration and coronary heart disease in women. Arch Intern Med. 2003;163:205–9.
18. Chen JC, Brunner RL, Ren H, et al. Sleep duration and risk of ischemic stroke in postmenopausal women. Stroke. 2008;39:3185–92.

19. Qureshi AI, Giles WH, Croft JB, Bliwise DL. Habitual sleep patterns and risk for stroke and coronary heart disease: a 10-year follow-up from NHANES I. Neurology. 1997;48:904–11.
20. Ikehara S, Iso H, Date C, et al. Association of sleep duration with mortality from cardiovascular disease and other causes for Japanese men and women: the JACC study. Sleep. 2009;32:295–301.
21. Ferrie JE, Shipley MJ, Cappuccio FP, et al. A prospective study of change in sleep duration: associations with mortality in the Whitehall II cohort. Sleep. 2007;30:1659–66.
22. Kronholm E, Laatikainen T, Peltonen M, Sippola R, Partonen T. Self-reported sleep duration, all-cause mortality, cardiovascular mortality and morbidity in Finland. Sleep Med. 2011;12:215–21.
23. Miller MA, Kandala NB, Kumari M, Marmot MG, Cappuccio FP. Relationships between sleep duration and von Willebrand factor, factor VII, and fibrinogen: Whitehall II study. Arterioscler Thromb Vasc Biol. 2010;30:2032–8.
24. Shearer WT, Reuben JM, Mullington JM, et al. Soluble TNF-alpha receptor 1 and IL-6 plasma levels in humans subjected to the sleep deprivation model of spaceflight. J Allergy Clin Immunol. 2001;107:165–70.
25. Patel SR, Zhu X, Storfer-Isser A, et al. Sleep duration and biomarkers of inflammation. Sleep. 2009;32:200–4.
26. King CR, Knutson KL, Rathouz PJ, Sidney S, Liu K, Lauderdale DS. Short sleep duration and incident coronary artery calcification. JAMA. 2008;300:2859–66.
27. Wolff B, Volzke H, Schwahn C, Robinson D, Kessler C, John U. Relation of self-reported sleep duration with carotid intima-media thickness in a general population sample. Atherosclerosis. 2008;196:727–32.
28. Meisinger C, Heier M, Lowel H, Schneider A, Doring A. Sleep duration and sleep complaints and risk of myocardial infarction in middle-aged men and women from the general population: the MONICA/KORA Augsburg cohort study. Sleep. 2007;30:1121–7.
29. Speidelhalder K, Scholtes C, Riemann D. The association between insomnia and cardiovascular diseases. Nat Sci Sleep. 2010;2:71–8.
30. Mallon L, Broman JE, Hetta J. High incidence of diabetes in men with sleep complaints or short sleep duration: a 12-year follow-up study of a middle-aged population. Diabetes Care. 2005;28:2762–7.
31. Nilsson PM, Nilsson JA, Hedblad B, Berglund G. Sleep disturbance in association with elevated pulse rate for prediction of mortality–consequences of mental strain? J Intern Med. 2001;250:521–9.
32. Vgontzas AN, Liao D, Bixler EO, Chrousos GP, Vela-Bueno A. Insomnia with objective short sleep duration is associated with a high risk for hypertension. Sleep. 2009;32:491–7.

Clinical Guidelines for the Evaluation of Adults with Obstructive Sleep Apnea

13

Lawrence J. Epstein

Abstract

Obstructive sleep apnea (OSA) is a common disorder with significant health and performance consequences. Effective therapies are available but require long-term management because of the chronic nature of the disorder. The low rates of treatment are due primarily to under-recognition of the disorder. Given the high prevalence and significant consequences of lack of treatment, an evaluation for OSA should become part of any general routine health maintenance evaluation. Herein, we discuss the evaluation of the patient with OSA.

Introduction

Obstructive sleep apnea (OSA) is a common disorder that produces daytime impairment and deleterious health consequences [1]. OSA is characterized by repetitive collapse of the upper airway, leading to drops in blood oxygen levels, repetitive arousals from sleep, and elevations in blood pressure [2]. The prevalence rate for OSA with symptoms in the 30- to 60-year-old general population has been estimated at 2–4%, with higher rates in high risk population such as diabetics, obese individuals, those with congestive heart failure, and certain population demographics [3]. However, the rate of abnormal breathing during sleep, defined as more than five obstructive episodes per hour of sleep, is much higher with a prevalence of 9% for women and 24% for men [3]. Even these values may underestimate the true prevalence as they were derived from studies performed at the start of the current obesity epidemic. Obesity rates have risen significantly over the last 20 years and, currently, over 30% of Americans are obese [4]. Since obesity is one of the major risk factors for the development of OSA, rates of OSA are likely higher as well.

The diagnosis of OSA is made by demonstrating obstructive events during sleep in individuals with symptoms and risk factors suggestive of OSA. Despite the high prevalence of the disorder, known risk factors and distinctive features of the disorder, OSA is still under-recognized and under-diagnosed. In one study of almost 5,000 adults in the general population, 93% of women and 82% of men with moderate-to-severe OSA had not been clinically diagnosed [5]. Reasons for this include a lack of awareness in the

L.J. Epstein, MD (✉)
Division of Sleep Medicine,
Brigham and Women's Hospital,
Harvard Medical School, Brighton, MA, USA
e-mail: lawrence_epstein@sleephealth.com

P.J. Shiromani et al. (eds.), *Sleep Loss and Obesity: Intersecting Epidemics*,
DOI 10.1007/978-1-4614-3492-4_13, © Springer Science+Business Media, LLC 2012

general population as well as lack of training in sleep disorders for physicians and other clinicians. Sleep medicine training is a postgraduate exercise, occurring after medical school for most physicians, with fewer than 5 h devoted to training in sleep or sleep disorders in most medical school curricula [6].

Programs to improve awareness of sleep disorders in the general population and the medical community have been instituted and are having success in raising recognition rates and increasing care programs. Fellowship programs in sleep medicine have been started to train sleep specialists; board certification is required to practice sleep medicine and sleep medicine has been formally recognized as a subspecialty field [7]. As a result, much study has been done on the appropriate use of diagnostic testing and treatments for sleep disorders such as OSA. To help clinicians sort through the large volume of available information, specialty organizations undertake evidence-based reviews and recommend clinical guidelines for evaluation and treatment of disorders. The American Academy of Sleep Medicine, the professional organization of sleep clinicians and sleep researchers, regularly publishes evidence-based Practice Parameters to assist clinicians in treating patients. Recently, a task force of the AASM pulled together all of the Practice Parameters to publish a comprehensive Clinical Guideline for the Evaluation, Management, and Long-term Care of Obstructive Sleep Apnea in Adults [8]. That Clinical Guideline is the basis for this discussion on the evaluation of the patient with OSA.

Diagnosis

The diagnosis of OSA is based on the clinical signs and symptoms determined during a comprehensive sleep evaluation, which should include a sleep-oriented history and physical examination and findings identified by sleep testing. The International Classification of Sleep Disorders defines OSA as the occurrence of symptoms in the presence of at least five obstructive respiratory events (apneas, hypopneas, or respiratory effort-related arousals) per hour of sleep. The presence of 15 or more obstructive respiratory events per hour of sleep in the absence of sleep-related symptoms is also sufficient for the diagnosis of OSA due to the greater association of this severity of obstruction with important consequences such as increased cardiovascular disease risk [9].

History and Physical Examination

A comprehensive sleep history should include an evaluation for symptoms of OSA: snoring, witnessed apneas, gasping/choking episodes, excessive sleepiness not explained by other factors (including assessment of sleepiness severity by the Epworth Sleepiness Scale [10]), total sleep amount, nocturia, morning headaches, sleep fragmentation/sleep maintenance insomnia, and decreased concentration and memory (see Table 13.1) [8]. In addition, there should be an investigation for conditions that either increase the risk of OSA or can be a consequence of OSA, including hypertension, stroke, cardiovascular disease, cor pulmonale, decreased daytime alertness, and motor vehicle crashes [8].

Findings on physical examination are not pathognomonic for OSA but can suggest an increased risk for having the disorder. The examination should include an evaluation of the respiratory, cardiovascular, and neurologic systems [11]. The person should be evaluated for signs of upper airway narrowing or the presence of other disorders that increase the risk of OSA. In particular, features suggestive of OSA include a neck circumference greater than 17" in men and 16" in women, a body mass index ≥30 kg/m^2, a modified Mallampati score of 3 or 4, or the presence of retrognathia, lateral peritonsillar narrowing, macroglossia, tonsillar hypertrophy, elongated/enlarged uvula, high arched high palate, and nasal abnormalities or overjet (see Table 13.1) [8, 12].

Objective Testing

Objective testing should be performed to confirm the clinical findings and determine the severity of

Table 13.1 Components of a comprehensive sleep evaluation for OSA

Symptoms on history	Signs on physical examination
Snoring	Neck circumference >17 inches in men, >16 inches in women
Witnessed apnea	Body mass index (BMI) ≥30 kg/m^2
Gasping/choking episodes	Modified Mallampati score of 3 or 4
Nocturia	Crowded oropharynx (tonsillar hypertrophy, elongated uvula, macroglossia)
Morning headaches	Nasal abnormalities (polyps, deviation, turbinate hypertrophy, valve abnormalities)
Fragmented sleep/sleep maintenance problems	Retro/micrognathia
Decreased concentration and memory	High arched palate
OSA risk factors and consequences	Overjet

OSA. Currently, no clinical model predicts the severity of OSA, and treatment decisions are based on the severity of the disorder [11, 13]. There are two recognized methods for objective testing, in-laboratory testing with a polysomnogram (PSG) and out of center testing (OCST) using portable monitors (PM) [8].

A PSG measures multiple physiologic parameters and allows a comprehensive analysis of sleep, including sleep stages, electroencephalography, respiratory events, snoring, cardiac rhythms, limb movement, and eye movements [14] (see Fig. 13.1). The PSG can be used not only to characterize sleep quality but also to detect the wide variety of sleep disorders. PSG is the reference standard for the diagnosis of OSA [11].

In-laboratory testing is time intensive, utilizes a trained technologist to monitor the patient throughout the sleep period, and requires the patient to sleep at the sleep center rather than at home. In an attempt to reduce cost and improve patient access to testing, OCST with PMs has been extensively studied as an alternative to in-laboratory PSG. The types of PMs available range from full PSG in the home to devices that measure only a few parameters (see Fig. 13.2). OCST with PMs is indicated only for the diagnosis of OSA because, compared to PSG, the devices typically measure fewer parameters, most frequently respiratory parameters (airflow and effort), oxyhemoglobin saturation, and heart rate (see Table 13.2). The current recommendation is that OCST with PMs may be used to diagnose OSA when utilized as part of a comprehensive sleep evaluation in patients with a high pretest likelihood of moderate to severe OSA [15].

Patient Education

The patient should review the results of the evaluation with the sleep specialist, including the sleep study findings. If a diagnosis of OSA is made, the patient should be educated on the pathophysiology of the disorder, disease severity, risk factors, natural history, and clinical consequences of OSA. General education on dealing with the diagnosis of OSA should include information on the impact of weight loss, sleeping position, alcohol avoidance, risk factor modification, and medication effects. The patient should receive specific counseling on the increased risk of drowsy driving due to OSA and how to prevent and manage problems while driving [8]. Specific reporting requirements for the counseling physician are state dependent, so the specialist should learn the appropriate regulations for their locale. Once the patient has an understanding of the impact and severity of their OSA, the patient should be presented with treatment options appropriate for their disease severity and situation. The decision on the type of treatment to use should be a collaboration between the patient, the patient's family, and the sleep specialist.

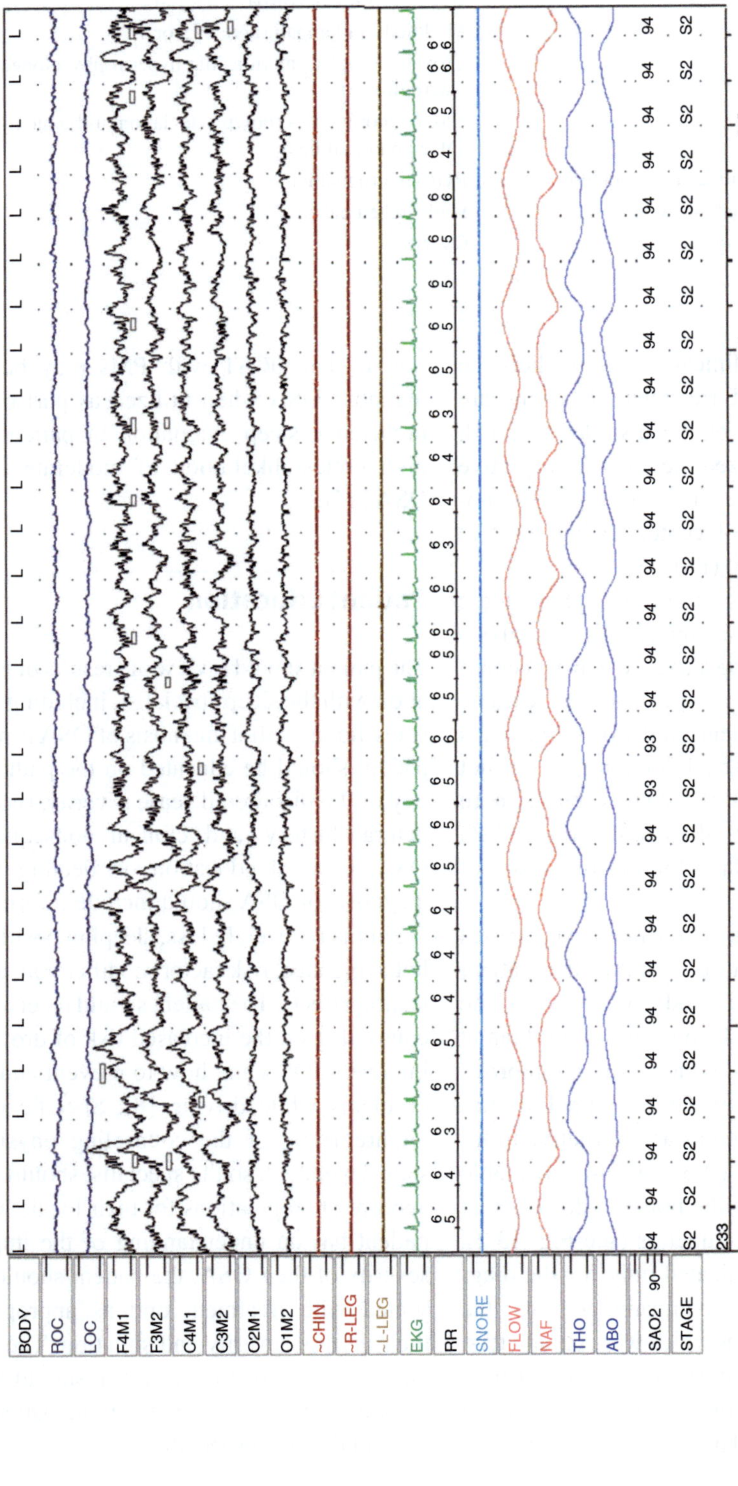

Fig. 13.1 Example of recording from in-laboratory PSG. Shown is a 30-s sample from an overnight diagnostic PSG. This tracing demonstrates normal sleep. BODY=body position, ROC=right oculogram, LOC=left oculogram, F4M1, F3M2, C4M1, C3M2, O2M1, O1M2=electroenephalograpy channels, CHIN=chin electromyogram, R LEG=right leg electromyogram, L LEG=left leg electromyogram, EKG=electrocardiogram, RR=heart rate, NAF=nasal air flow, THO=thoracic effort, ABD=abdominal effort, SAO2=oxyhemoglobin saturation, and STAGE=sleep stage

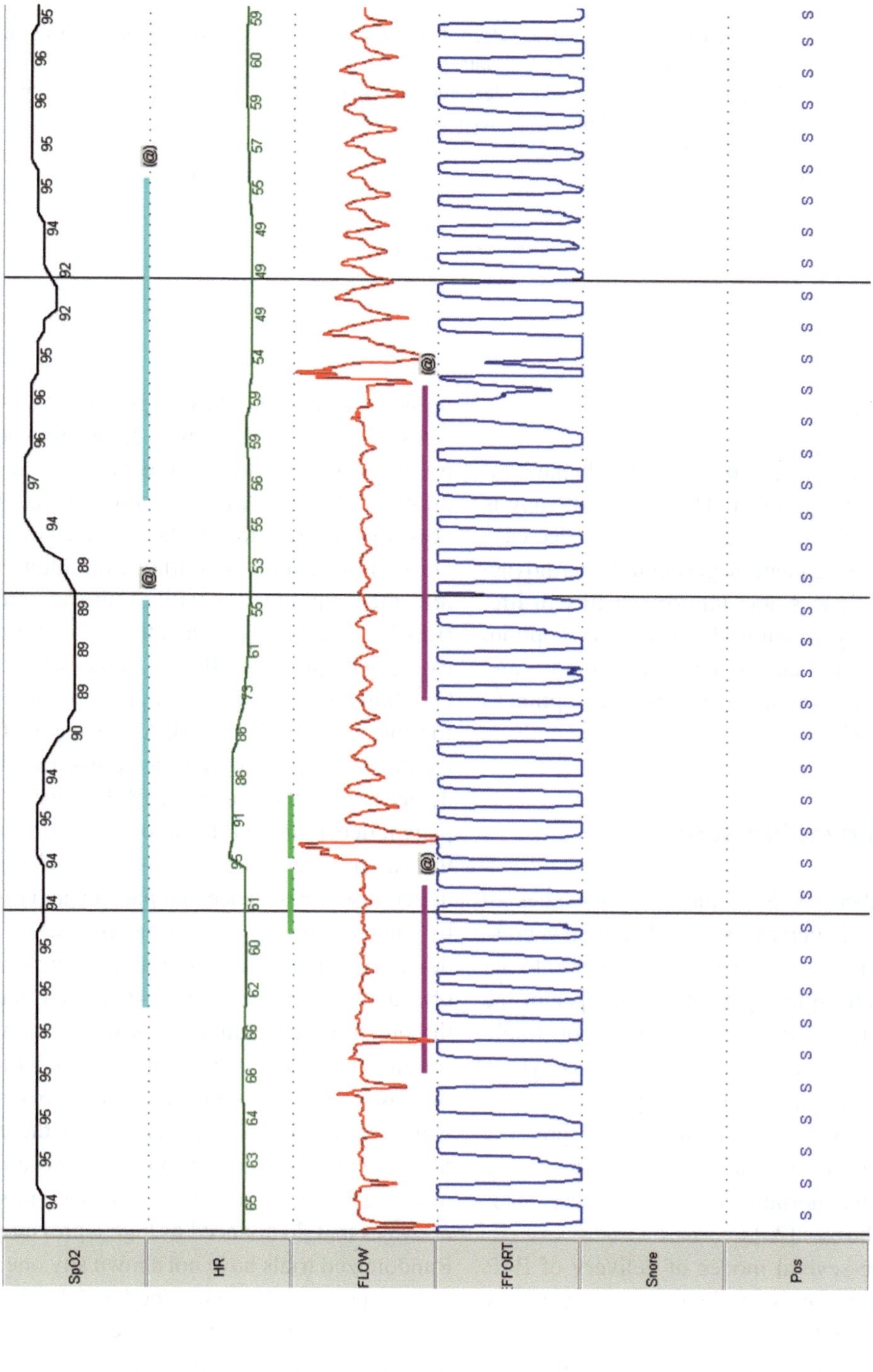

Fig. 13.2 Example of recording from OCST with PM. Shown is a 2-min sample from a 6-channel recorder obtained in the patient's home. The tracing shows recurrent obstructive events (underlined in the FLOW channel) accompanied by oxyhemoglobin desaturation. SpO2 = oxyhemoglobin saturation, HR = heart rate, and POS = body position

Table 13.2 Comparison of in-laboratory polysomnography with out of center sleep testing with portable monitors

	PSG	OCST with PMs
Site of testing	Sleep center	Patient's home
Parameters measured	12–19	1–19 possible, most often 4–7 parameters
Types of analysis performed	Comprehensive sleep analysis; sleep stages, respiratory, movement, electronencephalography	Respiratory analysis only
Sleep disorders evaluated	Sleep fragmentation disorders, sleep-related respiratory disorders, movement disorders, nocturnal seizure disorders, parasomnias	Obstructive sleep apnea
Technologist required during testing	Yes	No
Cost	+++	+

Treatment

There are medical, surgical, and behavioral options for treating OSA. The goal of therapy is to improve symptoms and performance, decrease deleterious consequences, particularly cardiovascular consequences, and improve quality of life. Positive airway pressure (PAP), a medical option, is the treatment of choice for mild, moderate, and severe OSA and should be offered as an option to all patients [8].

Positive Airway Pressure

First described by Sullivan and colleagues in 1981 [16], PAP therapy uses air delivered at pressure through the nose, or nose and mouth, to act as a pneumatic splint, preventing collapse of the upper airway. PAP quickly and demonstrably eliminates airway obstruction, improves daytime sleepiness, and improves quality of life [17]. Treatment of OSA can reduce blood pressure [18], decrease stroke and cardiovascular disease morbidity and mortality [19], and reduce automobile crash rates [20].

There are several modes of delivery of PAP. The most common is continuous positive airway pressure (CPAP), where the device delivers airflow sufficient to maintain a constant pressure throughout both inspiration and expiration. The pressure is set at the lowest pressure that eliminates all events. With bilevel positive airway pressure (BPAP), the pressure varies between the different phases of the respiratory cycle. Higher pressures are required to keep the airway open during inspiration than during expiration, allowing the expiratory positive airway pressure (EPAP) to be set lower than the inspiratory positive airway pressure (IPAP). Some patients find breathing out against a lower pressure more comfortable and will use BPAP but not CPAP. Autotitrating positive airway pressure (APAP) devices adjust the amount of PAP delivered to the patient depending on the amount needed to keep the airway open. The devices detect changes in upper airway resistance and patency and respond by increasing or decreasing pressure. These devices start at a low pressure, increase until obstructive events are eliminated, then maintain the pressure that eliminates obstruction. After a set time the pressure slowly drops in order to find the lowest pressure that maintains a patent airway. The mean airway pressure over the course of the night is lower than on CPAP, because the machine can drop the pressure in sleep positions or stages that do not need as high a pressure [21]. Randomized trials have not shown any one mode to be superior to the other modes [22, 23]; however, individual patients may be able to tolerate one type over another, so treatment should be tailored to the patients' tolerance and preference.

In-laboratory PSG is the preferred approach for determining the optimal CPAP or BPAP level but APAP devices may be used to determine a fixed CPAP level or as the treatment mode in an unattended fashion [17, 24]. Tolerance of PAP therapy may be enhanced by use of heated humidification and a systematic educational program and PAP usage should be objectively monitored with time and use meters to help assure utilization [17].

Adherence to therapy is the major limitation to treatment with PAP. Patients tend to under-report PAP use, so it is important to measure adherence objectively [25]. Long-term usage patterns are established early in the treatment period, within the first 1–3 months, so it is important to maximize the early experience with PAP [26, 27]. Close follow-up for problems, and monitoring PAP adherence, is especially important during the first few weeks of PAP and should be part of the treatment plan [17]. If PAP use is considered inadequate based on objective monitoring and the patient's report of symptom improvement, alternative therapies should be considered and implemented promptly [8].

Oral Appliances

Custom-made oral appliances (OA) may improve upper airway patency during sleep by enlarging the upper airway and/or by improving upper airway muscle tone and decreasing upper airway collapsibility [28]. The most commonly used OAs are mandibular repositioning appliances that cover the upper and lower teeth and hold the mandible in an advanced position. Tongue-retaining devices hold only the tongue in a forward position with respect to the resting position.

OAs are not as effective as PAP therapy but can eliminate obstructive events. They are indicated for the treatment of mild and moderate OSA and in patients who are unable to use PAP therapy or fail other treatment modalities [29]. Patients should undergo a thorough dental examination to assess candidacy for an OA and should be fitted by dental personnel who are trained in management of OSA as well as in the overall care of oral health, the temporomandibular joint, dental occlusion, and associated oral structures [8]. The goal of OA treatment includes the elimination of obstructive events, normalization of oxyhemoglobin saturation levels, and resolution of the clinical signs and symptoms of OSA. To ensure a satisfactory outcome has been achieved with OA therapy, patients should undergo objective sleep testing, either in-laboratory PSG or OCST with PMs, to demonstrate elimination of OSA [29].

Other Medical Therapies

A novel nasal expiratory positive airway pressure device was introduced recently. Small resistance valves are placed over each nostril with adhesive tape. The valves create resistance to expiratory airflow, increasing the pressure in the upper airway. The valves are disposable, small, and easily transportable. An initial study showed an average reduction of 43% in obstructive events, with some subjects having elimination of OSA [30]. The device was more effective in those with mild and moderate OSA than severe OSA. More study is required to understand the appropriate indications and the role for this therapy.

There are currently no effective pharmacotherapies for OSA except those that treat disorders that can contribute to the development of OSA, such as thyroid replacement therapy for hypothyroidism [31]. Oxygen and stimulant medications, such as modafinil, are adjunctive therapies that do not treat OSA directly but may treat some of the consequences of OSA, in particular, hypoxemia and daytime sleepiness [32].

Behavioral Therapies

This group of treatments is characterized by the patient modifying their behavior in a manner that improves OSA and includes weight loss and positional therapy. Weight gain narrows the size of the airway and is a major risk factor for the development of OSA. Weight loss, conversely, leads to an increase in airway diameter and reduction in the number of obstructive events [33]. With sufficient weight loss, OSA may be eliminated in some cases. Weight loss should be recommended for all overweight OSA patients. However, weight loss therapy should be combined with another primary treatment for OSA because of the low success rate of dietary programs and the low cure rate by dietary approach alone [32]. After substantial weight loss (i.e., 10% or more of body weight), a follow-up sleep study, either in-laboratory PSG or OCST with PM, is indicated to ascertain whether PAP therapy is still needed or whether adjustments in PAP level are necessary [8].

Sleep position also affects airway size and patency with a decrease in the area of the upper airway while in the supine position [34]. Positional therapy consists of methods to keep the persons from sleeping on their back. Before utilizing this form of therapy, objective testing, typically with an in-laboratory PSG, must demonstrate the person has a low frequency of obstructive events in the nonsupine position versus that in the supine position [32]. A positioning device (e.g., position alarm, pillow, backpack, tennis ball) should be used when initiating positional therapy. To establish the efficacy of a positioning device in the home, providers should consider use of an objective position monitor [8].

Surgical Therapy

The earliest treatments for OSA, prior to the introduction of CPAP, were surgical. The goal of surgical therapy is the resolution of the clinical signs and symptoms of OSA, elimination of obstructive events and oxyhemoglobin desaturation, and the normalization of sleep quality. The presence and severity of OSA must be determined by objective sleep testing before initiating surgical therapy [35]. Surgical therapy includes a variety of upper airway reconstructive or bypass procedures.

Tracheostomy is an effective single intervention for OSA and works by bypassing the site of upper airway obstruction. Though nearly uniformly successful, this procedure is typically considered only when other options are refused or are ineffective because of significant morbidity and poor patient acceptance [35].

Site-specific therapies are designed to increase the size of the airway and prevent obstruction by targeting the site of collapse. However, current diagnostic methods are not effective at identifying the site of collapse and most patients have multiple sites of collapse [36], making single site surgeries less successful. The most frequently performed OSA surgery is the uvulopalatopharyngoplasty (UPPP), but this procedure does not reliably eliminate OSA [35]. Because of the variable success rate, follow-up objective testing needs to be performed to prove efficacy. The variety of site-specific surgeries is listed in Table 13.3. Maxillo-mandibular advancement surgery has a higher success rate and is indicated for the surgical treatment of severe OSA in patients who cannot tolerate PAP therapy or OA therapy [35]. Surgical therapy is often performed as part of multi-level or stepwise surgery, in which surgical intervention progresses from site-specific to more involved maxillofacial procedures with efficacy checks between steps [37].

Radiofrequency tissue ablation and palatal implants are approved for use in snoring and OSA and are treatments for patients with mild-to-moderate OSA who cannot tolerate other treatment modalities [35]. Laser-assisted uvulopalatoplasty is not recommended as a treatment for OSA. Nasal reconstruction surgeries are used for symptomatic nasal airway blockage but typically do not resolve OSA [38].

A postoperative evaluation should be performed after an appropriate period of healing and should include clinical follow-up for symptom resolution and an objective sleep test to demonstrate elimination of OSA. If OSA has not been resolved, an alternative treatment should be instituted.

Table 13.3 Treatments for obstructive sleep apnea

Treatment category	Therapies
Medical	Continuous positive airway pressure (CPAP)
	Bilevel positive airway pressure (BPAP)
	Autotitrating positive airway pressure (APAP)
	Oral appliances
	Expiratory resistance valves
	Adjunctive therapies (oxygen, modafinil)
Behavioral	Weight loss
	Positional therapy
Surgical	
Upper airway bypass procedure	Tracheostomy
Site specific procedures	Nasal procedures (septoplasty, nasal valve surgery, turbinate reduction, polypectomy)
	Oropharyngeal procedures (UPPP, tonsillectomy +/– adenoidectomy, palatal implants)
	Hypopharyngeal procedures (tongue reduction, genioglossal advancement, hyoid suspension)
Global airway procedures	Maxilomandibular advancement
	Bariatric surgery

Bariatric surgery is considered an adjunctive surgery for the treatment of OSA. Bariatric surgery produces major weight loss and is currently indicated in individuals with a body mass index (BMI) ≥40 kg/m^2 or those with a BMI≥35 kg/m^2 with comorbidities related to obesity [39]. There is a large overlap between this population and those with OSA. The remission rate for OSA 2 years after bariatric surgery, related to the amount of weight lost, is 40%, emphasizing the need for ongoing clinical follow-up of these patients [40].

Long-Term Management

OSA is a chronic disease that requires long-term treatment unless there is modification of the underlying features contributing to the development of the disorder, either weight loss in cases related to obesity or surgical expansion of upper airway size. All patients with OSA should have ongoing, long-term management, education, and support. For those on therapy, they should have regular follow-up to check adherence to treatment, look for side effects, and conduct ongoing risk modification. Those with effective intervention that has eliminated OSA should be monitored for return of symptoms [8].

Conclusion

OSA is a common disorder with significant health and performance consequences. Effective therapies are available but require long-term management because of the chronic nature of the disorder. The low rates of treatment are due primarily to under-recognition of the disorder. Given the high prevalence and significant consequences of lack of treatment, an evaluation for OSA should become part of any general routine health maintenance evaluation (see Fig. 13.3).

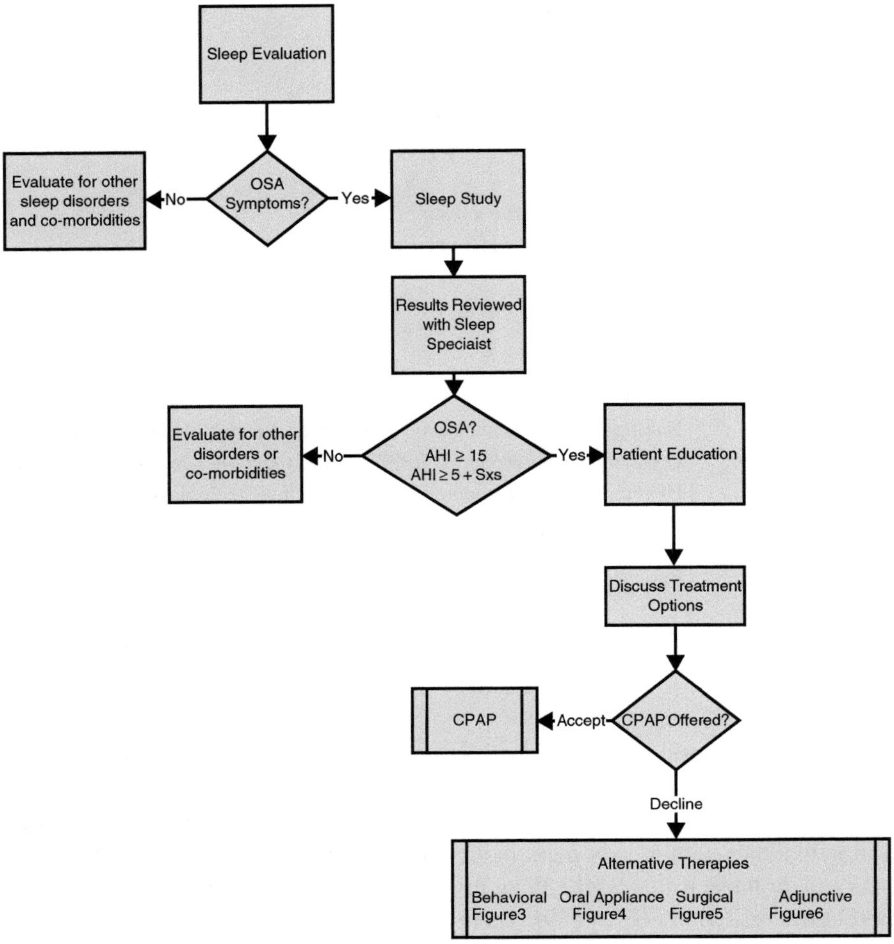

Fig. 13.3 Flow chart of evaluation of patients suspected of having OSA. Adapted with permission from [8]

References

1. Institute of Medicine of the National Academies, Committee on Sleep Medicine and Research. Sleep-disordered breathing. In: Colton HR, Altevogt BM, editors. Sleep disorders and sleep deprivation: an unmet public health problem. Washington, D.C.: The National Academies Press; 2006. p. 78–89
2. Epstein LJ, Weiss JW. Clinical consequences of obstructive sleep apnea. Semin Respir Crit Care Med. 1998;19:123–32.
3. Young T, Palta M, Dempsey J, Skatrud J, Weber S, Badr S. The occurrence of sleep-disordered breathing among middle-aged adults. N Engl J Med. 1993;329:1230–5.
4. Flegal KM, Carroll MD, Ogden CU, Curtin LR. Prevalence and trends in obesity among US adults, 1999–2008. JAMA. 2010;303(3):235–41.
5. Young T, Evans L, Finn L, Palta M. Estimation of the clinically diagnosed proportion of sleep apnea syndrome in middle-aged men and women. Sleep. 1997;20:705–6.
6. Rosen RC, Rosekind M, Rosevear C, Cole WE, Dement WC. Physician education in sleep and sleep disorders: a national survey of U.S. medical Schools. Sleep. 1993;16:249–54.
7. Epstein LJ. Milestone achieved: the specialty of sleep medicine. J Clin Sleep Med. 2005;1:407–9.
8. Epstein LJ, Kristo D, Strollo PJ, Friedman N, Malhotra A, Patil SP, Ramar K, Rogers R, Schwab RJ, Weaver EM, Weinstein MD. Clinical guidelines for the evaluation, management and long-term care of obstructive sleep apnea in adults. J Clin Sleep Med. 2009;5:263–76.
9. American Academy of Sleep Medicine. International classification of sleep disorders, 2nd Edition: Diagnostic and coding manual. Westchester, IL: American Academy of Sleep Medicine; 2005.
10. Johns MW. A new method for measuring daytime sleepiness: the Epworth sleepiness scale. Sleep. 1991;14:540–5.

11. Kushida CA, Littner MR, Morgenthaler T, et al. Practice parameters for the indications for polysomnography and related procedures: an update for 2005. Sleep. 2005;28:499–521.
12. Friedman M, Tanyeri H, La Rosa M, et al. Clinical predictors of obstructive sleep apnea. Laryngoscope. 1999;109:1901–7.
13. Kushida CA, Morgenthaler TI, Littner MR, et al. Practice parameters for the treatment of snoring and Obstructive Sleep Apnea with oral appliances: an update for 2005. Sleep. 2006;29:240–3.
14. Carskadon MA, Rechtschaffen A. Monitoring and staging human sleep. In: Kryger MH, Roth T, Dement W, editors. Principles and practice of sleep medicine. 4th ed. Philadelphia: Elsevier; 2005. p. 1359–77.
15. Collop NA, Anderson WM, Boehlecke B, Claman D, Goldberg R, Gottlieb DJ, Hudgel D, Sateia M, Schwab R. Clinical guidelines for the use of unattended portable monitors in the diagnosis of obstructive sleep apnea in adult patients. Portable Monitoring Task Force of the American Academy of Sleep Medicine. J Clin Sleep Med. 2007;3:737–47.
16. Sullivan CE, Issa FG, Berthon-Jones M, Eves L. Reversal of obstructive sleep apnoea by continuous positive airway pressure applied through the nares. Lancet. 1981;1:862–5.
17. Kushida CA, Littner MR, Hirshkowitz M, et al. Practice parameters for the use of continuous and bilevel positive airway pressure devices to treat adult patients with sleep-related breathing disorders. Sleep. 2006;29:375–80.
18. Haentjens P, Van Meerhaeghe A, Moscariello A, De Weerdt S, Poppe K, Dupont A, Velkeniers B. The impact of continuous positive airway pressure on blood pressure in patients with obstructive sleep apnea syndrome; evidence from a meta-analysis of placebo-controlled randomized trials. Arch Intern Med. 2007;167:757–65.
19. Marin JM, Carrizo SJ, Vicente E, Agusti AGN. Long-term cardiovascular outcomes in men with obstructive sleep apnoea-hypopnea with or without treatment with continuous positive airway pressure: an observational study. Lancet. 2005;365:1046–53.
20. Findley L, Smith C, hooper J, Dineen M, Suratt PM. Treatment with nasal CPAP decreases automobile accidents in patients with sleep apnea. Am J Respir Crit Care Med. 2000;161:857–9.
21. Rahangdale S, Epstein LJ. Positive airway pressure therapy: theory, indications, effectiveness, adverse effects. In: Butkov N, Lee-Chiong T, editors. Fundamentals of sleep technology. Lippincott: Williams & Wilkins. p. 417–29
22. Reeves-Hoché MK, Hudgel DW, Meck R, Witteman R, Ross A, Zwillich CW. Continuous versus bilevel positive airway pressure for obstructive sleep apnea. Am J Respir Crit Care Med. 1995;151:1443–9.
23. Damjanovic D, Fluck A, Bremer H, Müller-Quernheim J, Idzko M, Sorichter S. Compliance in sleep apnoea therapy: influence of home care support and pressure mode. Eur Resp J. 2009;33:804–11.
24. Morgenthaler TI, Aurora RN, Brown T, Zak R, Alessi C, Boehlecke B, Chesson Jr AL, Friedman L, Kapur V, Maganti R, Owens J, Pancer J, Swick TJ. Practice parameters for the use of autotitrating continuous positive airway pressure devices for titrating pressures and treating adult patients with obstructive sleep apnea syndrome: an update for 2007. An American Academy of Sleep Medicine report. Sleep. 2008;31:141–7.
25. Rauscher H, Formanek D, Popp W, Zwick H. Self-reported vs measured compliance with nasal CPAP for obstructive sleep apnea. Chest. 1993;103:1675–80.
26. Popescu G, Latham M, Allgar V, Elliot MW. Continuous positive airway pressure for sleep apnoea/hypopnea syndrome: usefulness of a 2 week trial to identify factors associated with long term use. Thorax. 2001;56:727–33.
27. Mc Ardle N, Devereux G, Heidarnejad H, Engleman HM, Mackay TW, Douglas NJ. Long-term use of CPAP therapy for sleep apnea/hypopnea syndrome. Am J Resp Crit Care Med. 1999;159:1108–14.
28. Ferguson KA, Cartwright R, Rogers R, Schmidt-Nowara W. Oral appliances for snoring and obstructive sleep apnea: a review. Sleep. 2006;29:244–62.
29. Kushida CA, Morgenthaler TI, Littner MR, Alessi CA, Bailey D, Coleman J, Friedman L, Hirshkowitz M, Kapen S, Kramer M, Lee-Chiong T, Owens J, Pancer JP. Practice parameters for the treatment of snoring and Obstructive Sleep Apnea with oral appliances: an update for 2005. Sleep. 2006;29:240–3.
30. Berry RB, Kryger MH, Massie CA. A novel nasal expiratory positive airway pressure (EPAP) device for the treatment of obstructive sleep apnea: a randomized controlled trial. Sleep. 2011;34(4):479–85.
31. Veasey SC, Guilleminault C, Strohl KP, Sanders MH, Ballard RD, Magalang UJ. Medical therapy for obstructive sleep apnea: a review. Sleep. 2006;29:1036–44.
32. Morgenthaler TI, Kapen S, Lee-Chiong T, Alessi C, Boehlecke B, Brown T, Coleman J, Friedman L, Kapur V, Owens J, Pancer J, Swick TJ. Practice parameters for the medical therapy of obstructive sleep apnea. Sleep. 2006;29:1031–5.
33. Welch KC, Foster GD, Ritter CT, Thomas BA, Wadden A, Arens R, Maislin G, Schwab RJ. A novel volumetric magnetic resonance imaging paradigm to study upper airway anatomy. Sleep. 2002;25:530–40.
34. Pevernagie DA, Stanson AW, Sheedy 2nd PF, Daniels BK, Shepard Jr JW. Effects of body position on the upper airway of patients with obstructive sleep apnea. Am J Respir Crit Care Med. 1995;152:179–85.
35. Aurora RN, Casey KR, Auerbach S, Bista SR, Chowdhuri S, Karippot A, Lamm C, Ramar K, Zak R, Morgenthaler TI. Practice parameters for the surgical modifications of the upper airway for obstructive sleep apnea in adults. Sleep. 2010;33:1408–13.
36. Morrison DL, Launois SH, Isono S, Feroah TR, Whitelaw WA, Remmers JE. Pharyngeal narrowing

and closing pressures in patients with obstructive sleep apnea. Am Rev Respir Dis. 1993;148:606–11.
37. Riley RW, Powell NB, Guilleminault C. Obstructive sleep apnea syndrome: a review of 306 consecutively treated surgical patients. Otolaryngol Head Neck Surg. 1993;108:117–25.
38. Powell NB, Riley RW, Guilleminault C. Surgical management of sleep-disordered breathing. In: Kryger M, Roth T, Dement WC, editors. Principles and practice of sleep medicine. 4th ed. Philadelphia: Elsevier; 2005. p. 1085–97.
39. SAGES Guidelines Committee. Guidelines for clinical application of laparoscopic bariatric surgery: Society of American Gastrointestinal and Endoscopic Surgeons; 2008.
40. Mechanick JI, Kushner RF, Sugerman HJ, et al. American Association of Clinical Endocrinologists, The Obesity Society, and American Society for Metabolic & Bariatric Surgery Medical guidelines for clinical practice for the perioperative nutritional, metabolic, and nonsurgical support of the bariatric surgery patient. Endocr Pract. 2008;14 Suppl 1:1–83.

Pharmacological Treatment of Obesity

14

Dhiraj G. Kabra, Uma D. Kabra, Matthias H. Tschöp, and Susanna Hofmann

Abstract

Obesity is a chronic disease resulting in major morbidity and premature death globally. Current pharmacological treatments are not fully acceptable because of their poor safety and efficacy. Obesity is also associated with other serious complications such as diabetes mellitus, hypertension, hyperlipidemia, hypercholesterolemia, cardiovascular disease etc. Because of its complex nature there is a need for safe and efficacious long-term pharmacological treatment. One treatment approach is not sufficient to manage complex pathological circuitry of obesity. Recently, several novel targets have been proposed that control energy homeostasis and prevent obesity. Although, newer drugs are years away from clinical use, the hope for research investments made to date is translation into safe and effective pharmacological treatment in the future. The goal of this chapter is to describe the latest strategies of pharmacological treatment that are under development, which may finally be used in future.

Introduction

Obesity is an epidemic condition affecting the population worldwide. The prevalence of obesity has increased tremendously over the past five decades [1]. In the USA, obesity is increasing not only in adults, but especially among children and adolescents [2]. As many as one-third of all Americans are obese. In 1997, The World Health Organization (WHO) recommended a standard classification of adult overweight and obesity based on BMI criteria: a BMI of 25.0–29.9 kg/m^2 is defined as overweight; a BMI of 30.0 kg/m^2 or more is defined as obesity [1].

Obesity means abnormal accumulation of body fat, usually 20% or more over an individual's ideal body weight. Increasing evidence suggest that obesity is not a simple problem but a complex disorder involving appetite regulation and energy metabolism. It is associated with a variety of comorbid conditions such as hypertension,

D.G. Kabra, PhD • U.D. Kabra, MPharm
M.H. Tschöp, MD (✉) • S. Hofmann, MD
Cincinnati Diabetes and Obesity Centre,
Division of Internal Medicine, University of Cincinnati,
Cincinnati, OH, USA
e-mail: tschoemh@ucmail.uc.edu

P.J. Shiromani et al. (eds.), *Sleep Loss and Obesity: Intersecting Epidemics*,
DOI 10.1007/978-1-4614-3492-4_14,

diabetes mellitus, hypercholesterolemia, sleep apnea, stroke, arthritis, gallstones, and certain types of cancer [3, 4]. The etiology of obesity is still not completely established but genetic, metabolic, biochemical, cultural, and psychosocial factors contribute to obesity [5, 6]. The goal of obesity treatment is to achieve and maintain healthier weight and enhance quality of life. There is no specific strategy or medication recommended for obesity on a routine basis. However, obesity management requires a comprehensive approach involving diet and nutrition, regular physical activity and behavioral changes, with an emphasis on long-term weight management with pharmacological agents [7].

Pharmacological treatment is always seen as an additional tool and should always be used in conjunction with a weight loss program. There are some practical guidelines for the pharmacological treatment of obesity which can be summarized as follows: (1) if lifestyle changes do not promote weight loss equivalent to 10% of initial body weight over 6 month then pharmacotherapy may be considered; (2) Pharmacological drug treatment should be given to people with BMI > 30 if there are no obesity related risk factors; or to people with BMI > 27 if there are risk factors such as hypertension; (3) pharmacological treatment is not recommended for children since there is not enough evidence on the effect at this age group; (4) failure to lose >2 kg after 4–8 weeks of drug treatment should be considered as a treatment failure and then concern should be given to dose management or substituting different medication; (5) the occurrence of plateau in body weight after several months of treatment does not mean that the drug is ineffective; (6) treatment should be maintained only if it is safe and effective for the patient; (7) the drug should be considered effective if it induces and maintains a loss of 5–10% of baseline body weight [8–12].

The high incidence of obesity and lack of safe pharmaceutical agents have lead to an increase in antiobesity drug-related research. A useful weight loss drug must: (1) be orally active; (2) be tolerable with no or few side effects; (3) not be addictive; (4) be effective in producing loss of body mass without significant loss of lean body mass; (5) be cheap; (6) be long acting with known mechanism of action [13, 14]. An optimal agent does not exist. A pharmacologic agent effective against treating obesity can either decrease availability of energy or increase energy output. A decrease in energy availability can be accomplished by decreasing appetite or by decreasing absorption of nutrients. An increase in energy output can be accomplished by increasing metabolic rate or thermogenesis. Information presented here offers an overview of medication used in the past that are no longer available, current therapies, and outlines targets for future drug development.

Weight Loss Drugs: A Short History

Thyroid extract was the first drug tried and used as early as 1893 [15]. However, the doses required to produce the desired effects lead to hyperthyroidism, lean mass loss, osteoporosis, myocardial toxicity, and sudden death. Weight was also rapidly regained after discontinuation of therapy [16–18]. In view of these complications, the use of thyroid hormone in the treatment of obesity was abandoned. During the 1920s, various laxatives were used for obesity treatment. They stimulated the large bowel to empty, which occurs only after the food and calories have been absorbed via the small intestines, but they do not act on the small intestine. Prolonged use leads to electrolyte disorder, cramps, dehydration followed by water retention and bloating. The most serious consequences of excessive laxative use are dependency, severe constipation, rectal bleeding, and osteomalacia. In 1933, a derivative of aniline dinitrophenol that uncouples mitochondrial oxidative phosphorylation was used alone or in combination with thyroid hormone, and induced rapid weight loss accompanied by sweating and elevated body temperature. Severe toxicities of dinitrophenol, including hepatotoxicity, dermatitis, visual impairment, and death, led to discontinuation of its use [19–21].

Amphetamine, which was introduced in 1937, rapidly replaced dinitrophenol for the treatment of obesity [22]. Amphetamine and related drugs, such as methamphetamine, phenmetrazine, are centrally acting sympathomimetics that increase levels of norepinephrine, serotonin, and dopamine in the brain. Use of these drugs was discontinued because of severe addictive potential and strong stimulatory effects [23]. The "Rainbow Pills," consisting of amphetamine, thyroid hormone, digitalis, and a diuretic, were prescribed during 1940s to 1960s. This drug regimen caused severe addiction, cardiac toxicities, and deaths in 1967 [23–27]. An amphetamine derivative with anorectic property called aminorex was withdrawn from the European market in 1971 as it was found to be associated with pulmonary hypertension [28]. Acupuncture technique, which was introduced for obesity treatment in 1975, had no clinical benefits of weight loss [29]. In 1978, several deaths were reported with the use of very low calorie diets, containing collagen [30]. A few years later, number of drugs were developed and marketed to replace amphetamines as appetite suppressants. These anorectic drugs include phendimetrazine, diethylpropion, mazindol, and, phentermine, which produced some amphetamine-like effects but with reduced stimulatory properties [31].

Fenfluramine and dexfenfluramine were used during 1980s and 1990s for obesity treatment. They enhanced the release and also inhibited the reuptake of neurotransmitter serotonin without addiction and CNS stimulant activity, promoting reduced appetite and weight. In 1992, an effective therapy began with the combination of anorectic serotoninergic drug fenfluramine and sympathomimetic agent phentermine (fen–phen) [32]. Both these drugs acted synergistically and produced significant weight loss with fewer side effects. However, the drugs fenfluramine, dexfenfluramine, and the combination fen–phen were withdrawn from the market in September 1997 after the reports of cardiac valvular disease and primary pulmonary hypertension in patients undergoing the treatment [33].

Other drugs that have been used to treat obesity, without a specific license, include phenylpropanolamine (PPA) and ephedrine, the latter in combination with caffeine or aspirin. PPA is a sympathomimetic drug which acts as an appetite suppressant by stimulating the release of noradrenaline at α and β receptors. A study reported that when PPA is used as an antiobesity medication, it increases the risk of hemorrhagic stroke and is not considered safe for over-the-counter use. Therefore, the drug was removed from the US market in 2001 [34]. Ephedrine is the most common ingredient in herbal dietary supplements used for weight loss. It stimulates sympathetic nervous system and elevates circulating catecholamine levels which in turn activates β-1, β-2, and, at high levels, β-3 receptors in adipocytes. This results in increased lipolysis and thermogenesis in brown adipose tissue and skeletal muscle [35]. Thus, it can slightly suppress the appetite; however, no studies have shown that it is effective as a weight loss agent. Moreover, ephedrine is also associated with number of adverse effects such as cardiac arrhythmias, angina pectoris, vasoconstriction with hypertension, and tremors [36]. It can also interact with many prescription and over-the-counter medications. Due to high incidence of abuse, in February 2004, the FDA officially banned the sale of ephedrine in USA. Caffeine, a component of coffee is a thermogenic agent that stimulates lipolysis, restricts accumulation of fats. Ephedrine is often combined with caffeine or aspirin to produce weight loss with fewer side effects. This combination had both anorectic and thermogenic actions, thereby assisting weight loss and weight maintenance by affecting both aspects of energy balance [35, 37]. It was approved for sale in Denmark in 1990. Licensed under the trade name of Letigen. In 2002, the marketing license was suspended, after a number of reports had suggested a safety issue [38]. Table 14.1 represents a historical tabulation of several treatments that have been tried along with their unintended consequences.

Table 14.1 History of drug treatments for obesity

Year	Drug	Side effects
1893	Thyroid extract	Hyperthyroidism, lean mass loss, myocardial toxicity
1920	Laxatives	Electrolyte imbalance, rectal bleeding, osteomalacia
1933	Dinitrophenol	Liver and retinal toxicity
1937	Amphetamine	Addiction
1967	Rainbow pills	Addiction and cardiac toxicity
1971	Aminorex	Pulmonary hypertension
1975	Acupuncture technique	No clinical benefits
1978	Collagen-based VLCD	Death
1997	Fenfluramine/dexfenfluramine	Heart valvular insufficiency
1997	Fen-phen	Cardiac valvular disease, primary pulmonary hypertension
2000	Phenylpropanolamine	Hemorrhagic stroke
2004	Ephedrine and caffeine	Insomnia, palpitation, nausea

Weight Loss Drugs: Currently Prescribed

Past forms of pharmacotherapy for obesity were generally misguided and caused unacceptable morbidity and mortality. Previously, the weight-loss drugs were not subjected to long-term studies. That is, there was an underlying assumption that obesity can be cured. This has been proven wrong, and obesity is now viewed as another chronic disease that requires long-term treatment. Viewing obesity as a chronic disease necessitates the development of safer drugs with longer efficacy. This need is supported by the fact that many patients using traditional anti-obesity drugs regain weight after initial loss.

Keeping this in mind, the pharmacological treatment for obesity has undergone radical changes over the past few years leading to the development of new products and treatment proposals (see Table 14.2). Most of the weight-loss medications approved by The Food and Drug Administration (FDA) for short-term use, generally 12 weeks, are appetite suppressants. These include phentermine, phendimetrazine, benzphetamine, and diethylpropion. Of these, phentermine is one of the most commonly prescribed noradrenergic appetite suppressants in the USA. Generally, noradrenergic drugs produce a weight loss of 3–8% compared with placebo and may cause increased blood pressure and heart rate, sleeplessness, dry mouth, dizziness, headache, and nervousness. Phendimetrazine have greater abuse potential than phentermine and diethylpropion.

At present, sibutramine and orlistat are the only weight-loss medications approved by FDA for long-term use in patients who are significantly obese. The safety and effectiveness of these drugs have not been established for use beyond 2 years.

Orlistat

Orlistat is the hydrogenated derivative of lipstatin produced by the bacteria *Streptococcus toxytricini* [39]. It is a potent inhibitor of gastrointestinal pancreatic lipase, which is inactive in its folded state. Such enzymes catalyze the hydrolysis of triglycerides to free fatty acids in the lumen of gut. Binding of the enzyme to triglycerides is facilitated by co-lipase in the presence of bile salts. Due to this interaction the active site is exposed. Orlistat binds irreversibly to the lipase active site, preventing the digestion and absorption of dietary fats by approximately 30% [40]. The unabsorbed fat is excreted in the stool. Orlistat does not inhibit other intestinal enzymes or systemic lipases due to its very small absorption [41]. The FDA approved orlistat as a prescription weight loss agent in 1999. In 2010, the FDA approved a revised label for orlistat that

Table 14.2 Currently used drugs for treating obesity

Drug	FDA approval	Side effects
Phentermine	Yes; short term	Increased blood pressure and heart rate, dizziness, headache, sleeplessness, nervousness
Diethylpropion	Yes; short term	Same as phentermine
Phendimetrazine	Yes; short term	Same as phentermine
Sibutramine	Yes; long term	Increased blood pressure, dry mouth, constipation
Orlistat	Yes; long term	GI effects: oily spotting, flatulence
Orlistat OTC	Yes; short term	Same as orlistat
Bupropion	No	Dry mouth
Fluoxetine/sertaline	No	Weight regain, sleep disorder, amnesia
Topiramate	No	Difficultly with memory, paresthesia
Zonisamide	No	Reduction in blood pressure, kidney stones
Metformin	No	Headache, diarrhea

includes new safety information about cases of severe liver injury that have been reported rarely with the use of this medication.

A 6 month, multicenter, dose ranging study was carried out at 30, 60, 120, and 240 mg of orlistat thrice daily in obese patients together with a low fat diet. Orlistat produced a dose-dependent significant difference in weight loss from 60 mg doses and reached a plateau at 120 mg doses. Higher doses did not increase weight loss [42]. In a 2-year clinical trial study, orlistat-treated subjects lost 8.76 kg of initial body weight compared to 5.81 kg in placebo-treated subjects. During the second year, subjects treated with orlistat 120 mg three times a day regained 35.2% of lost weight compared to 63.4% regain with placebo-treated subjects [43].

The most common GI side effects of orlistat include oily and frequent bowel movements, bowel urgency, and flatulence with discharge; these side effects are usually short term and tend to decrease considerably after the first weeks of treatment [44]. One study demonstrated that these side effects can be minimized by reducing fat in the diet and by using natural fiber [45]. Orlistat causes significant reductions in total and low density lipoprotein cholesterol levels and in systolic and diastolic blood pressure [46, 47]. In nondiabetic obese patients, the use of orlistat with a calorie and fat restricted diet is associated with significant reduction in insulin and plasma glucose levels [43]. Orlistat may reduce absorption of some fat soluble vitamins; therefore, a multivitamin supplement is recommended to prevent nutritional deficiencies [48]. Orlistat also interferes with the absorption of many drugs, such as warfarin, cyclosporine, and levothyroxine [49–51]. It is contraindicated in patients with history of oxalate kidney stones and cholestasis [50, 52].

Orlistat OTC

Orlistat OTC was approved by the FDA in February 2007 for the treatment of obesity. The OTC strength is lower (60 mg, thrice daily) than its prescription counterpart. It blocks the fat absorption and promotes weight reduction. It is used in conjunction with a reduced-calorie diet and weight maintenance program to treat obesity in people with certain risk factors such as diabetes, hypertension, and high cholesterol or triglycerides [53]. It is contraindicated in patients who have food absorption problems, have received organ transplants, and in children younger than 18 years of age. Other parameters are the same as one would observe for the prescription drug orlistat.

Sibutramine

Sibutramine is a beta-phenethylamine, initially developed as an antidepressant agent [54]. It was approved by the FDA in November 1997 as a weight loss agent. Sibutramine is a selective presynaptic reuptake inhibitor of both norephinephrine and serotonin, thereby potentiating anorectic effect and stimulating thermogenesis [55]. In contrast to fenfluramine and dexfenfluramine, it does not stimulate the release of norephinephrine and serotonin; so the risk of pulmonary hypertension or cardiac valvular abnormalities is negligible [56]. Sibutramine is well absorbed in the GIT and reaches peak plasma level in the first two hours of treatment. It is metabolized in the liver by cytochrome P450 and produces an active metabolite with a long half-life [57].

A double-blind, randomized, placebo-controlled trial of sibutramine has demonstrated a significant dose-dependent effect on body weight in obese patients, with doses of 10, 15, 20, and 30 mg being more effective than placebo. This study suggested that sibutramine is a valuable drug for obesity treatment at a recommended dosage of 10–20 mg/day [58]. In a number of studies lasting up to 1 year, weight loss with a hypocaloric diet and 10–20 mg/d ranged from 4.7% to 7.3% of initial body weight [59–63].

Consistent noradrenergic side effects of the drug include a 1–3 mm Hg increase in diastolic blood pressure, a 0.3–3 mm Hg increase in systolic blood pressure, and a 2–5 beat/min increase in resting heart rate [62]. Other side effects include headache, dry mouth, insomnia, and constipation. One study found that frequency of these adverse effects could be reduced if sibutramine was given intermittently, 12 weeks of active drug alternated wtih 7 weeks of placebo over 44-week period [62]. Weight loss with sibutramine produced a decrease in fasting triglycerides and low-density lipoprotein cholesterol and an increase in high-density lipoprotein cholesterol [60]. Sibutramine is contraindicated in patients with history of seizure, cholestasis, glaucoma, and eating disorders. The drug should be used with caution in patients receiving monoamine oxidase or other serotonergic agents due to the risk of serotonin syndrome [64].

Some of the drugs from the antiepileptic, antidiabetic, and antidepressant categories are not approved by the FDA for the treatment of obesity, but they have been shown to promote short-term weight loss in clinical studies and are currently prescribed off-label.

Bupropion

Bupropion is currently used as an antidepressant and smoking cessation aid [65]. It acts by inhibiting the reuptake of the neurotransmitters dopamine and norepinephrine. In a double-blind, placebo-controlled trial, it was found that bupropion SR 300 and 400 mg/day were well tolerated by obese adults and were associated with a 24-week weight losses of 7.2% and 10.1%, respectively, and sustained weight losses at 48 weeks [66]. A recent meta-analysis confirmed the efficacy of bupropion given at 400 mg/day for treating obesity. Results demonstrated that weight loss in the bupropion group (4.4 kg) was significantly greater than in the placebo group (1.7 kg) [67]. The most common side effect associated with the drug is dry mouth. Bupropion is contraindicated in patients with a history of bulimia, anorexia, or seizure [65]. Combinations of bupropion with other agents are currently being studied for treatment of obesity.

Fluoxetine and Sertaline

Both fluoxetine and sertaline are selective serotonin reuptake inhibitors (SSRIs) that are approved for use as antidepressants, but not for obesity treatment [68, 69]. Unlike fenfluramine, fluoxetine is an effective anorectic agent that promotes weight loss in obese patient without producing pulmonary hypertension or cardiac valvular abnormalities [70]. In a short term, double blind, placebo controlled study fluoxetine (60 mg/day) produced a weight loss of 0.23 kg/week more than placebo [71]. A study in which sertaline was used as a weight loss agent showed no significant difference between the experimental and placebo groups [72]. Both fluoxetine and sertaline have a weak and non-sustained weight-reducing effect when given at doses higher than those normally used in the treatment of depression [70]. The principle problem with fluoxetine as an antiobesity agent is weight regain, as observed in long-term clinical trials [70, 73]. SSRIs are not effective as antiobesity agents, although they may be useful for depressed obese patients

Topiramate

Topiramate is an antiepileptic drug associated with weight loss. It is a GABA agonist and glutamenergic antagonist. Its appetite suppressant property might be due to its antagonistic effect on glutamate [65]. A short-term, randomized, double-blind, placebo-controlled, dose-ranging trial suggested that topiramate produced significantly greater weight loss than placebo at all doses (ranging from 64 to 384 mg/day). The two lower doses (64 mg/day and 96 mg/day) produced similar weight loss, which was less than that produced by the two higher doses (192 mg/day and 384 mg/day). The two lower doses were better tolerated by the subjects than the higher doses [74]. Therefore, lower doses were found to be more clinically effective in producing weight loss and require further clinical evaluation for the long-term treatment of obesity. The most frequent adverse events were related to the central or peripheral nervous system, including paresthesia, somnolence, and difficulty with memory, concentration, and attention. Weight loss with topiramate produced reductions in blood pressure and modest changes in total cholesterol, low density lipoprotein, and triglycerides [74]. As a result of these unfavorable effects, phase III clinical trials were discontinued [65]; however, topiramate is now being developed in combination with other agents for obesity treatment.

Zonisamide

Zonisamide is marketed as an antiepileptic drug [75]. In short-term clinical trials in epileptic patients the drug caused weight loss, which was an adverse side effect. Based on this zonisamide could be one of the therapeutic drugs for obesity treatment. It has both serotonergic and dopaminergic activity, responsible for its anorectic effects. In a 16-week, randomized, double-blind, placebo-controlled trial, the mean weight observed for the zonisamide-treated group changed from 97.2 kg at baseline to 90.9 kg, and for the placebo-treated group, the corresponding change was 97.6–96.5 kg [76]. Thus, in this short-term preliminary trial, zonisamide and hypocaloric diet resulted in more weight loss than placebo. In addition to this, there was a significant reduction waist circumference with zonisamide compared with placebo. The drug was tolerated well, with a few adverse effects such as somnolence, psychomotor slowing, cognitive impairment, and fatigue and depression [77]. Further studies are warranted in order to recommend zonisamide as an antiobesity drug.

Metformin

Metformin is an antidiabetic drug, but several studies have shown that it also helps in weight reduction in nondiabetics patients [78–80]. It acts either by reducing gastrointestinal absorption of carbohydrates and fats or by inducing anorectic and lipolytic effect. Clinical trial studies provided evidence that metformin has a beneficial effect on obesity in children and adolescent patients [81–83]. These studies have also shown positive effects on waist circumference, fasting insulin and glucose levels, and triglycerides. Metformin also promotes weight loss in obese patients with non-insulin-dependent diabetes mellitus (NIDDM) [84]. It is not recommended in patients with kidney or heart failure or any medical condition that could make blood acidic. Gastrointestinal effects were the most frequently reported adverse effects [82, 83]. Long-term, large-scale, randomized, placebo-controlled clinical trials are still required to firmly establish the role of metformin as a weight loss agent.

Over-the-Counter Treatments

Many over the counter are also used for obesity treatment including diet pills and combination of herbals and vitamins. Herbal preparations are not recommended as part of a weight loss program as these preparations have unpredictable amounts of active ingredients and potentially harmful effects. Many obese people use dietary supplements for a weight loss. To date, there is little clinical evidence to support their use. More data are necessary to determine the efficacy and safety of these supplements.

Conjugated Linoleic Acid

Conjugated linoleic acid (CLA) is formed in the first biohydrogenation step of linoleic acid by the action of linoleic acid isomerase of the bacterium *Butyrivibrio fibrisolvens*. It reduces fat deposition, possibly by increasing fat oxidation and decreasing triglyceride uptake in adipose tissue. Several randomized, double-blind, clinical trials showed that CLA reduces body fat mass in overweight and moderately obese, healthy volunteers. The reduction of body fat mass within the groups was significant for the 3.4 and 6.8 g CLA group [85–88]. The data also suggested that CLA may reduce body fat mass in humans and that no additional effect is achieved with doses >3.4 g CLA/day. However, adverse GI effects are frequent. Further clinical studies are still required to confirm the optimal dose and duration of treatment of CLA as anti-obesity agent.

Chitosan

Chitosan is a dietary fiber derived from chitin, a starch found in the skeleton of shellfish [89]. Chitosan is a nonabsorbable polymer; therefore, it passes through the intestinal tract without adding any calories. It binds with intestinal triacylglycerol or its hydrolysis products and removes the fat from the body. A randomized, double-blind, placebo-controlled trial by Pittler et al. [90] has shown that chitosan treatment did not result in significant weight loss over placebo. However, recent studies demonstrated that chitosan is well tolerated and may be useful as a weight loss agent with minor gastrointestinal side effects [91–93]. Further clinical studies are still required to recommend it for weight loss.

Yerba Mate

Like caffeine, yerba mate is also a strong central nervous stimulant. To date, it has not been proven as weight loss agent. The most frequent side effect include excessive stimulation and high blood pressure [94, 95].

Guar Gum

Guar gum is a dietary fiber obtained from the Indian cluster bean *Cyamopsis tetragonolobus*. It is commonly used in small amounts as a thickening agent for foods and medications. It decreases appetite by providing a "feeling of fullness." However, guar gum causes obstruction of the esophagus. Research studies show that it has no significant effect on weight loss [96].

Dandelion

Dandelion (*Taraxacum officinale*) is a natural diuretic. *Taraxacum officinalle* exhibits strong inhibitory activity against pancreatic lipase in vitro and in vivo, thereby inhibiting fat absorption [97]. Studies need to be carried out to identify and elucidate the active component of *T. officinalle* responsible for its pancreatic lipase inhibitory activity.

Garcinia

Garcinia is a fruit containing hydroxycitric acid. Hydroxycitric acid inhibits the mitochondrial enzyme citrate lyase, involved in the fatty acid synthesis. Randomized clinical trials have shown that garcinia is effective as a weight loss agent [98–100]. However, the methodology used was poor, and so long-term clinical trials are required to ascertain the role of garcinia as a weight loss agent.

Glucomannan

Glucomannan is derived from the roots of *Amorphophallus konjac*. It contributes to weight loss by delaying the absorption of glucose from the intestines. It is effective in producing modest weight loss by providing a "feeling of fullness." However, esophageal obstruction has been reported in people taking glucomannan [101].

Chromium

Chromium plays an important role in carbohydrate and lipid metabolism, potentially influencing body composition, including reducing fat mass and increasing lean body mass. Because the supplement is absorbed better than dietary chromium, most studies have focused on the use of chromium picolinate as weight loss agent [102]. However, random clinical trials have not shown any differences in weight loss between the treatment and placebo groups [103]. Because of the lack of large, long-term, well-designed studies, the efficacy and safety of chromium picolinate as a weight loss medication remains uncertain.

Green Tea

Green tea is a "non-fermented" tea that contains catechins. Catechins are strong antioxidants. The cathechins in green tea increase fat oxidation and thermogenesis [104]. Recent human studies suggested that green tea promotes weight reduction. While all the evidence from research is promising, future studies are necessary to fully understand the role of green tea in obesity [105]. Harmful effects of overconsumption of green tea can cause oxidative DNA damage of pancreas and liver [106].

Pyruvate

Pyruvate is formed in the body during digestion of carbohydrates and protein from food. According to some clinical studies, pyruvate effectively promotes weight loss. It is generally well tolerated with minimal gastrointestinal adverse effects [107, 108]. However, studies are still required to prove the role of pyruvate as a weight loss agent.

Weight Loss Drugs: New Approaches

The maximum weight loss achieved by currently available pharmacological drugs for treating obesity varies with individuals but appears to be not more than 10% of initial body weight when combined with dietary, behavioral, and exercise therapy. Once this threshold is achieved physiological mechanism in the body cause a progressive increase in appetite and decrease in energy expenditure. These regulatory responses prevent further weight loss and make maintenance of achieved weight loss difficult. Although this degree of weight loss may have a beneficial effect, there is need for more effective and better tolerated antiobesity drugs.

The shortcomings of available therapies have led to massive interest in finding new and innovative strategies for treating obesity. Our growing knowledge of peripheral signals and CNS pathways involved in the control and regulation of body adiposity have resulted in the development of a large number of new targets, some of which have yielded promising data in clinical trials for weight loss. The peptide therapy for obesity has become better validated in recent studies, and this may allow more rapid exploitation of novel targets. In addition to this, several new drugs are in the pipeline for approval or are in the research and developmental stage. Many medications approved for other conditions are now being studied for their use in obesity management.

Drugs Affecting Peripheral Mechanisms

Cetilistat

Cetilistat (ATL-962) is a novel gastrointestinal lipase inhibitor under development. In a phase II, randomized, double-blinded, clinical trial study, the drug showed significant weight loss of 4.1 kg compared to 2.4 kg with placebo [109]. Moreover, it has also shown a significant reduction in total serum and low-density lipoprotein cholesterol levels. Cetilistat is generally well tolerated with fewer GI side effects and has been shown to have a weight loss profile similar to orlistat [110]. It has also shown increase in dietary fat excretion in healthy human volunteers. Currently, the drug is undergoing phase III clinical trial.

Fatostatin

Fatostatin, presently in early-stage research, inhibits sterol regulatory element-binding proteins (SREBPs) activation. SREBPs are transcription factors involved in the regulation of genes required for cholesterol and fatty acid synthesis. Pharmacological studies with fatostatin show that it promotes reduction of bodyweight, hepatic fat accumulation, and blood glucose in ob/ob mice [111]. However, these results are very preliminary and further studies are warranted.

Beta-3-Adrenoceptor Agonists

β3-adrenoreceptors are highly expressed in adipose tissue and skeletal muscle, playing an important role in the regulation of energy balance and glucose homeostasis. Activation of β3-adrenoreceptors stimulates lipolysis and thermogenesis in white and brown adipose tissue [112]. The β3-adrenoceptor agonist CL.316243 increases fat oxidation and decreases carbohydrate oxidation in humans [113]. A double-blind, randomized, placebo-controlled, clinical study revealed that a new, potent, and highly selective β3-adrenoceptor agonist TAK-677 has no effect on fat oxidation and, at high doses, produced a slight increase in thermogenesis coupled with an increase in heart rate [114]. The study also found large variabilities in plasma concentrations of tak-677 in subjects, indicating some possible problems with bioavailability and therefore efficacy. To date, no β3-adrenoceptor agoinst has progressed beyond phase II clinical trials. Thus, the role of β3-adrenoreceptors agonists in the treatment of obesity remains unclear, and further studies are still needed to assess their potential clinical efficacy.

Drugs Affecting Central Mechanism

Tesofensine

Tesofensine, developed by neurosearch, is a novel pharmacological agent, that inhibits the presynaptic reuptake of the neurotransmitters noradrenaline, dopamine, and serotonin. Phase II clinical trials in obese subjects revealed that the tesofensine group lost 6.7–12.8 kg of initial body weight compared with a 2.2-kg loss in the placebo group [115]. It has also shown dose-related reductions in body weight, body fat, and waist circumference as well as improvements in other obesity-related endocrine factors. Side effects such as nausea, dry mouth, mood disturbance, and insomnia were reported with the drug treatment. Currently, the drug is under phase III trials to study and carefully monitor the above-mentioned adverse effects.

Bupropion–Naltrexone (Contrave)

Contrave is a combination of bupropion, a dopamine and noradrenaline reuptake inhibitor, and naltrexone, an opioid antagonist. These two agents are reported to act synergistically by stimulating central melanocortin pathways, resulting in increased energy expenditure and reduced appetite. Individually, both the drugs have been shown to reduce appetite and body-weight in humans [116]. Recently phase III clinical trials with contrive revealed loss of 9.2–11.4 kg of initial body weight compared with 7.3 kg loss in the placebo group [117]. Nausea, Headache, constipation, dizziness, vomiting, and dry mouth were the most frequent adverse events reported with the use of contrave.

Phentermine–Topiramate (Qnexa)

Qnexa, developed by vivus pharmaceutical, combines immediate-release phentermine and controlled-release topiramate for long-term obesity management. Phentermine is an appetite suppressant used for weight loss, whereas topiramate is a GABA agonist and glutamenergic antagonist approved for the treatment of seizures and migraines. The dose-ranging study demonstrated that topiramate promotes weight loss in a dose-dependent manner [74]. The phase III clinical trial study shows that subjects taking full-dose and mid-dose of qnexa achieved an average weight loss of 9.2% and 8.5%, respectively, as compared with 1.7% for the placebo group [Vivus inc].[1] Side effects include paraesthesia, dry mouth, altered taste, increased heart rate, possible birth defects, and psychiatric problems (depression, suicidal thoughts, impaired memory, and concentration). In July 2010, an FDA advisory committee agreed that the phentermine/topiramate combination was effective in reducing weight; however, the committee refused to approve a recommendation for the obesity treatment due to safety concerns.

Bupropion–Zonisamide (Empatic)

Empatic is a combination of long-acting bupropion (antidepressant) and long-acting zonisamide (antiepileptic). It has already been shown that taken individually these drugs promote weight

[1] VIVUS announces positive results from two phase 3 studies; Obese patients on Qnexa achieve average weight loss up to 14.7% and significant improvements in comorbidities. Results of EQUIP and CONQUER phase 3 studies exceed FDA benchmarks for obesity treatments, demonstrate positive safety profile. September 2009.

loss. In a preliminary, randomized, open-label study the combinatorial therapy of empatic resulted in greater weight loss (8.1 kg) than with zonisamide alone (3.0 kg) in obese women [77]. Nausea, headache, and insomnia are the most frequent side effects associated with combined therapy. Currently, the drug is under phase III clinical trial to evaluate the efficacy and safety of the combination.

Lorcaserin

We know that different serotonin receptors are responsible for different effects. The 5-HT2C receptor is responsible for many physiological functions including mood anxiety and feeding behavior. These receptors are widely distributed throughout the CNS. It has been suggested that targeting the specific 5-HT2C receptor can help the patient lose weight effectively in a safe manner [118]. Lorcaserin (Arena Pharmaceuticals) is a selective serotonin 2 C receptor agonist having potential to treat obesity with a safer side effect profile. It has been shown to produce dose-dependent reduction in food intake and body weight in rats when administered chronically [119]. In phase III clinical study known as Behavioral Modification and Lorcaserin for overweight and Obesity Management (BLOOM), the obese or overweight patients were randomized to lorcaserin 10 mg twice daily or placebo for 52 weeks [120]. All patients also underwent diet and exercise counseling. At 1 year, individuals treated with lorcaserin lost 5.8 kg compared with 2.2 kg lost in the placebo-treated patients. At week 52, patients in the placebo group continued to receive placebo but patients in the lorcaserin group were randomly reassigned to receive either placebo or lorcaserin. Primary outcomes were weight loss at 1 year and maintenance of weight loss at 2 years. The most frequently reported side effects with drug were headache, dizziness, and nausea. Regarding the more serious concern of cardiac valvulopathy, the BLOOM investigators reported that at 1 year FDA-defined valvulopathy developed in 2.3% of patients in the placebo and 2.7% of patients in the lorcaserin group. At 2 years, the rates were similar. The Behavioral Modification and Lorcaserin Second Study for Obesity Management (BLOSSOM) trial showed that lorcaserin did not increase the risk of cardiac valvulopathy or worsen valve problems, including in some patients with preexisting mild to greater aortic or mitral regurgitation [120]. Thus, from the recent data lorcaserin was found to be safe and effective in reducing weight as compared with placebo, but the FDA denied its approval for use as an antiobesity agent due to safety concerns.

Endocannabinoid Receptor Blockers (Rimonabant, Taranabant)

Rimonabant

Rimonabant, developed by Sanofi-Aventis, is a selective endocannabinoid receptor-1 (CB1) antagonist indicated for the treatment of obesity. It acts centrally on the brain, thus decreasing appetite by blocking the "munchie receptor" that stimulates hunger. It may also act peripherally by increasing thermogenesis and thereby increasing energy expenditure [121, 122]. In a 1 year randomized, double-blind, clinical trial with doses of 5 or 20 mg, rimonabant was found to produce significant reduction in weight and waist circumference as well as improved lipid and glycemic profiles compared with placebo [123]. Generally, the drug was well tolerated with mild and transient side effects such as nausea, dizziness, diarrhea, and insomnia [124]. In February 2006, the FDA issued an approvable letter for rimonabant for obesity treatment. The European commission also approved the sale of rimonabant in Europe for the treatment of obesity as an adjunct to diet and exercise in July 2006. However, in 2007 the FDA's Endocrine and Metabolic Drugs Advisory Committee (EMDAC) voted against recommending rimonabant for obesity treatment since Sanofi-Aventis failed to demonstrate the safety of the drug [124]. In October 2008 European Medicines Agency recommended the suspension of the sale of rimonabant in UK due to the risk of serious psychiatric problems and even suicide. Presently, the drug is under phase III clinical trial in the USA and the UK to evaluate the safety of the drug. Results from the phase III trial suggested that rimonabant was effective in reducing weight; however, it indicated an increased risk of suicide and other psychiatric and neurological effects in patients [125].

Taranabant

The CB1 receptor inverse agonist Taranabant has been evaluated for the treatment of obesity. Taranabant promotes weight loss by reducing appetite and increasing energy expenditure. A 12-week study with the drug demonstrated a significant weight loss in obese subjects compared to placebo. Incidence of side effects with taranabant treatment increases with increasing dose, including mild to moderate GIT and psychiatric effects [126]. A clinical trial, low dose study demonstrated that taranabant produces weight loss in a dose-dependent fashion. Mean weight loss achieved after 1 year of drug treatment was 5.0 kg (0.5 mg dose), 5.2 kg (1 mg dose), and 6.4 kg (2 mg dose) compared to 1.4 kg for placebo [127]. Recently, a study using higher doses of taranabant (2, 4, and 6 mg) found the drug to be effective in achieving clinically significant weight loss at 1 year of treatment; this weight loss persisted to 2 years but was found to be associated with dose-related increase in adverse events particularly serious psychiatric disorders [128]. On the basis of these clinical trials, the overall safety and efficacy profile of taranabant did not support its further development in the treatment of obesity [128, 129].

Gut Hormones

Leptin

Leptin is a protein secreted by fat cells and the circulating concentrations of leptin are directly proportional to the total amount of body fat or adipose tissue. Leptin deficiency is associated with severe obesity in rodents and humans [130]. Leptin has a number of actions that result in the inhibition of food intake: (a) it works by inhibiting the activity of neuropeptide Y (NPY) and agouti-related peptide (AgRP), which stimulates appetite; (b) it enhances the effect of alpha-melanocyte stimulating hormone (α-MSH), which is an appetite suppressant; (c) it raises the body temperature, resulting in greater caloric reduction [131–134]. Subcutaneous daily injections of recombinant leptin along with a weight reduction diet led to a mean weight loss of 15 pounds in comparison to 3.75 pounds with placebo in a 24-week treatment period. More than 95% of the lost weight was from body fat [135]. Additional human studies are required to determine the therapeutic potential of leptin and its analogue for the obesity treatment. Leptin treatment was considered as the cure for obesity, but it failed to achieve the expected results as patients became resistant to the hormone and regained the lost weight [136]. Recently, it was discovered that pretreating mice with Buphenyl (4-PBA) and Tauroursodeoxycholic acid (TUDCA) increases leptin sensitivity by almost tenfold, and the mice showed a significant weight loss even when fed with a high-fat diet [137]. If this effect can be reproduced in humans, then it could provide an effective treatment for obesity. Large and frequent doses of leptin provide only moderate weight loss because of its short circulating half-life, low potency, and poor solubility [138]. Furthermore, high levels of leptin are inflammatory and studies have shown that excessive inflammation can lead to heart disease, diabetes, and many other chronic diseases [139]. Currently, leptin is under phase III clinical trials to check the availability of leptin analogues that are more soluble with longer serum half-life. A form of leptin, Fc-leptin, is highly soluble, more biologically potent, and possesses a much longer serum half-life. Fc-leptin was found to treat obesity in both leptin-deficient and normal mice, although studies have not been undertaken on human subjects [138]. Thus, Fc-leptin could be a potential treatment for obesity in humans after more extensive research. Although the latest results are promising, more studies are needed to completely understand leptin's role in weight loss.

Amylin Analogues (Pramlintide, Davalintide)

Pramlintide

Pramlintide (Amylin Pharmaceuticals) is a synthetic, stable, subcutaneously delivered analogue of the hormone amylin, which is secreted by the pancreas in response to eating. Amylin potentially inhibits glucagon release and slows gastric emptying. Pramlintide is already licensed in the USA to be used along with insulin for the treatment of both type 1 and type 2 diabetes. Pramlintide seems to mediate its anorectic effects by delaying GIT motility [140, 141]. In a 16-week, phase II, randomized, placebo-controlled dose escalation trial, pramlintide (240 μg) demonstrated 3.7% mean weight loss when given as a subcutaneous (SC) injection compared to placebo [142]. These results supported continued evaluation of pramlintide as a potential treatment for obesity. Pramlintide treatment achieved greater weight loss and enhanced long-term maintenance of weight loss, when used over 12 months as an adjunct to lifestyle intervention in obese subject [143]. Recently, preliminary trials have assessed modest weight loss in obese patients without diabetes and demonstrated a weight loss of up to 8 kg after 1 year with the use of pramlintide. Nausea was the most common adverse event [144]. However, current trials are limited; wider-scale clinical testing of the drug is anticipated.

Davalintide

Another amylin analogue, davalintide, induced dose-dependent weight loss up to 8 weeks and maintained metabolic rate during active weight loss [145]. Currently, the drug has entered phase II clinical trials to study its safety, tolerability, and efficacy as an antiobesity agent.

Pramlintide–Metreleptin

Recently, Amylin Pharmaceuticals has developed a combination treatment containing pramlintide and metreleptin (recombinant human leptin agoinst) following the observation that concurrent peripheral administration of amylin and leptin elicits synergistic weight loss in leptin-resistant, diet-induced obese rats [146]. In a phase II clinical trial study, pramlintide–metreleptin combination treatment led to significantly more weight loss (11.5 kg) from baseline as compared to metreleptin (7.4 kg) and pramlintide (7.9 kg) alone [147]. The most common adverse effects were injection site events and nausea. Collectively, these findings support further development of pramlintide/metreleptin as a novel, integrated, neurohormonal approach to obesity pharmacotherapy.

Glucagon-Like Peptide-1 Analogues (Exenatide, Liraglutide)

Exenatide

Glucagon-like peptide-1 (GLP-1), a gut peptide, decreases glucagon secretion from the pancreas, increases beta cell mass, and increases insulin secretion in a glucose-dependent manner. It also reduces gastric emptying rate, thereby increasing satiety, decreasing hunger, and reducing food intake with an ensuing weight loss in both lean and overweight humans [148, 149]. Exenatide (synthetic exendin-4) is a long-acting GLP-1 receptor agonist, developed by Amylin Pharmaceuticals in conjunction with Eli Lilly. It has already been approved for the treatment of type 2 diabetes. Several studies show that exenatide not only lowers blood sugar but also results in weight loss in people with type 2 diabetes [150, 151]. Exenatide in conjunction with diet and exercise decreased calorie intake and resulted in weight loss of 5.06 kg as compared to loss of 1.61 kg in the placebo group in nondiabetic patients. Only exenatide-treated subjects (9.6%) lost more than 10% of their body weight [152]. The most common side effects include mild to moderate nausea and diarrhea in some patients, especially when therapy is initiated; these effects are mitigated with continued use. Cases of acute pancreatitis have been reported in patients taking exenatide. Currently, the drug is administered subcutaneously twice daily; once-a-week injection is in late stage clinical trials [153].

Liraglutide

Liraglutide is a recently introduced another GLP-1 analogue, 97% identical to native GLP-1. It has been approved in Europe for the treatment of type 2 diabetes [151] and is under evaluation in the USA by the FDA. Like exenatide, it also increases the secretion of leptin, resulting in suppressed appetite, energy intake, and a delay in gastric emptying. It is also associated with improved glycemic control and a weight loss profile similar to exenatide [151, 154]. A High-dose study with liraglutide revealed a weight loss of 5.5–6.0 kg compared to placebo and a total of 75% of subjects receiving high-dose liraglutide lost more than 5% of their body weight, whereas 35% lost over 10% [155]. Side effects, including nausea, were comparable to those seen with exenatide use. There was no increased incidence of pancreatitis in trial participants. However, a slightly increased incidence of medullary thyroid tumors has been reported in two nonhuman species in clinical trials.

Recently, preclinical studies have shown that a novel peptide agonist of glucagon and GLP-1 receptors produces substantial amount of body fat reduction with fewer adverse effects [156]. This peptide based therapy might be a potential breakthrough in the treatment of obesity. Clinical trials need to be carried out to test the peptide's efficacy in human subjects.

Oxyntomodulin

Oxyntomodulin (OXM), secreted by intestinal L-cells, is released in proportion to calories ingested. It is known to inhibit gastric secretion, gastric emptying, and pancreatic exocrine secretion [157]. OXM also inhibits food intake and reduces bodyweight in rodents [158, 159]. Coadministration of a GLP-1 antagonist blocks the anorectic effect of OXM. Intravenous infusion in lean human volunteers reduces appetite without increasing nausea or affecting food palatability [160]. In a 4-week, double-blind, randomized, controlled trial in overweight and obese OXM administration resulted in a 2.3-kg weight loss compared with 0.5-kg weight reduction in placebo-treated controls [161]. Subsequently, it was shown that thrice daily subcutaneous administration of oxyntomodulin in obese volunteers induced a 26% rise in energy expenditure over 4 days, in addition to a reduced food intake [162]. In detail, long-term clinical trials are required to predict whether OXM administered in this way would achieve desired weight loss or meet other criteria for regulatory approval.

Cholecystokinin

Cholecystokinin (CCK) is widely distributed in the gut, secreted mainly by entero-endocrine I cells in the small intestine. It is released rapidly postprandially in response to fat and protein [163]. Its actions include: inhibition of food intake, delayed gastric emptying, stimulation of pancreatic enzyme secretion, and stimulation of gall bladder contraction. These effects are mediated via binding to CCK receptors on the vagus nerve [164]. Intravenous infusion of CCK decreases food consumption by reducing meal size and duration in rats and human; this effect is similar for lean and obese subjects. CCK administration to humans and animals inhibits food intake by reducing meal size and duration [165–166]. In rodents, continuous intraperitoneal infusion produces rapid and reversible tolerance after 24 h [168]. Recent studies suggested that daily CCK injections produces synergistic weight loss effect in rats in combination with leptin [169, 170], possibly by increasing the rate of leptin transport across the blood–brain barrier. Further studies are required to investigate whether coadministration with leptin or other hormones increases the effect of CCK receptor agonists in the treatment of human obesity. Recently, nausea and taste aversion have been detected with CCK treatment at high doses, making it unlikely candidate for an antiobesity treatment [171]

Ghrelin

Ghrelin is a peptide hormone principally secreted by cells in oxyntic gland of stomach [172]. It stimulates appetite as well as regulates energy homeostasis following central and systemic administration. A posttranslational octanoylation of ghrelin's third serine residue by the enzyme ghrelin-*O*-acyltransferase (GOAT) is essential for its receptor binding [173]. Ghrelin exerts its action by binding specifically to receptors in the arcuate nucleus that include the orexigenic neuropeptide Y (NPY) neurons and agouti-related protein (AgRP) [174]. Ablation of the ARC eliminates ghrelin's orexigenic effects [175]. Plasma ghrelin levels rise prior to meals, ultimately acting as a meal initiator, and fall immediately after a meal [176, 177]. In general, plasma ghrelin levels are inversely related with bodyweight [178–180]. Chronic administration of ghrelin potentially increases food intake and leads to weight gain in rats and healthy lean humans [181, 182]. Similarly, subchronic administration of ghrelin agonists BIM-28125 or BIM-28131 resulted in increased food consumption and body weight [183]. These data suggest that suppression of ghrelin action could be an effective target for treatment of obesity. This could be achieved by developing GHS-R1a (ghrelin receptor) antagonists. The selective ghrelin receptor antagonist BIM-28163 unexpectedly leads to an increase in food intake and bodyweight in vivo [184]. NOX-B11, a spiegelmers, stable L-enantiomer RNA-based aptamers, is highly specific bioactive, *n*-octanoylated form of ghrelin which efficiently suppresses ghrelin-induced growth hormone release and food intake in rats [185]. The related NOX-B11–2 also promotes weight loss when administered chronically to diet-induced obese mice [186, 187]. Although the current results are promising, further clinical studies are still required to establish NOX-B11 as a weight loss agent in obese human. Recently, Pfizer has been granted license to take forward development of NOX-B11. Ghrelin may be implicated in at least a part of the effectiveness of the Roux-en-Y gastric bypass. It has been suggested that this particular procedure is also associated with a substantial decrease in the levels of circulating ghrelin, but this conclusion remains controversial [188].

Melanin-Concentrating Hormone Antagonists

Melanin-concentrating hormone (MCH) is a potent orexigenic neuropeptide produced by neurons of the lateral hypothalamic area (LHA) [189]. Mice in which the MCH gene has been ablated are hypophagic and lean [190]. Chronic infusion of MCH significantly increases food intake, body weight, white adipose tissue (WAT) mass, and liver mass in mice on a moderately high-fat diet [191]. This suggested that overexpression of MCH might result in obesity [192]. Two receptors for MCH have been identified: MCH1-R and MCH2-R [193]. MCH1R receptor in mice plays an important role in regulation of energy homeostasis through multiple actions on locomotor activity, metabolism, appetite, and neuroendocrine function [194]. MCH1R antagonists has demonstrated efficacy for the treatment of obesity and related comorbidities in preclinical models, suggesting the potential for improved therapy in humans [195]. SNAP-7941, a selective, high-affinity MCH1 receptor (MCH1-R) antagonist, inhibited food intake, reduced consumption of palatable food, and, after chronic administration to rats with diet-induced obesity, resulted in a marked, sustained decrease in body weight [196]. MCH-1 receptor antagonism with T-226296 decreases food intake in DIO rats by selectively reducing meal size [197]. Chronic administration of SCH-A to DIO mice decreased food intake, body weight, and adiposity and plasma leptin and free fatty acids [198]. Recently, S38151 antagonizes food intake when injected intra-cerebroventricularly in the rats [199]. Several MCH-1 receptor antagonists are under development and characterization for the treatment of obesity.

Melanocortin Receptor Selective (MC4R) Agonists: Much effort is currently underway to develop melanocortin receptor selective compounds for clinical use, with a focus on the development of selective MC4R agonists to treat obesity. Melanocortins are the peptide derived from pro-opiomelanocortin (POMC) in arcuate neurons and project to many sites in and beyond the hypothalamus, where various types of melanocortin receptor are found [200]. Recent work on melanocortin system has provided evidence that hypothalamic MC3/4 receptors are involved in the regulation of energy homeostasis as revealed by genetic and pharmacological evidence. α-MSH is an agonist of MC3/4 receptors that acts mainly on the melanocortin-4 receptor (MC4-R) to inhibit feeding, and increase energy expenditure [201]. A synthetic analogue of α-MSH, melanoton II (MT II) exhibits agonistic properties at both MC3/4 receptors resulting in decreased food intake and body weight in rodents [202, 203]. The drug was well tolerated in clinical trials but was associated with mild to moderate side effects that include flushing, somnolence, nausea, vomiting, headache, taste disturbances, and hypertension. Similarly, long-term administration of MT II in both normal and obese rats was found to be associated with increased blood pressure and heart rate [204]. For these reasons, the FDA has expressed concerns about safety and further trials have been halted. Recently, a non-peptide MC4R agonist, MK-0493, exhibited increase energy expenditure and weight loss in diet-induced obese rats. When administered to human subjects at well-tolerated doses, only a marginally significant increase in energy expenditure was observed [205]. These results suggest that agonists for MC3/4 receptor may be an important perspective on pharmacology, physiology, and obesity drug discovery.

Neuropeptide Y

NPY is one of the most potent stimulators of food intake and has been shown to act in the perifornical hypothalamus [206]. NPY or its analogues stimulated the food intake in rat and mice leading to obesity, when injected into the paraventricular neurons or the adjacent perifornical region of the lateral hypothalamic area (PF-LHA) [207]. This property of NPY makes it an obvious choice as an antiobesity drug. Most evidence indicates that the Y1 and Y5 receptors are critical to the effects of NPY on ingestive behavior [208]. A highly selective Y5 antagonist CGP 71683A, when given intraperitoneally, antagonized NPY-induced feeding and inhibited spontaneous food intake in various rats models, and this effect was progressively attenuated over time [209]. Although with prolonged NPY Y5 receptor blockade food intake returned to normal but body weight remained low. These data suggest that NPY Y5 antagonists have a role to play in the treatment of obesity. Another recently studied Y5 antagonist, MK-0557, failed to induce significant amount of weight loss compared to placebo [210]. Thus, solely targeting NPY 5R in the future is unlikely to produce desired therapeutic efficacy. Y1 antagonists also represent an obvious target for obesity treatment. BW1229U91 is a potent Y1 antagonist that inhibits feeding in the rats [211]. With significant support from both preclinical pharmacological and molecular genetic studies, further human trials are warranted.

Conclusion

The neurobiology of obesity is extremely complex with many overlapping and redundant pathways. This complexity decreases the probability that any one treatment approach is superior in the management of obesity. However, pharmacological interventions in addition to lifestyle change (diet and physical activity) and in some cases behavioral modifications show promise. A more detailed understanding of the physiological control of energy balance and the pathophysiology of obesity will continue to inform the development of highly selective, better-targeted compounds. Further, gut hormones are likely to be an ongoing area of intense research in the field of antiobesity drug development. Future advances in this field will include the development of long-acting peptide hormone analogues to overcome the problem of the very short plasma half-life of endogenous peptide hormones, and the introduction of new forms of delivery: slow-release injectable depots, protease-resistant oral formulations, inhalation devices, and transdermal patches. Although newer drugs are years away from clinical use, the hope for research investments made to date is translated into safe and effective antiobesity drugs in the future. The search for novel drug treatments for obesity is both legitimate and necessary.

References

1. Prevention and management of the global epidemic of obesity: report of the WHO consultation on obesity 1997, 3–5 June; Geneva, Switzerland.
2. Ogden CL, Carroll MD, Flegal KM. High body mass index for age among US children and adolescents, 2003–2006. JAMA. 2008;299(20):2401–5.
3. Haslam DW, James WP. Obesity. Lancet. 2005; 366(9492):1197–209.

4. Kopelman PG. Obesity as a medical problem. Nature. 2000;404(6778):635–43.
5. Bouchard C. Current understanding of the etiology of obesity: genetic and nongenetic factors. Am J Clin Nutr. 1991;53(6):1561S–5S.
6. Goran MI, Weinsier RL. Role of environmental vs. metabolic factors in the etiology of obesity: time to focus on the environment. Obes Res. 2000;8(5):407–9.
7. Weinsier RL, Hunter GR, Heini AF, et al. The etiology of obesity: relative contribution of metabolic factors, diet, and physical activity. Am J Med. 1998;105(2): 145–50.
8. Guidelines for the approval and use of drugs to treat obesity. A position paper of The North American Association for the Study of Obesity. Obes Res 1995;3(5):473–8.
9. Atkinson RL. Use of drugs in the treatment of obesity. Annu Rev Nutr. 1997;17:383–403.
10. Bray GA, DeLany J. Opinions of obesity experts on the causes and treatment of obesity–a new survey. Obes Res. 1995;3 Suppl 4:419S–23S.
11. Bray GA, Ryan DH. Drug treatment of the overweight patient. Gastroenterology. 2007;132(6): 2239–52.
12. Finer N. Present and future pharmacological approaches. Br Med Bull. 1997;53(2):409–32.
13. Bray GA. Evaluation of drugs for treating obesity. Obes Res. 1995;3 Suppl 4:425S–34S.
14. Guy-Grand B. Pharmacological approaches to intervention. Int J Obes Relat Metab Disord. 1997;21 Suppl 1:S22–4.
15. Henry JP. A case of myxoedema cured by thyroid extract. Br Med J. 1893;1(1684):737–8.
16. Rivlin RS. Therapy of obesity with hormones. N Engl J Med. 1975;292(1):26–9.
17. Bhasin S, Wallace W, Lawrence JB, Lesch M. Sudden death associated with thyroid hormone abuse. Am J Med. 1981;71(5):887–90.
18. Mittleman RE, Goldberg RB, Nadji M. Severe thyroid atrophy due to prolonged ingestion of thyroid hormone for treatment of obesity. South Med J. 1984;77(2):268–70.
19. Boardman WW. Rapidly developing cataracts after dinitrophenol. Cal West Med. 1935;43(2):118–9.
20. Cutting WC, Rytand DA, Tainter ML. Relationship between blood cholesterol and increased metabolism from dinitrophenol and thyroid. J Clin Invest. 1934; 13(4):547–52.
21. Koch RA, Lee RC, Tainter ML. Dinitrophenol on liver function. Cal West Med. 1935;43(5):337–9.
22. Lesses MF, Myerson A. Human autonomic pharmacology. XVI. Benzedrine sulfate as an aid in the treatment of obesity. 1938. Obes Res. 1994;2(3): 286–92.
23. Mayer J. Genetic, traumatic and environmental factors in the etiology of obesity. Physiol Rev. 1953;33(4):472–508.
24. Jelliffe RW, Hill D, Tatter D, Lewis Jr E. Death from weight-control pills. A case report with objective postmortem confirmation. JAMA. 1969;208(10): 1843–7.
25. Kaplan NM, Jose A. Thyroid as an adjuvant to amphetamine therapy of obesity. A controlled double-blind study. Am J Med Sci. 1970;260(2):105–11.
26. Kattus Jr AA, Biscoe BW, Dashe AM, Davis JH. Spurious heart disease induced by digitalis-containing reducing pills. Arch Intern Med. 1968;122(4):298–304.
27. Smith HJ, Roche AH, Jausch MF, Herdson PB. Cardiomyopathy associated with amphetamine administration. Am Heart J. 1976;91(6):792–7.
28. Follath F, Burkart F, Schweizer W. Drug-induced pulmonary hypertension? Br Med J. 1971;1(5743): 265–6.
29. Soong YS. The treatment of exogenous obesity employing auricular acupuncture. Am J Chin Med (Gard City N Y). 1975;3(3):285–7.
30. Van Itallie TB, Yang MU. Cardiac dysfunction in obese dieters: a potentially lethal complication of rapid, massive weight loss. Am J Clin Nutr. 1984; 39(5):695–702.
31. Silverstone T. Appetite suppressants. A review. Drugs. 1992;43(6):820–36.
32. Weintraub M. Long-term weight control study: conclusions. Clin Pharmacol Ther. 1992;51(5):642–6.
33. Connolly HM, Crary JL, McGoon MD, et al. Valvular heart disease associated with fenfluramine-phentermine. N Engl J Med. 1997;337(9):581–8.
34. Kernan WN, Viscoli CM, Brass LM, et al. Phenylpropanolamine and the risk of hemorrhagic stroke. N Engl J Med. 2000;343(25):1826–32.
35. Astrup A, Breum L, Toubro S. Pharmacological and clinical studies of ephedrine and other thermogenic agonists. Obes Res. 1995;3 Suppl 4:537S–40S.
36. Haller CA, Benowitz NL. Adverse cardiovascular and central nervous system events associated with dietary supplements containing ephedra alkaloids. N Engl J Med. 2000;343(25):1833–8.
37. Daly PA, Krieger DR, Dulloo AG, et al. Ephedrine, caffeine and aspirin: safety and efficacy for treatment of human obesity. Int J Obes Relat Metab Disord. 1993;17 Suppl 1:S73–8.
38. Hallas J, Bjerrum L, Stovring H, Andersen M. Use of a prescribed ephedrine/caffeine combination and the risk of serious cardiovascular events: a registry-based case-crossover study. Am J Epidemiol. 2008;168(8):966–73.
39. Hadvary P, Lengsfeld H, Wolfer H. Inhibition of pancreatic lipase in vitro by the covalent inhibitor tetrahydrolipstatin. Biochem J. 1988;256(2):357–61.
40. Guerciolini R. Mode of action of orlistat. Int J Obes Relat Metab Disord. 1997;21 Suppl 3:S12–23.
41. Zhi J, Melia AT, Guerciolini R, et al. Retrospective population-based analysis of the dose-response (fecal fat excretion) relationship of orlistat in normal and obese volunteers. Clin Pharmacol Ther. 1994; 56(1):82–5.
42. Drent ML, van der Veen EA. Lipase inhibition: a novel concept in the treatment of obesity. Int J Obes Relat Metab Disord. 1993;17(4):241–4.
43. Davidson MH, Hauptman J, DiGirolamo M, et al. Weight control and risk factor reduction in obese

subjects treated for 2 years with orlistat: a randomized controlled trial. JAMA. 1999;281(3):235–42.
44. Drent ML, Larsson I, William-Olsson T, et al. Orlistat (Ro 18-0647), a lipase inhibitor, in the treatment of human obesity: a multiple dose study. Int J Obes Relat Metab Disord. 1995;19(4):221–6.
45. Cavaliere H, Floriano I, Medeiros-Neto G. Gastrointestinal side effects of orlistat may be prevented by concomitant prescription of natural fibers (psyllium mucilloid). Int J Obes Relat Metab Disord. 2001;25(7):1095–9.
46. Halpern A, Mancini MC, Suplicy H, et al. Latin-American trial of orlistat for weight loss and improvement in glycaemic profile in obese diabetic patients. Diabetes Obes Metab. 2003;5(3):180–8.
47. Zhi J, Melia AT, Eggers H, et al. Review of limited systemic absorption of orlistat, a lipase inhibitor, in healthy human volunteers. J Clin Pharmacol. 1995; 35(11):1103–8.
48. McDuffie JR, Calis KA, Booth SL, et al. Effects of orlistat on fat-soluble vitamins in obese adolescents. Pharmacotherapy. 2002;22(7):814–22.
49. Colman E, Fossler M. Reduction in blood cyclosporine concentrations by orlistat. N Engl J Med. 2000;342(15):1141–2.
50. Filippatos TD, Derdemezis CS, Gazi IF, et al. Orlistat-associated adverse effects and drug interactions: a critical review. Drug Saf. 2008;31(1): 53–65.
51. MacWalter RS, Fraser HW, Armstrong KM. Orlistat enhances warfarin effect. Ann Pharmacother. 2003; 37(4):510–2.
52. Ahmed MH. Orlistat and calcium oxalate crystalluria: an association that needs consideration. Ren Fail. 2010;32(8):1019–21.
53. Schwartz SM, Bansal VP, Hale C, et al. Compliance, behavior change, and weight loss with orlistat in an over-the-counter setting. Obesity (Silver Spring). 2008;16(3):623–9.
54. King DJ, Devaney N. Clinical pharmacology of sibutramine hydrochloride (BTS 54524), a new anti-depressant, in healthy volunteers. Br J Clin Pharmacol. 1988;26(5):607–11.
55. Stock MJ. Sibutramine: a review of the pharmacology of a novel anti-obesity agent. Int J Obes Relat Metab Disord. 1997;21 Suppl 1:S25–9.
56. Bach DS, Rissanen AM, Mendel CM, et al. Absence of cardiac valve dysfunction in obese patients treated with sibutramine. Obes Res. 1999;7(4):363–9.
57. Bae SK, Cao S, Seo KA, et al. Cytochrome P450 2B6 catalyzes the formation of pharmacologically active sibutramine (N-{1-[1-(4-chlorophenyl) cyclobutyl]-3-methylbutyl}-N, N-dimethylamine) metabolites in human liver microsomes. Drug Metab Dispos. 2008;36(8):1679–88.
58. Bray GA, Ryan DH, Gordon D, et al. A double-blind randomized placebo-controlled trial of sibutramine. Obes Res. 1996;4(3):263–70.
59. Apfelbaum M, Vague P, Ziegler O, et al. Long-term maintenance of weight loss after a very-low-calorie diet: a randomized blinded trial of the efficacy and tolerability of sibutramine. Am J Med. 1999;106(2): 179–84.
60. Dujovne CA, Zavoral JH, Rowe E, Mendel CM. Effects of sibutramine on body weight and serum lipids: a double-blind, randomized, placebo-controlled study in 322 overweight and obese patients with dyslipidemia. Am Heart J. 2001;142(3): 489–97.
61. Hazenberg BP. Randomized, double-blind, placebo-controlled, multicenter study of sibutramine in obese hypertensive patients. Cardiology. 2000;94(3): 152–8.
62. McMahon FG, Fujioka K, Singh BN, et al. Efficacy and safety of sibutramine in obese white and African American patients with hypertension: a 1-year, double-blind, placebo-controlled, multicenter trial. Arch Intern Med. 2000;160(14):2185–91.
63. Smith IG, Goulder MA. Randomized placebo-controlled trial of long-term treatment with sibutramine in mild to moderate obesity. J Fam Pract. 2001; 50(6):505–12.
64. Florentin M, Liberopoulos EN, Elisaf MS. Sibutramine-associated adverse effects: a practical guide for its safe use. Obes Rev. 2008;9(4):378–87.
65. Ioannides-Demos LL, Proietto J, McNeil JJ. Pharmacotherapy for obesity. Drugs. 2005;65(10): 1391–418.
66. Anderson JW, Greenway FL, Fujioka K, et al. Bupropion SR enhances weight loss: a 48-week double-blind, placebo- controlled trial. Obes Res. 2002;10(7):633–41.
67. Li Z, Maglione M, Tu W, et al. Meta-analysis: pharmacologic treatment of obesity. Ann Intern Med. 2005;142(7):532–46.
68. Fuller RW, Wong DT. Serotonin uptake and serotonin uptake inhibition. Ann N Y Acad Sci. 1990;600:68–78. discussion 9–80.
69. Koe BK, Weissman A, Welch WM, Browne RG. Sertraline, 1S,4S-N-methyl-4-(3,4-dichlorophenyl)-1,2,3,4-tetrahydro-1-naphthylamine, a new uptake inhibitor with selectivity for serotonin. J Pharmacol Exp Ther. 1983;226(3):686–700.
70. Goldstein DJ, Rampey Jr AH, Roback PJ, et al. Efficacy and safety of long-term fluoxetine treatment of obesity–maximizing success. Obes Res. 1995;3 Suppl 4:481S–90S.
71. Wise SD. Clinical studies with fluoxetine in obesity. Am J Clin Nutr. 1992;55(1 Suppl):181S–4S.
72. Wadden TA, Bartlett SJ, Foster GD, et al. Sertraline and relapse prevention training following treatment by very-low-calorie diet: a controlled clinical trial. Obes Res. 1995;3(6):549–57.
73. Darga LL, Carroll-Michals L, Botsford SJ, Lucas CP. Fluoxetine's effect on weight loss in obese subjects. Am J Clin Nutr. 1991;54(2):321–5.
74. Bray GA, Hollander P, Klein S, et al. A 6-month randomized, placebo-controlled, dose-ranging trial of topiramate for weight loss in obesity. Obes Res. 2003;11(6):722–33.

75. Wilensky AJ, Friel PN, Ojemann LM, et al. Zonisamide in epilepsy: a pilot study. Epilepsia. 1985;26(3):212–20.
76. Gadde KM, Franciscy DM, Wagner 2nd HR, Krishnan KR. Zonisamide for weight loss in obese adults: a randomized controlled trial. JAMA. 2003;289(14):1820–5.
77. Gadde KM, Yonish GM, Foust MS, Wagner HR. Combination therapy of zonisamide and bupropion for weight reduction in obese women: a preliminary, randomized, open-label study. J Clin Psychiatry. 2007;68(8):1226–9.
78. Munro JF, MacCuish AC, Marshall A, et al. Weight-reducing effect of diguanides in obese non-diabetic women. Br Med J. 1969;2(5648):13–5.
79. Fendri S, Debussche X, Puy H, et al. Metformin effects on peripheral sensitivity to insulin in non diabetic obese subjects. Diabete Metab. 1993;19(2):245–9.
80. Baysal B, Batukan M, Batukan C. Biochemical and body weight changes with metformin in polycystic ovary syndrome. Clin Exp Obstet Gynecol. 2001;28(4):212–4.
81. Atabek ME, Pirgon O. Use of metformin in obese adolescents with hyperinsulinemia: a 6-month, randomized, double-blind, placebo-controlled clinical trial. J Pediatr Endocrinol Metab. 2008;21(4):339–48.
82. Freemark M, Bursey D. The effects of metformin on body mass index and glucose tolerance in obese adolescents with fasting hyperinsulinemia and a family history of type 2 diabetes. Pediatrics. 2001;107(4):E55.
83. Love-Osborne K, Sheeder J, Zeitler P. Addition of metformin to a lifestyle modification program in adolescents with insulin resistance. J Pediatr. 2008;152(6):817–22.
84. Lee A, Morley JE. Metformin decreases food consumption and induces weight loss in subjects with obesity with type II non-insulin-dependent diabetes. Obes Res. 1998;6(1):47–53.
85. Blankson H, Stakkestad JA, Fagertun H, et al. Conjugated linoleic acid reduces body fat mass in overweight and obese humans. J Nutr. 2000;130(12):2943–8.
86. Gaullier JM, Halse J, Hoye K, et al. Conjugated linoleic acid supplementation for 1 y reduces body fat mass in healthy overweight humans. Am J Clin Nutr. 2004;79(6):1118–25.
87. Larsen TM, Toubro S, Gudmundsen O, Astrup A. Conjugated linoleic acid supplementation for 1 y does not prevent weight or body fat regain. Am J Clin Nutr. 2006;83(3):606–12.
88. Gaullier JM, Halse J, Hoye K, et al. Supplementation with conjugated linoleic acid for 24 months is well tolerated by and reduces body fat mass in healthy, overweight humans. J Nutr. 2005;135(4):778–84.
89. Kafetzopoulos D, Martinou A, Bouriotis V. Bioconversion of chitin to chitosan: purification and characterization of chitin deacetylase from Mucor rouxii. Proc Natl Acad Sci USA. 1993;90(7):2564–8.
90. Pittler MH, Abbot NC, Harkness EF, Ernst E. Randomized, double-blind trial of chitosan for body weight reduction. Eur J Clin Nutr. 1999;53(5):379–81.
91. Mhurchu CN, Poppitt SD, McGill AT, et al. The effect of the dietary supplement, Chitosan, on body weight: a randomised controlled trial in 250 overweight and obese adults. Int J Obes Relat Metab Disord. 2004;28(9):1149–56.
92. Zahorska-Markiewicz B, Krotkiewski M, Olszanecka-Glinianowicz M, Zurakowski A. Effect of chitosan in complex management of obesity. Pol Merkur Lekarski. 2002;13(74):129–32.
93. Kaats GR, Michalek JE, Preuss HG. Evaluating efficacy of a chitosan product using a double-blinded, placebo-controlled protocol. J Am Coll Nutr. 2006;25(5):389–94.
94. Pittler MH, Schmidt K, Ernst E. Adverse events of herbal food supplements for body weight reduction: systematic review. Obes Rev. 2005;6(2):93–111.
95. Andersen T, Fogh J. Weight loss and delayed gastric emptying following a South American herbal preparation in overweight patients. J Hum Nutr Diet. 2001;14(3):243–50.
96. Pittler MH, Ernst E. Guar gum for body weight reduction: meta-analysis of randomized trials. Am J Med. 2001;110(9):724–30.
97. Zhang J, Kang MJ, Kim MJ, et al. Pancreatic lipase inhibitory activity of taraxacum officinale in vitro and in vivo. Nutr Res Pract. 2008;2(4):200–3.
98. Heymsfield SB, Allison DB, Vasselli JR, et al. Garcinia cambogia (hydroxycitric acid) as a potential antiobesity agent: a randomized controlled trial. JAMA. 1998;280(18):1596–600.
99. Preuss HG, Bagchi D, Bagchi M, et al. Effects of a natural extract of (-)-hydroxycitric acid (HCA-SX) and a combination of HCA-SX plus niacin-bound chromium and Gymnema sylvestre extract on weight loss. Diabetes Obes Metab. 2004;6(3):171–80.
100. Vasques CA, Rossetto S, Halmenschlager G, et al. Evaluation of the pharmacotherapeutic efficacy of Garcinia cambogia plus Amorphophallus konjac for the treatment of obesity. Phytother Res. 2008;22(9):1135–40.
101. Keithley J, Swanson B. Glucomannan and obesity: a critical review. Altern Ther Health Med. 2005;11(6):30–4.
102. Anderson RA. Effects of chromium on body composition and weight loss. Nutr Rev. 1998;56(9):266–70.
103. Yazaki Y, Faridi Z, Ma Y, et al. A pilot study of chromium picolinate for weight loss. J Altern Complement Med. 2010;16(3):291–9.
104. Kao YH, Chang HH, Lee MJ, Chen CL. Tea, obesity, and diabetes. Mol Nutr Food Res. 2006;50(2):188–210.

105. Boschmann M, Thielecke F. The effects of epigallocatechin-3-gallate on thermogenesis and fat oxidation in obese men: a pilot study. J Am Coll Nutr. 2007;26(4):389S–95S.
106. Takabayashi F, Tahara S, Kaneko T, Harada N. Effect of green tea catechins on oxidative DNA damage of hamster pancreas and liver induced by N-Nitrosobis (2-oxopropyl)amine and/or oxidized soybean oil. Biofactors. 2004;21(1–4):335–7.
107. Kalman D, Colker CM, Wilets I, et al. The effects of pyruvate supplementation on body composition in overweight individuals. Nutrition. 1999;15(5): 337–40.
108. Stanko RT, Tietze DL, Arch JE. Body composition, energy utilization, and nitrogen metabolism with a 4.25-MJ/d low-energy diet supplemented with pyruvate. Am J Clin Nutr. 1992;56(4):630–5.
109. Kopelman P, Bryson A, Hickling R, et al. Cetilistat (ATL-962), a novel lipase inhibitor: a 12-week randomized, placebo-controlled study of weight reduction in obese patients. Int J Obes (Lond). 2007;31(3):494–9.
110. Bryson A, de la Motte S, Dunk C. Reduction of dietary fat absorption by the novel gastrointestinal lipase inhibitor cetilistat in healthy volunteers. Br J Clin Pharmacol. 2009;67(3):309–15.
111. Kamisuki S, Mao Q, Abu-Elheiga L, et al. A small molecule that blocks fat synthesis by inhibiting the activation of SREBP. Chem Biol. 2009;16(8): 882–92.
112. Arch JR, Ainsworth AT, Cawthorne MA, et al. Atypical beta-adrenoceptor on brown adipocytes as target for anti-obesity drugs. Nature. 1984; 309(5964):163–5.
113. Weyer C, Tataranni PA, Snitker S, et al. Increase in insulin action and fat oxidation after treatment with CL 316,243, a highly selective beta3-adrenoceptor agonist in humans. Diabetes. 1998;47(10):1555–61.
114. Redman LM, de Jonge L, Fang X, et al. Lack of an effect of a novel beta3-adrenoceptor agonist, TAK-677, on energy metabolism in obese individuals: a double-blind, placebo-controlled randomized study. J Clin Endocrinol Metab. 2007;92(2):527–31.
115. Astrup A, Madsbad S, Breum L, et al. Effect of tesofensine on bodyweight loss, body composition, and quality of life in obese patients: a randomised, double-blind, placebo-controlled trial. Lancet. 2008;372(9653):1906–13.
116. Greenway FL, Whitehouse MJ, Guttadauria M, et al. Rational design of a combination medication for the treatment of obesity. Obesity (Silver Spring). 2009; 17(1):30–9.
117. Greenway FL, Fujioka K, Plodkowski RA, et al. Effect of naltrexone plus bupropion on weight loss in overweight and obese adults (COR-I): a multicentre, randomised, double-blind, placebo-controlled, phase 3 trial. Lancet. 2010;376(9741):595–605.
118. Roth BL, Willins DL, Kristiansen K, Kroeze WK. 5-Hydroxytryptamine2-family receptors (5-hydroxytryptamine2A, 5-hydroxytryptamine2B, 5-hydroxytryptamine2C): where structure meets function. Pharmacol Ther. 1998;79(3):231–57.
119. Thomsen WJ, Grottick AJ, Menzaghi F, et al. Lorcaserin, a novel selective human 5-hydroxytryptamine2C agonist: in vitro and in vivo pharmacological characterization. J Pharmacol Exp Ther. 2008;325(2):577–87.
120. Smith SR, Weissman NJ, Anderson CM, et al. Multicenter, placebo-controlled trial of lorcaserin for weight management. N Engl J Med. 2010; 363(3):245–56.
121. Carai MA, Colombo G, Gessa GL. Rimonabant: the first therapeutically relevant cannabinoid antagonist. Life Sci. 2005;77(19):2339–50.
122. Padwal RS, Majumdar SR. Drug treatments for obesity: orlistat, sibutramine, and rimonabant. Lancet. 2007;369(9555):71–7.
123. Van Gaal L, Pi-Sunyer X, Despres JP, et al. Efficacy and safety of rimonabant for improvement of multiple cardiometabolic risk factors in overweight/obese patients: pooled 1-year data from the Rimonabant in Obesity (RIO) program. Diabetes Care. 2008;31 Suppl 2:S229–40.
124. Christensen R, Kristensen PK, Bartels EM, et al. Efficacy and safety of the weight-loss drug rimonabant: a meta-analysis of randomised trials. Lancet. 2007;370(9600):1706–13.
125. Christopoulou FD, Kiortsis DN. An overview of the metabolic effects of rimonabant in randomized controlled trials: potential for other cannabinoid 1 receptor blockers in obesity. J Clin Pharm Ther. 2011;36(1):10–8.
126. Addy C, Wright H, Van Laere K, et al. The acyclic CB1R inverse agonist taranabant mediates weight loss by increasing energy expenditure and decreasing caloric intake. Cell Metab. 2008;7(1): 68–78.
127. Proietto J, Rissanen A, Harp JB, et al. A clinical trial assessing the safety and efficacy of the CB1R inverse agonist taranabant in obese and overweight patients: low-dose study. Int J Obes (Lond). 2010;34(8): 1243–54.
128. Aronne LJ, Tonstad S, Moreno M, et al. A clinical trial assessing the safety and efficacy of taranabant, a CB1R inverse agonist, in obese and overweight patients: a high-dose study. Int J Obes (Lond). 2010;34(5):919–35.
129. Koch L. Obesity: taranabant no longer developed as an antiobesity agent. Nat Rev Endocrinol. 2010; 6(6):300.
130. Maffei M, Halaas J, Ravussin E, et al. Leptin levels in human and rodent: measurement of plasma leptin and ob RNA in obese and weight-reduced subjects. Nat Med. 1995;1(11):1155–61.
131. Ahima RS, Saper CB, Flier JS, Elmquist JK. Leptin regulation of neuroendocrine systems. Front Neuroendocrinol. 2000;21(3):263–307.
132. Cone RD. Anatomy and regulation of the central melanocortin system. Nat Neurosci. 2005;8(5): 571–8.

133. Cowley MA, Smart JL, Rubinstein M, et al. Leptin activates anorexigenic POMC neurons through a neural network in the arcuate nucleus. Nature. 2001;411(6836):480–4.
134. Schwartz MW, Woods SC, Porte Jr D, et al. Central nervous system control of food intake. Nature. 2000;404(6778):661–71.
135. Heymsfield SB, Greenberg AS, Fujioka K, et al. Recombinant leptin for weight loss in obese and lean adults: a randomized, controlled, dose-escalation trial. JAMA. 1999;282(16):1568–75.
136. Van Heek M, Compton DS, France CF, et al. Diet-induced obese mice develop peripheral, but not central, resistance to leptin. J Clin Invest. 1997;99(3): 385–90.
137. Ozcan L, Ergin AS, Lu A, et al. Endoplasmic reticulum stress plays a central role in development of leptin resistance. Cell Metab. 2009;9(1):35–51.
138. Lo KM, Zhang J, Sun Y, et al. Engineering a pharmacologically superior form of leptin for the treatment of obesity. Protein Eng Des Sel. 2005; 18(1):1–10.
139. Wannamethee SG, Tchernova J, Whincup P, et al. Plasma leptin: associations with metabolic, inflammatory and haemostatic risk factors for cardiovascular disease. Atherosclerosis. 2007;191(2): 418–26.
140. Uwaifo GI, Ratner RE. Novel pharmacologic agents for type 2 diabetes. Endocrinol Metab Clin North Am. 2005;34(1):155–97.
141. Hoogwerf BJ, Doshi KB, Diab D. Pramlintide, the synthetic analogue of amylin: physiology, pathophysiology, and effects on glycemic control, body weight, and selected biomarkers of vascular risk. Vasc Health Risk Manag. 2008;4(2):355–62.
142. Aronne L, Fujioka K, Aroda V, et al. Progressive reduction in body weight after treatment with the amylin analog pramlintide in obese subjects: a phase 2, randomized, placebo-controlled, dose-escalation study. J Clin Endocrinol Metab. 2007;92(8):2977–83.
143. Smith SR, Aronne LJ, Burns CM, et al. Sustained weight loss following 12-month pramlintide treatment as an adjunct to lifestyle intervention in obesity. Diabetes Care. 2008;31(9):1816–23.
144. Dunican KC, Adams NM, Desilets AR. The role of pramlintide for weight loss. Ann Pharmacother. 2010;44(3):538–45.
145. Mack CM, Soares CJ, Wilson JK, et al. Davalintide (AC2307), a novel amylin-mimetic peptide: enhanced pharmacological properties over native amylin to reduce food intake and body weight. Int J Obes (Lond). 2010;34(2):385–95.
146. Roth JD, Roland BL, Cole RL, et al. Leptin responsiveness restored by amylin agonism in diet-induced obesity: evidence from nonclinical and clinical studies. Proc Natl Acad Sci USA. 2008;105(20): 7257–62.
147 Ravussin E, Smith SR, Mitchell JA, et al. Enhanced weight loss with pramlintide/metreleptin: an integrated neurohormonal approach to obesity pharmacotherapy. Obesity (Silver Spring). 2009;17(9): 1736–43.
148. Naslund E, King N, Mansten S, et al. Prandial subcutaneous injections of glucagon-like peptide-1 cause weight loss in obese human subjects. Br J Nutr. 2004;91(3):439–46.
149. Verdich C, Flint A, Gutzwiller JP, et al. A meta-analysis of the effect of glucagon-like peptide-1 (7-36) amide on ad libitum energy intake in humans. J Clin Endocrinol Metab. 2001;86(9):4382–9.
150. Poon T, Nelson P, Shen L, et al. Exenatide improves glycemic control and reduces body weight in subjects with type 2 diabetes: a dose-ranging study. Diabetes Technol Ther. 2005;7(3):467–77.
151. Vilsboll T, Zdravkovic M, Le-Thi T, et al. Liraglutide, a long-acting human glucagon-like peptide-1 analog, given as monotherapy significantly improves glycemic control and lowers body weight without risk of hypoglycemia in patients with type 2 diabetes. Diabetes Care. 2007;30(6):1608–10.
152. Rosenstock J, Klaff LJ, Schwartz S, et al. Effects of exenatide and lifestyle modification on body weight and glucose tolerance in obese subjects with and without pre-diabetes. Diabetes Care. 2010;33(6):1173–5.
153. Madsbad S, Kielgast U, Asmar M, et al. An overview of once-weekly GLP-1 receptor agonists - available efficacy and safety data and perspectives for the future. Diabetes Obes Metab. 2011;13: 394–407.
154. Nauck M, Frid A, Hermansen K, et al. Efficacy and safety comparison of liraglutide, glimepiride, and placebo, all in combination with metformin, in type 2 diabetes: the LEAD (liraglutide effect and action in diabetes)-2 study. Diabetes Care. 2009;32(1):84–90.
155. Astrup A, Rossner S, Van Gaal L, et al. Effects of liraglutide in the treatment of obesity: a randomised, double-blind, placebo-controlled study. Lancet. 2009;374(9701):1606–16.
156. Day JW, Ottaway N, Patterson JT, et al. A new glucagon and GLP-1 co-agonist eliminates obesity in rodents. Nat Chem Biol. 2009;5(10):749–57.
157. Stanley S, Wynne K, Bloom S. Gastrointestinal satiety signals III. Glucagon-like peptide 1, oxyntomodulin, peptide YY, and pancreatic polypeptide. Am J Physiol Gastrointest Liver Physiol. 2004;286(5): G693–7.
158. Dakin CL, Small CJ, Batterham RL, et al. Peripheral oxyntomodulin reduces food intake and body weight gain in rats. Endocrinology. 2004;145(6):2687–95.
159. Dakin CL, Gunn I, Small CJ, et al. Oxyntomodulin inhibits food intake in the rat. Endocrinology. 2001;142(10):4244–50.
160. Cohen MA, Ellis SM, Le Roux CW, et al. Oxyntomodulin suppresses appetite and reduces food intake in humans. J Clin Endocrinol Metab. 2003;88(10):4696–701.
161. Wynne K, Park AJ, Small CJ, et al. Subcutaneous oxyntomodulin reduces body weight in overweight and obese subjects: a double-blind, randomized, controlled trial. Diabetes. 2005;54(8):2390–5.

162. Wynne K, Park AJ, Small CJ, et al. Oxyntomodulin increases energy expenditure in addition to decreasing energy intake in overweight and obese humans: a randomised controlled trial. Int J Obes (Lond). 2006;30(12):1729–36.
163. Polak JM, Bloom SR, Rayford PL, et al. Identification of cholecystokinin-secreting cells. Lancet. 1975; 2(7943):1016–8.
164. Dufresne M, Seva C, Fourmy D. Cholecystokinin and gastrin receptors. Physiol Rev. 2006;86(3): 805–47.
165. Gibbs J, Young RC, Smith GP. Cholecystokinin decreases food intake in rats. J Comp Physiol Psychol. 1973;84(3):488–95.
166. Lieverse RJ, Jansen JB, Masclee AA, Lamers CB. Satiety effects of a physiological dose of cholecystokinin in humans. Gut. 1995;36(2):176–9.
167. Pi-Sunyer X, Kissileff HR, Thornton J, Smith GP. C-terminal octapeptide of cholecystokinin decreases food intake in obese men. Physiol Behav. 1982; 29(4):627–30.
168. Crawley JN, Beinfeld MC. Rapid development of tolerance to the behavioural actions of cholecystokinin. Nature. 1983;302(5910):703–6.
169. Matson CA, Reid DF, Ritter RC. Daily CCK injection enhances reduction of body weight by chronic intracerebroventricular leptin infusion. Am J Physiol Regul Integr Comp Physiol. 2002;282(5): R1368–73.
170. Wang L, Barachina MD, Martinez V, et al. Synergistic interaction between CCK and leptin to regulate food intake. Regul Pept. 2000;92(1–3):79–85.
171. Zac-Varghese S, Tan T, Bloom SR. Hormonal interactions between gut and brain. Discov Med. 2010;10(55):543–52.
172. Kojima M, Hosoda H, Date Y, et al. Ghrelin is a growth-hormone-releasing acylated peptide from stomach. Nature. 1999;402(6762):656–60.
173. Yang J, Brown MS, Liang G, et al. Identification of the acyltransferase that octanoylates ghrelin, an appetite-stimulating peptide hormone. Cell. 2008;132(3):387–96.
174. Kamegai J, Tamura H, Shimizu T, et al. Chronic central infusion of ghrelin increases hypothalamic neuropeptide Y and Agouti-related protein mRNA levels and body weight in rats. Diabetes. 2001;50(11): 2438–43.
175. Tamura H, Kamegai J, Shimizu T, et al. Ghrelin stimulates GH but not food intake in arcuate nucleus ablated rats. Endocrinology. 2002;143(9):3268–75.
176. Cummings DE, Purnell JQ, Frayo RS, et al. A preprandial rise in plasma ghrelin levels suggests a role in meal initiation in humans. Diabetes. 2001;50(8): 1714–9.
177. Tschop M, Wawarta R, Riepl RL, et al. Post-prandial decrease of circulating human ghrelin levels. J Endocrinol Invest. 2001;24(6):RC19–21.
178. Cummings DE, Frayo RS, Marmonier C, et al. Plasma ghrelin levels and hunger scores in humans initiating meals voluntarily without time- and food-related cues. Am J Physiol Endocrinol Metab. 2004;287(2):E297–304.
179. Salbe AD, Tschop MH, DelParigi A, et al. Negative relationship between fasting plasma ghrelin concentrations and ad libitum food intake. J Clin Endocrinol Metab. 2004;89(6):2951–6.
180. Tschop M, Weyer C, Tataranni PA, et al. Circulating ghrelin levels are decreased in human obesity. Diabetes. 2001;50(4):707–9.
181. Tschop M, Smiley DL, Heiman ML. Ghrelin induces adiposity in rodents. Nature. 2000;407(6806): 908–13.
182. Wren AM, Seal LJ, Cohen MA, et al. Ghrelin enhances appetite and increases food intake in humans. J Clin Endocrinol Metab. 2001;86(12):5992.
183. Strassburg S, Anker SD, Castaneda TR, et al. Long-term effects of ghrelin and ghrelin receptor agonists on energy balance in rats. Am J Physiol Endocrinol Metab. 2008;295(1):E78–84.
184. Halem HA, Taylor JE, Dong JZ, et al. A novel growth hormone secretagogue-1a receptor antagonist that blocks ghrelin-induced growth hormone secretion but induces increased body weight gain. Neuroendocrinology. 2005;81(5):339–49.
185. Helmling S, Maasch C, Eulberg D, et al. Inhibition of ghrelin action in vitro and in vivo by an RNA-Spiegelmer. Proc Natl Acad Sci USA. 2004; 101(36):13174–9.
186. Kobelt P, Helmling S, Stengel A, et al. Anti-ghrelin Spiegelmer NOX-B11 inhibits neurostimulatory and orexigenic effects of peripheral ghrelin in rats. Gut. 2006;55(6):788–92.
187. Shearman LP, Wang SP, Helmling S, et al. Ghrelin neutralization by a ribonucleic acid-SPM ameliorates obesity in diet-induced obese mice. Endocrinology. 2006;147(3):1517–26.
188. Diniz Mde F, Azeredo Passos VM, Diniz MT. Bariatric surgery and the gut-brain communication–the state of the art three years later. Nutrition. 2010;26(10):925–31.
189. Tritos NA, Maratos-Flier E. Two important systems in energy homeostasis: melanocortins and melanin-concentrating hormone. Neuropeptides. 1999;33(5): 339–49.
190. Shimada M, Tritos NA, Lowell BB, et al. Mice lacking melanin-concentrating hormone are hypophagic and lean. Nature. 1998;396(6712):670–4.
191. Ito M, Gomori A, Ishihara A, et al. Characterization of MCH-mediated obesity in mice. Am J Physiol Endocrinol Metab. 2003;284(5):E940–5.
192. Ludwig DS, Tritos NA, Mastaitis JW, et al. Melanin-concentrating hormone overexpression in transgenic mice leads to obesity and insulin resistance. J Clin Invest. 2001;107(3):379–86.
193. An S, Cutler G, Zhao JJ, et al. Identification and characterization of a melanin-concentrating hormone receptor. Proc Natl Acad Sci USA. 2001; 98(13):7576–81.
194. Marsh DJ, Weingarth DT, Novi DE, et al. Melanin-concentrating hormone 1 receptor-deficient mice are

lean, hyperactive, and hyperphagic and have altered metabolism. Proc Natl Acad Sci USA. 2002;99(5): 3240–5.
195. Kowalski TJ, McBriar MD. Therapeutic potential of melanin-concentrating hormone-1 receptor antagonists for the treatment of obesity. Expert Opin Investig Drugs. 2004;13(9):1113–22.
196. Borowsky B, Durkin MM, Ogozalek K, et al. Antidepressant, anxiolytic and anorectic effects of a melanin-concentrating hormone-1 receptor antagonist. Nat Med. 2002;8(8):825–30.
197. Ito M, Ishihara A, Gomori A, et al. Mechanism of the anti-obesity effects induced by a novel melanin-concentrating hormone 1-receptor antagonist in mice. Br J Pharmacol. 2010;159(2):374–83.
198. Kowalski TJ, Spar BD, Weig B, et al. Effects of a selective melanin-concentrating hormone 1 receptor antagonist on food intake and energy homeostasis in diet-induced obese mice. Eur J Pharmacol. 2006;535(1–3):182–91.
199. Audinot V, Zuana OD, Fabry N, et al. S38151 [p-guanidinobenzoyl-[Des-Gly(10)]-MCH(7-17)] is a potent and selective antagonist at the MCH(1) receptor and has anti-feeding properties in vivo. Peptides. 2009;30(11):1997–2007.
200. Mountjoy KG, Mortrud MT, Low MJ, et al. Localization of the melanocortin-4 receptor (MC4-R) in neuroendocrine and autonomic control circuits in the brain. Mol Endocrinol. 1994;8(10):1298–308.
201. Roselli-Rehfuss L, Mountjoy KG, Robbins LS, et al. Identification of a receptor for gamma melanotropin and other proopiomelanocortin peptides in the hypothalamus and limbic system. Proc Natl Acad Sci USA. 1993;90(19):8856–60.
202. Murphy B, Nunes CN, Ronan JJ, et al. Melanocortin mediated inhibition of feeding behavior in rats. Neuropeptides. 1998;32(6):491–7.
203. Trivedi P, Jiang M, Tamvakopoulos CC, et al. Exploring the site of anorectic action of peripherally administered synthetic melanocortin peptide MT-II in rats. Brain Res. 2003;977(2):221–30.
204. Kuo JJ, Silva AA, Hall JE. Hypothalamic melanocortin receptors and chronic regulation of arterial pressure and renal function. Hypertension. 2003; 41(3 Pt 2):768–74.
205. Krishna R, Gumbiner B, Stevens C, et al. Potent and selective agonism of the melanocortin receptor 4 with MK-0493 does not induce weight loss in obese human subjects: energy intake predicts lack of weight loss efficacy. Clin Pharmacol Ther. 2009;86(6):659–66.
206. Stanley BG, Magdalin W, Seirafi A, et al. The perifornical area: the major focus of (a) patchily distributed hypothalamic neuropeptide Y-sensitive feeding system(s). Brain Res. 1993;604(1–2):304–17.
207. Stanley BG, Kyrkouli SE, Lampert S, Leibowitz SF. Neuropeptide Y chronically injected into the hypothalamus: a powerful neurochemical inducer of hyperphagia and obesity. Peptides. 1986;7(6): 1189–92.
208. Gerald C, Walker MW, Criscione L, et al. A receptor subtype involved in neuropeptide-Y-induced food intake. Nature. 1996;382(6587):168–71.
209. Criscione L, Rigollier P, Batzl-Hartmann C, et al. Food intake in free-feeding and energy-deprived lean rats is mediated by the neuropeptide Y5 receptor. J Clin Invest. 1998;102(12): 2136–45.
210. Erondu N, Gantz I, Musser B, et al. Neuropeptide Y5 receptor antagonism does not induce clinically meaningful weight loss in overweight and obese adults. Cell Metab. 2006;4(4):275–82.
211. Widdowson PS, Henderson L, Pickavance L, et al. Hypothalamic NPY status during positive energy balance and the effects of the NPY antagonist, BW 1229U91, on the consumption of highly palatable energy-rich diet. Peptides. 1999;20(3): 367–72.

Bariatric Surgery for Treatment of Obesity

15

Mark J. Perna, Thomas Karl Byrne, and Chitharanja C. Pullattrana

Abstract

Obesity and obstructive sleep apnea (OSA) are tightly interconnected. Both diseases are rapidly increasing in prevalence and adding tremendous cost to health-care systems worldwide. A multitude of studies have demonstrated bariatric surgery, especially Roux-en-Y gastric bypass, an invaluable solution for many patients. Bariatric surgery not only provides profound weight loss but also offers resolution or significant improvements in obesity-related comorbidities including OSA. After surgical weight loss occurs, the bariatric patient can expect fewer apnea/hypopnea events, improved sleep efficiency, improved or resolved CPAP requirements, better sleepiness score, and improved quality of life. Therefore, these transformations will translate into more rested, more productive, overall safer, and healthier population.

The prevalence of obesity is increasing worldwide and is associated with unfavorable and serious health outcomes including type II diabetes mellitus, hypertension, hyperlipidemia, coronary artery disease, gastroesophageal reflux disease, obstructive sleep apnea (OSA), and early death. As the global obesity epidemic unfolds, so too does an epidemic of complicated and costly chronic medical conditions. In this review, bariatric surgery is evaluated as a valuable tool in the treatment of obesity.

Surgical Procedures for Obesity and Obesity-Related Disorders

Until the 1950s, obesity was not a surgical problem as morbid obesity affected less than <1% of the US adults. Unfortunately, in 2008, morbid obesity rates had ballooned to 6% of US adults affecting approximately 18 million people [1]. With an evolving epidemic of obesity and obesity-related comorbidities, a demand for viable and safe surgical options developed as a result of this mounting problem. Therefore, new

M.J. Perna, MD (✉) • T.K. Byrne, MD • C.C. Pullattrana
Department of Surgery, Medical University of South Carolina, 96th Jonathan Lucas Avenue, Charleston, SC 29403, USA
e-mail: perna@musc.edu

P.J. Shiromani et al. (eds.), *Sleep Loss and Obesity: Intersecting Epidemics*,
DOI 10.1007/978-1-4614-3492-4_15,

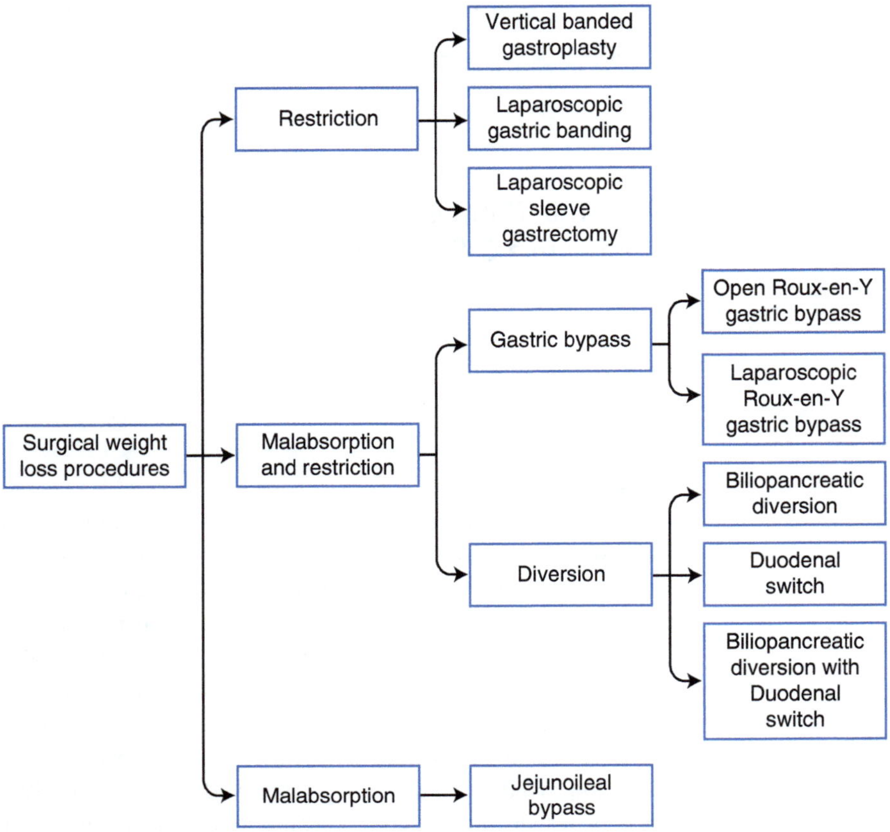

Fig. 15.1 The genealogy of bariatric surgery including current standard procedures as well as early weight loss procedures

surgical procedures were revolutionized to meet this demand and eventually the multibillion-dollar surgical subspecialty of metabolic and bariatric surgery was spawned. Here, a brief review of malabsorptive, restrictive, and combination bariatric surgical procedures will be discussed (Fig. 15.1).

In the 1950s, the first surgical weight loss procedures were performed at the University of Minnesota [2]. The jejunoileal bypass was the first procedure developed specifically for weight loss. Mimicking intestinal "short gut" syndrome, the intestinal bypass operation was solely a malabsorptive procedure; connecting the proximal jejunum to the distal ileum or colon (Fig. 15.2). The rerouted nutrients through the gastrointestinal tract underwent severely altered overall intestinal absorption causing massive weight loss. This procedure offered substantial weight loss, but at the cost of high mortalities and unacceptable late complications. The jejunoileal bypass was completed through large open incisions with long recovery times. The bypass also left a long blind limb of small bowel which was prone to stasis. In addition to weight loss, many patients developed uncontrollable diarrhea, severe electrolyte imbalances, and severe vitamin deficiencies, such as Vitamin A, leading to night blindness. Also, liver failure developed in many of these patients as a result of bacterial overgrowth in the blind jejunal limb [3]. These procedures have since been abandoned and are no longer recognized as a recommended bariatric procedure. Many of these patients were eventually reversed or converted to other operations. However, the concepts of intestinal malabsorption for weight loss paved the way for the future viable weight loss surgery.

The gastric bypass was the next surgical weight loss procedure developed in the 1960s by

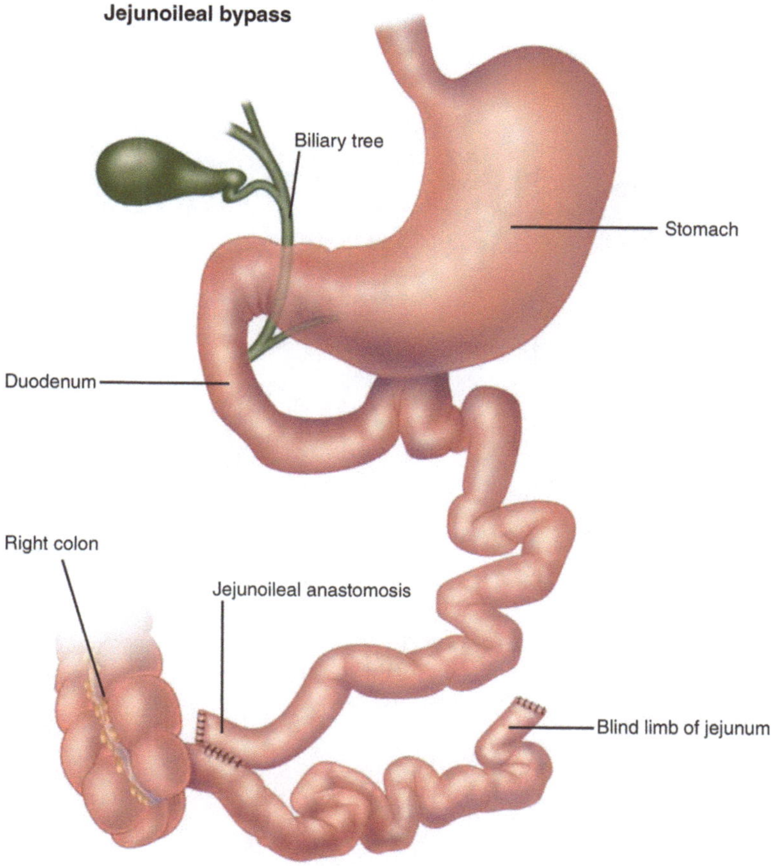

Fig. 15.2 The jejunoileal bypass was the first bariatric surgery and was primarily a malabsorptive procedure. Due to the unacceptably high complication rates and excessive mortality which plagued the jejunoileal bypass, it is now an abandoned procedure

Mason and Ito [4]. Gastric bypass provided two mechanisms for weight loss; gastric restriction via creation of small stomach pouch and intestinal malabsorption via a proximal jejunal bypass approximately 75–150 cm in length (Fig. 15.3). Initially, this procedure also carried relatively high surgical mortality rate as well as list a short- and long-term morbidities. However, gastric bypass offered substantial weight loss and vast improvements in comorbidities for morbidly obese patients. The initial bypass reconstruction performed was a looped gastrojejunal anastomosis [4], which carried an extra dilemma of bile reflux. This procedure eventually evolved into the Roux-en-Y gastric bypass, the preferred reconstruction today. Due to proximal bypass, intestinal malabsorption was less severe in gastric bypass than the jejunoilieal bypass. However, these patients were prone to nutritional deficiencies including folate, B_{12}, vitamin D, and iron.

Unsatisfied by associated late malabsorptive complications of the jejunoileal and gastric bypass, the gastroplasty was developed in the 1970s by Mason and others [5]. This technique focused solely on gastric restriction with normal absorption. One particular attraction to gastroplasty was the relatively low postoperative morbidity and mortality. The failures of horizontal gastroplasty eventually gave rise to vertical banded gastroplasty, which saw its peak in the 1980s (Fig. 15.4). Early weight loss results were promising. However, late weight regain was common particularly in “sweet eaters.” Staple line

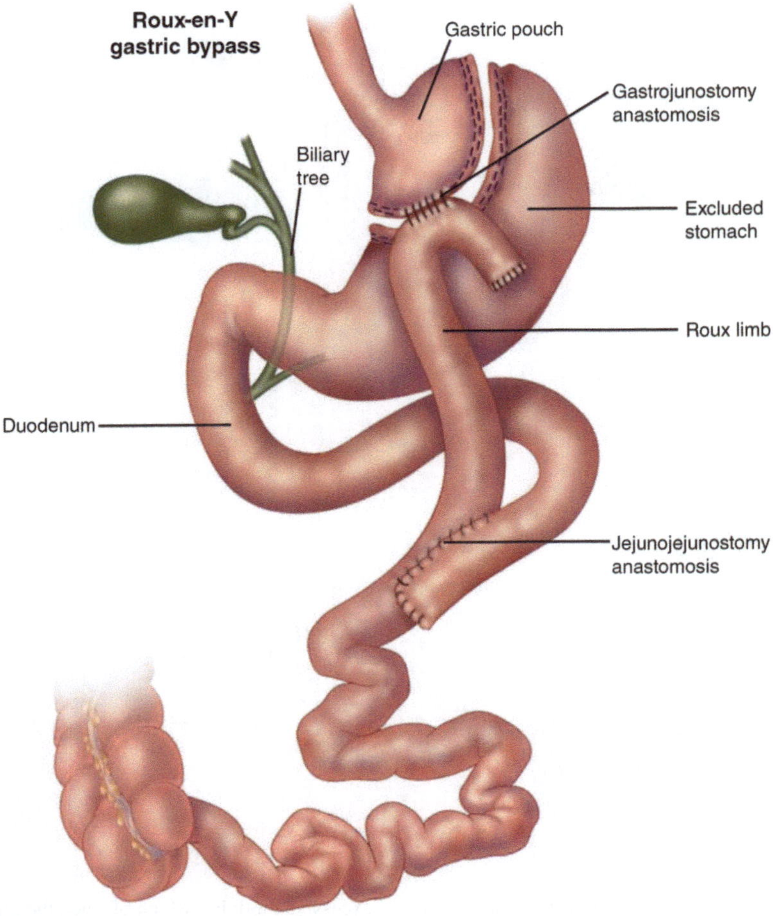

Fig. 15.3 The gastric bypass combined caloric restriction and malabsorption in the same procedure to improve weight loss. This procedure has been refined over many years and currently provides an excellent option in bariatric surgery. It is the most widely performed bariatric procedure

dehiscence occurred frequently, which compromised gastric restriction, and thus weight loss. Finally, band erosions and band migrations also occurred destroying the integrity pouch requiring reoperative surgery. Many of these procedures were eventually converted to Roux-en-Y gastric bypass or other procedures. Therefore, this procedure was largely abandoned as a result of poor long-term results.

Current Bariatric Procedures

Biliopancreatic diversion with duodenal switch was first described in 1988 by Hess and Hess [6], which was a combination of work described by Scopinaro [7] and DeMeester [8]. Like the gastric bypass, this procedure again combines gastric restriction and intestinal malabsorption to synergistically improve weight loss. In brief, the stomach is made into a gastric sleeve, while the duodenum is divided and then connected proximally to the ileum. An ileoileal anastomosis is created reestablishing intestinal continuity (Fig. 15.5). Patients lose between 70% to 90% of excess body weight in some studies and comorbidities often are resolved [9]. This procedure is technically more complex with longer operating times, but it can be done laparoscopically as well as robotically [10, 11]. Although weight loss and comorbidities results are strong, the complications can be devastating. Staple line hemorrhage,

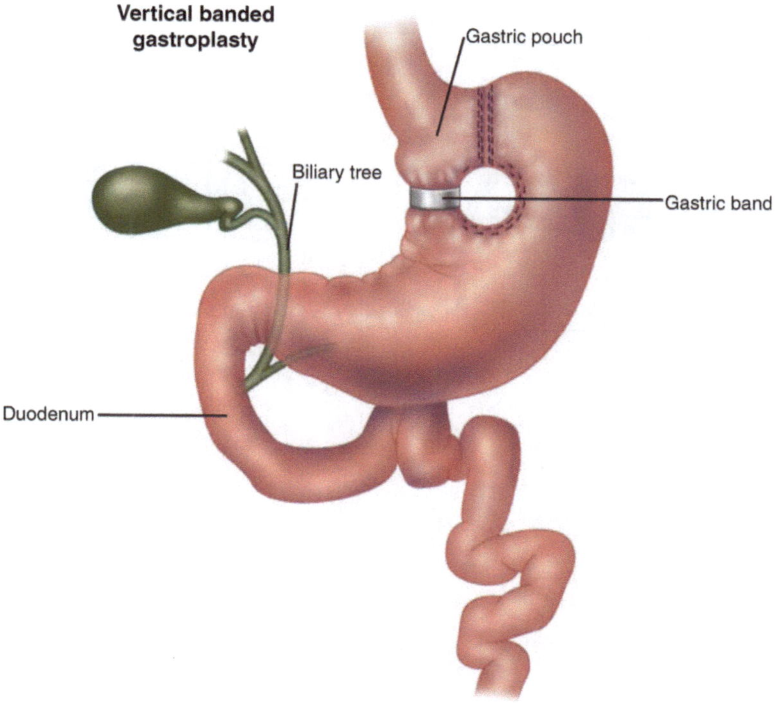

Fig. 15.4 The vertical banded gastroplasty was purely a restrictive procedure. The benefits were the early safety profile with low complication rates. However, the long-term weight regain and "pouch" failures were often seen. This procedure has been largely abandoned; however, it paved the way for other purely restrictive procedures

anastomotic leak particularly at the duodenojejunal connection, and an often fatal duodenal stump leak are among a few of the serious complications of the procedure [9]. Nonetheless, this procedure is currently performed at a few select centers in the US (Table 15.1).

Laparoscopic gastric banding is a second-generation restrictive procedure in which a band is wrapped around the incisura of the stomach (Fig. 15.6). The strengths of this operation are that it is quick, easy, and a generally safe procedure with few serious short-term complications. Also, the current bands are adjustable by increasing or decreasing the amount of gastric restriction experienced by the patient. Finally, patients are less prone to long-term nutritional deficiencies compared to the malabsorptive procedures. However, this procedure does require close follow-up for repeated band fills to optimize subsequent weight loss. Also, the weight loss and comorbidity improvement appears to be less drastic than gastric bypass [12]. Bands may be prone to high rates of late conversions to other procedures due to either poor weight loss or band-related long-term complications. For instance, band slips and band erosions can occur as a late complication requiring a reoperation and can be life threatening [13] (Table 15.1). The laparoscopic gastric band has increased though the 2000s, but its popularity may have already peaked.

Laparoscopic sleeve gastrectomy is a relatively new weight loss procedure described in late 1990s. This procedure converts the stomach from a reservoir to a sleeve-like tube (Fig. 15.7). The sleeve gastrectomy is again a pure restrictive procedure, which appears to have an acceptable safety profile. Weight loss and comorbidity resolution appear better than gastric banding, though not quite as strong as gastric bypass [12]. A particular strength of this operation is that there are no anastomoses obviating the concern for stomal ulceration or internal hernia. Also, this procedure may work well for patients who have had prior

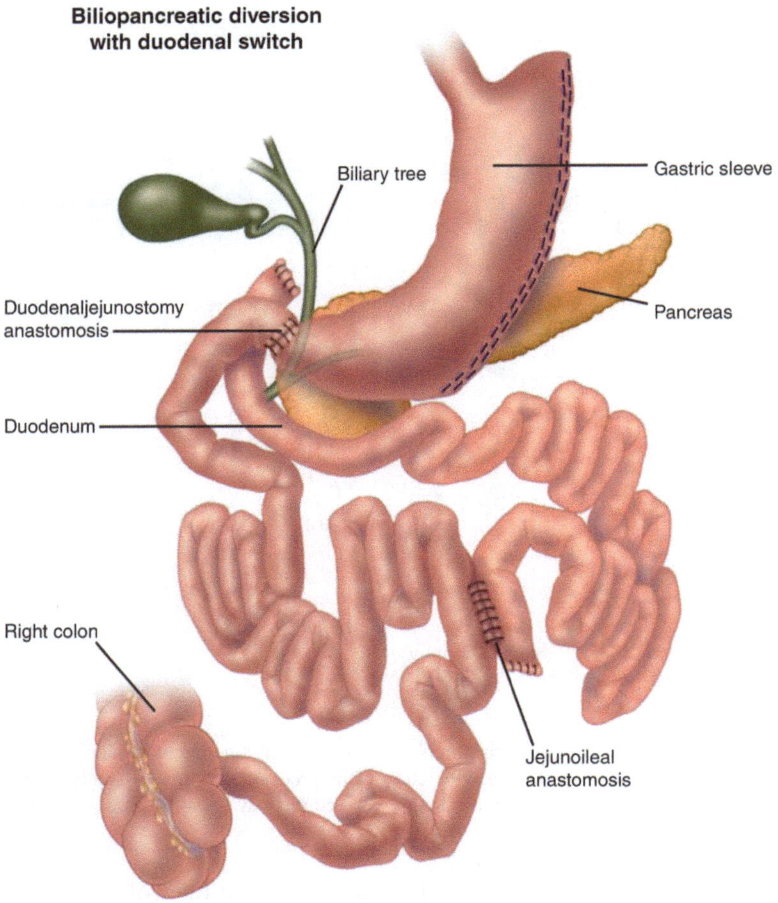

Fig. 15.5 The biliopancreatic diversion with duodenal switch combines caloric restriction and malabsorption to synergistically improve weight loss results. Although this procedure can be done laparoscopically, it is technically challenging and operating times are long. As a result, it has yet to be accepted as a mainstream bariatric procedure

small bowel procedures or extensive pelvic surgeries. Finally, another attractive component of this operation is its ability to be converted to gastric bypass for further weight loss. However, due to the long vertical sleeve, complications do occur. For instance, the staple line can leak causing sepsis or intraabdominal abscess and the staple line can bleed postoperatively (Table 15.1). Recently, the laparoscopic sleeve gastrectomy has increasingly gained popularity as a viable surgical weight loss option in the last decade.

Currently, the Roux-en-Y gastric bypass is the most common bariatric procedure performed in the US. This procedure balances an improving patient safety record with excellent sustained weight loss and comorbidity results. In the mid-1990s, laparoscopic Roux-en-Y gastric bypass was described, which is now the preferred technique. Currently, greater than 90% Roux-en-Y gastric bypasses are performed laparoscopically with relative ease, which improves postoperative pain and recovery while decreasing wound infections [14–16]. Although Roux-en-Y gastric bypass is now safer than ever, it is still plagued by several feared early complications. Anastomotic leak occurs in 1–3% of cases, and can lead to severe sepsis quite promptly. Also, patients are a setup for pulmonary embolism, which can lead to profound respiratory insufficiency in patients with already reduced cardiopulmonary reserve.

Table 15.1 Current bariatric procedures compared

Procedure	30-day mortality (%)	Morbidity (%)	1-year weight loss (−kg/m2)	Approximate excess weight loss (%)	Improvement of resolution of DM	Early complications	Late complications
Laparoscopic gastric banding (LAGB)	0.05	1.4	−7.0 kg/m^2	35–50	44	Few	Band slip, band erosion
Laparoscopic sleeve gastrectomy (LSG)	0.11	5.6	−11.9 kg/m^2	50–69	55	Staple line leak, staple line hemorrhage	Sleeve stricture
Open Roux-en-Y gastric bypass (RYGB)	0.71	15.0	−15.3 kg/m^2	65–75	68	Wound infection, anastomotic leak, staple line hemorrhage, DVT/PE	Internal hernia, stomal stenosis, stomal ulcer, incisional hernia, nutritional hernias
Laparoscopic Roux-en-Y gastric bypass (LRYGB)	0.14	5.9	−15.3 kg/m^2	65–75	83	Anastomotic leak, staple line hemorrhage, DVT/PE	Internal hernia, stomal stenosis, stomal ulcer, port site hernia, nutritional deficiencies
Biliopancreatic diversion with duodenal switch (BPD-DS)	0.57	6.0	~−20 kg/m^2	70–90	~>90	Duodenal stump leak, anastomotic leak, staple line hemorrhage, DVT/PE	Stomal stricture, internal hernia, nutritional deficiencies

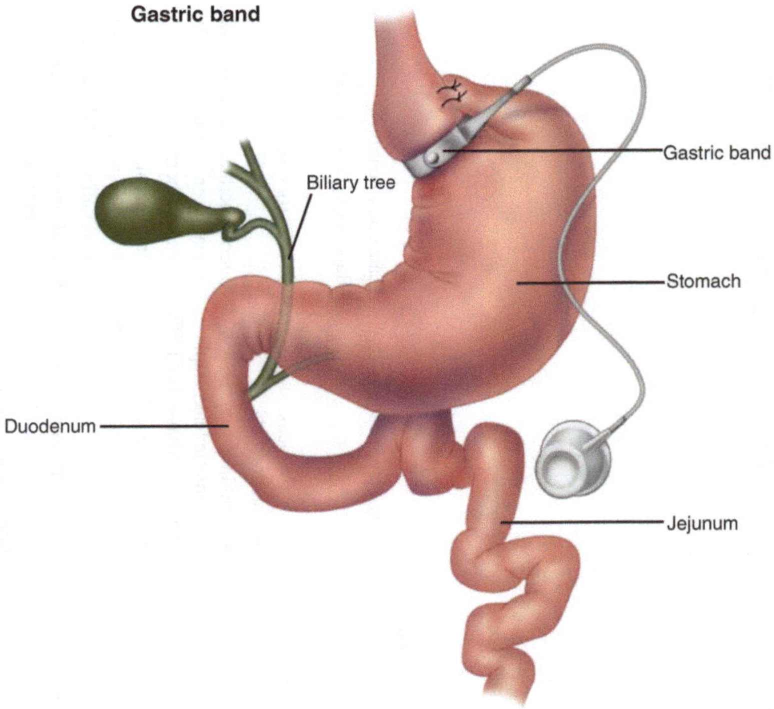

Fig. 15.6 The gastric band is a second-generation purely restrictive procedure and demonstrates an excellent early safety profile; however, long-term safety results are less robust compared to the combined restrictive and malabsorptive procedures

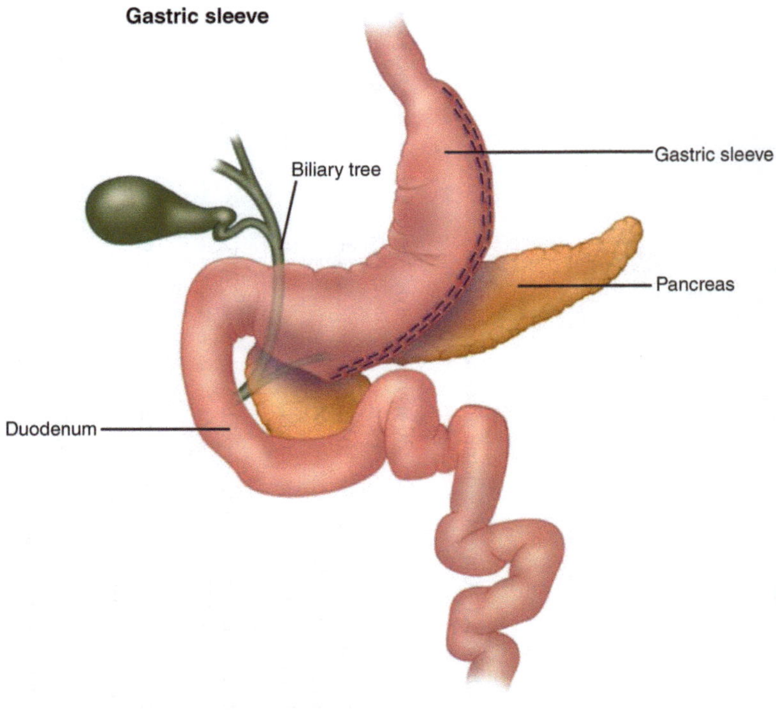

Fig. 15.7 The gastric sleeve is a relatively new gastric restriction procedure, which is quickly gaining popularity as a bariatric surgery. Recent weight loss data following gastric sleeve are promising and may be a great option for patients with previous small bowel surgery

Late complications also occur. The gastrojejunal anastomosis is especially prone to long-term complications such as stomal stenosis and stomal ulcers. Finally, internal hernia can occur at any time period postoperatively, potentially leading to disastrous outcomes such as massive small bowel necrosis (Table 15.1).

Bariatric and metabolic surgery has boomed particularly in the last 20 years. The number of registered ASMBS bariatric surgeons has increase by ninefold in 10 years [17]. In 2009, 220,000 bariatric procedures were performed. However, this was approximately only 1% of the population, which may benefit from these procedures [18]. Currently, the American Society for Metabolic and Bariatric Surgery (ASMBS) indications for bariatric surgery includes [18]:

BMI≥40 kg/m^2
BMI≥35 kg/m^2 with an obesity-related chronic disease

Surgical weight loss (i.e., the weight loss as a direct result of altered anatomy and physiology) mostly occurs in the first year postoperatively. Unfortunately, weight gain can occur after these procedures. This dilemma has led to the multidisciplinary team approach seen in many bariatric clinics often consisting of clinicians, dietitians, nursing staff, and psychologists. Potential bariatric patients should undergo extensive preoperative counseling and education. Also, there needs to be a particular emphasis on thorough long-term follow-up to insure safety and success [19]. It is often stressed to bariatric patients that surgical procedures offer merely a tool for weight loss, and that most optimal results occur with strict adherence to the dietary and lifestyle recommendations (i.e., consuming small but frequent meals, consuming foods high in protein and low in sugars and fats, and regular exercise and activity). Finally, bariatric surgery may influence family members. Recent evidence suggests a patient's lifestyle changes after gastric bypass may alter lifestyle choices in the family. For instance, family members of gastric bypass patients were more likely to increase activity, improve eating habits, and lose some weight after their family member's surgical intervention [20].

In summary, bariatric and metabolic surgery has evolved since the 1950s to meet the demands of a global obesity epidemic. Malabsorptive procedures were developed first, followed by combination malabsorptive and restrictive procedures, and then finally the purely restrictive procedures. The combination procedures likely offer the most significant weight loss and resolution of comorbidities, with the slightly increased risk of serious morbidity and mortality. In contrast, pure restrictive procedures are relatively safer early on, but have less drastic long-term weight loss results. Today, most bariatric procedures including the gastric band, sleeve gastrectomy, Roux-en-Y gastric bypass, and biliopancreatic diversion with duodenal switch can be performed with laparoscopy improving postoperative recovery. Overall, these procedures have been refined and improved offering viable options for treating morbid obesity when medical weight loss has failed.

Obesity as a Risk Factor for Obstructive Sleep Apnea

OSA is an important obesity-related comorbidity which has a multitude of negative health sequelae. OSA is characterized by intermittent upper-airway collapse which leads to impaired ventilation and disrupted sleep (see also Chaps. 9, 11 and 13). In US adults, OSA affects 4–7% of males and 1–2% of females of general population [20]. Therefore, in 2010 some 6–11 million US adults suffer from OSA [21]. Independent risks factors associated with OSA are age (Odds Ratio (OR)—1.52 per increased decade), sex (OR—1.51 for males), obesity (OR—1.55 per 5.3 kg/m^2), and neck size (OR—1.44 per 1.7-inch increment) in addition to breathing pauses, snoring frequency, and snoring loudness [22].

OSA leads to a constellation of poor health outcomes. In addition to excessive daytime sleepiness and diminished quality of life, it is also associated with coronary artery disease, cardiac arrhythmia, congestive heart failure, hypertension, stroke, pulmonary hypertension, insulin resistance, diabetes mellitus, and nonalcoholic fatty liver disease [20, 22–27]. As a result of these complicated health outcomes, OSA increases health-care expenditures compared to age- and gender-matched controls by approximately $1,400 per patient ($p<0.01$) [28]. Also, it is estimated undiagnosed untreated sleep apnea may add 3.4 billion dollars per year in additional health-care costs in the US [28].

The societal impacts of OSA are costly as well as hazardous. OSA affects workers productivity by adding to missed work days and lost work. In a recent study, workers with OSA and excessive daytime sleepiness were 13 times more likely to report recent work disability than those without OSA and excessive daytime sleepiness [29]. Also, excessive daytime sleepiness from OSA is dangerous for the US highways and roads. OSA-related crashes is believed to be responsible for approximately 800,000 crashes/year, 1,400 deaths/year, and cost an estimated 15.9 billion dollars in the USA/year [30]. Effective treatment may save lives and reduce costs [30].

OSA is exceptionally common in morbidly obese bariatric populations. Approximately 70–77% of bariatric patients screened with polysomnography meet American Academy of Sleep Medicine criteria for OSA by apnea/hypopnea index>5 events/h [31–34]. However, unfortunately many of morbidly obese patients have undiagnosed OSA prior to surgical consultation [32].

Bariatric Surgery and OSA

In large longitudinal studies of the general non-bariatric population, small increments of weight gain and loss are associated with changes in apnea/hypopnea index, an objective OSA measure of the number of breathing pauses per hour. There is a proportional relationship between weight loss and a decrease in apnea/hypopnea index. Also, this relationship appears to be more pronounced in males than their female counterparts [35]. Therefore, small decreases in weight appeared to be an important modifiable risk factor for apnea/hypopnea events [35, 36].

Drastic weight loss after bariatric surgery improves comorbidities including diabetes mellitus, hypertension, hyperlipidemia, and gastroesophageal reflux disease [37, 38]. Weight loss and bariatric surgery also appears to be beneficial for OSA. In a meta-analysis of all bariatric procedures including, Roux-en-Y gastric bypass, gastric banding, vertical banded gastroplasty, biliopancreatic diversion or duodenal switch, OSA was resolved in 85.7% (95% CI, 79.2–92.2%). Also, each individual procedure was individually associated with resolution of OSA. In addition, the authors demonstrated from four of the analyzed studies, which included primarily gastric bypass data, apnea/hypopnea index decreased from a baseline by 33.85/h (95% CI, 17.47–50.23/h) [39]. In multiple other studies including Roux-en-Y gastric bypass and vertical banded gastroplasty, weight loss was confirmed to be substantial following these procedures. Additionally, there appeared to be corresponding improvements in apnea/hypopnea index by at least 50% from preoperative levels [40–48]. However, as weight gain often occurs after vertical banded gastroplasty and the majority of patients (54–64%) require eventual revisional bariatric surgery, the gastroplasty is no longer a widely performed bariatric procedure. Therefore, grouping the gastroplasty and gastric bypass results together may skew these results negatively [49, 50].

Specifically, the Roux-en-Y gastric bypass is an effective operation with an improved safety profile. Many studies again demonstrate not only dramatic sustained reductions in excess weight, but also improvement and resolution of comorbidities including diabetes mellitus, hypertension, hyperlipidemia, and gastroesophageal reflux disease. Additionally, Roux-en-Y gastric bypass is also quite therapeutic for OSA. For example, Dhahuwala et al. found weight loss after Roux-en-Y gastric bypass improved multiple comorbidities including OSA. They reported on a series of 157 bariatric patients of which 12 (7.6%) were diagnosed with OSA preoperatively. Postoperatively, 11/12 reported resolution and 1/12 patients with improvement in OSA. However, this small study grossly underestimated the true incidence of OSA in the bariatric population. Additionally, the authors failed to report precise objective measures of OSA [51].

Guardiano et al. presented another small series of 8 bariatric patients which underwent both pre- and postoperative polysomnography at an average of greater than 2 years from Roux-en-Y gastric bypass surgery. They reported a reduction in apnea/hypopnea index after surgery (55±31 vs. 14±17 events/h, $p<0.01$). In this series, 4/8 of patients returned to a normal apnea/hypopnea index, less than 5 events/h. Also, 5/8 of patients required no CPAP postoperatively. Also, their results were blunted by one outlier patient with baseline BMI of 73 kg/m^2, only marginal weight loss postoperatively, and apnea/hypopnea index remained unchanged [52]. Kalra et al. demonstrated Roux-en-Y gastric bypass improved OSA in morbidly obese adolescents. In their series of 10 bariatric adolescents with average age of approximately 17, patients BMI decreased following surgery (60.8±11.1kg/m^2 vs. 41.6±9.5kg/m^2, $p<0.01$). Furthermore, the authors found a dramatic decrease in apnea/hypopnea index(9.1 vs. 0.65 events/h, $p<0.01$) and improvement in lowest asleep 0_2 saturation (82.9±5.7% vs. 91.7±3.6%, $p<0.01$) [53]. This study suggests perhaps younger patients after Roux-en-Y gastric bypass may have superb results in terms of OSA parameters.

Further studies confirm improvement in OSA measures after Roux-en-Y gastric bypass. For instance, Rasheid et al. reported follow-up data from pre- and postoperative polysomnography after Roux-en-Y gastric bypass. In a series of 100 bariatric patients with snoring history, they demonstrated an incidence of OSA was 87%. The 11 patients with repeat polysomnography demonstrated improved objective OSA parameters. For example, Epworth Sleep Scale, a measure of daytime sleepiness, (14±2 vs. 3±1, $p<0.001$), apnea/hypopnea index (56±13 vs. 23±7, $p<0.041$), and REM Latency (261±34 vs.118±12, $p<0.001$) were all decreased significantly postoperatively. Also, the authors noted the lowest asleep O_2 saturation improved (77±5 vs. 86±2, $p<0.001$) as did sleep efficiency (65±5 vs. 85±2, $p<0.001$) [54]. From the same group, Haines et al. reported further follow-up data of 101 patients who underwent both pre- and postoperative polysomnography. The authors again found quite similar improvements in OSA with an average follow-up of 11 months. In addition, they found of the 96/101(95%) of Roux-en-Y gastric bypass patients who required preoperative CPAP/BiPAP, only 31/101(30%) patients required it postoperatively [55].

Roux-en-Y gastric bypass improves symptoms of OSA and pulmonary function tests. Varela et al. demonstrated Roux-en-Y gastric bypass improved sleepiness symptoms and noninvasive ventilation requirements in their series of 56 morbidly obese patients with OSA. At 12 months postoperatively, they report Epworth Sleep Scale recovered from an elevated baseline score to within a normal range (13.7±5.5 to 3.4±1.1, $p<0.05$). Additionally, the percent of patients requiring CPAP decreased from 52% to 0% at 9 months postoperatively [56]. Also, Marti-Varli et al. evaluated 30 patients which required noninvasive ventilation preoperatively. These patients underwent a series of tests, which included room air blood gases, pulmonary function tests, and polysomnography preoperatively. The patients underwent the same battery of testing again at 12 months after surgery. The authors noted improvements in A-a 0_2 gradients (18.42±8.47 mmHg vs. 10.7±6.7 mmHg), improvements in pulmonary function including FEV_1 and FVC back to near 100% of predicted, and again a decrease in apnea/hypopnea index (63±38 vs. 17±16 events/h, $p<0.004$) [57].

The current bariatric restrictive procedures likely benefit OSA postoperatively. For instance, gastric banding as a weight loss procedure has gained much popularity during the last decade and has been evaluated in the OSA parameters of morbidly obese patients. Several studies have reported significant improvements in the apnea/hypopnea index following gastric banding [58–60]. For example, Lettierri et al. found apnea/hypopnea index decreased postoperatively after gastric banding (47.9±33.8 vs. 24.5±18.1 events/h, $p<0.001$). Moreover, the authors demonstrated an improved EDS (100% vs. 47.1%, $p<0.001$), an improved Epworth Sleep Scale score (15.0±4.9 vs. 10.6±4.0, $p<0.001$), and an improved lowest asleep O_2 saturation postoperatively (76.5±12.1 vs. 84.5±5.8,

$p < 0.004$). However, according to the authors, only 4% of patients demonstrated complete resolution of OSA after gastric banding [59]. This result may be consistent with other studies demonstrating gastric banding to be less effective than Roux-en-Y gastric bypass in weight loss and resolution of comorbidities [61–64]. The laparoscopic sleeve gastrectomy has shown promise in OSA treatment. According to Hutter et al., the laparoscopic sleeve gastrectomy appeared to be as effective as the laparoscopic Roux-en-Y gastric bypass and more effective than the laparoscopic gastric band at resolving OSA. However, these results were limited by small numbers of sleeve gastrectomy patients with OSA at one year. Further studies will be needed to confirm the efficacy of laparoscopic sleeve gastrectomy in treating OSA [12].

Recent studies have questioned true efficacy of bariatric surgery for "resolving" OSA. For instance, in a recent meta-analysis, Greenburg et al. reviewed multiple studies which again demonstrated vast improvements in apnea/hypopnea index (54.7 ± 5.6 vs. 15.8 ± 3.2 events/h, $p < 0.001$) after bariatric surgery. According to their analysis, only 25% of patients were cured of OSA by apnea/hypopnea index criteria (apnea/hypopnea index <5), and only 44% of patients obtained apnea/hypopnea index <10. The authors also found age and a follow-up weight of less than 100 kg to be independent predictors of OSA resolution. However, in critique, this meta-analysis demonstrated a high heterogeneity. Additionally, this study included procedures such as vertical banded gastroplasty, now obsolete, and gastric banding, both perhaps negatively affecting results. Finally, as few as 43% of this study population included underwent Roux-en-Y gastric bypass, generally the procedure of choice with the most robust results [65].

Many studies have demonstrated short-term follow-up in general only out to 2 years postoperatively. Few studies have reported on follow-up as far out as 10 years after surgery. In a recent series, Higa et al. reported on 51 patients underwent Roux-en-Y gastric bypass. At 10 years, 79% of patients demonstrated improvement or resolution in OSA. Unfortunately, the authors did not report more objective sleep apnea measures in their study [66]. Additionally, they found sustained results for diabetes mellitus, hypertension, hyperlipidemia, and gastroesophageal reflux disease. This study hinted at the longevity of the Roux-en-Y gastric bypass procedure at resolving or improving comorbidities long-term.

Unfortunately, many people with indications for bariatric surgery will not undergo a therapeutic operation. Harakeh et al. followed patients who were denied Roux-en-Y gastric bypass for insurance-related issues. The denied patients were more likely to report new onset of OSA as well as diabetes mellitus, hypertension, and hyperlipidemia ($p < 0.001$). Furthermore, the denied patients did not lose any weight [67]. Also, Adams et al. studied Roux-en-Y gastric bypass patients compared two well-matched cohorts; one group interested in bariatric surgery but denied by insurance and the other group not interested Roux-en-Y gastric bypass surgery. The Roux-en-Y gastric bypass group experienced strong improvements in apnea/hypopnea index from baseline compared to the control groups (46% vs. 3.6% and 5.4% respectively, $p < 0.0001$). The authors also revealed improved mean nighttime SpO_2 and less percentage nighttime $Sp0_2 < 90\%$ in the Roux-en-Y gastric bypass group compared to the nonoperative groups [68]. Therefore, morbidly obese patients turned away from surgical options pose potentially costly dilemma for already strained health- care systems as they accumulate chronic medical conditions.

Conclusion

In summary, morbid obesity and OSA are tightly interconnected. Both diseases are rapidly increasing in prevalence and adding tremendous cost to health-care systems worldwide. A multitude of studies have demonstrated bariatric surgery, especially Roux-en-Y gastric bypass, an invaluable solution for many patients. Bariatric surgery not only provides profound weight loss but also offers resolution or significant improvements in obesity-related comorbidities including OSA.

After surgical weight loss occurs, the bariatric patient can expect fewer apnea/hypopnea events, improved sleep efficiency, improved or resolved CPAP requirements, better sleepiness score, and improved quality of life. Therefore, these transformations will translate into more rested, more productive, more safe, and overall healthier population.

References

1. Ogden CL, Carroll MD. Prevalence of overweight, obesity, and extreme obesity among adults: United States, trends 1960–1962 through 2007–2008. 2010. http://www.cdc.gov/NCHS/data/hestat/obesity_adult_07_08/obesity_adult_07_08.pdf. Accessed 26 Oct 2011.
2. Buchwald H, Rucker RD. The rise and fall of the jejunoileal bypass. In: Nelson RL, Nyhus LM, editors. Surgery of the small instines. Norwalk, CT: Appleton Century Croft; 1987.
3. Buchwald H, Buchwald JN. Evolution of operative procedures for the management of morbid obesity 1950-2000. Obes Surg. 2002;12(5):705–17.
4. Mason EE, Ito C. Gastric bypass in obesity. Surg Clin North Am. 1967;47(6):1345–51.
5. Mason EE. Development and future of gastroplasties for morbid obesity. Arch Surg. 2003;138(4):361–6.
6. Hess DS, Hess DW. Biliopancreatic diversion with a duodenal switch. Obes Surg. 1998;8(3):267–82.
7. Scopinaro N, Gianetta E, Civalleri D, Bonalumi U, Bachi V. Bilio-pancreatic bypass for obesity: II. Initial experience in man. Br J Surg Sep. 1979;66(9): 618–20.
8. DeMeester TR, Fuchs KH, Ball CS, Albertucci M, Smyrk TC, Marcus JN. Experimental and clinical results with proximal end-to-end duodenojejunostomy for pathologic duodenogastric reflux. Ann Surg. 1987;206(4):414–26.
9. Nelson D, Beekley A, Carter P, Kjorstad R, Sebesta J, Martin M. Early results after introduction of biliopancreatic diversion/duodenal switch at a military bariatric center. Am J Surg. 2011;201(5):678–84.
10. Baltasar A, Bou R, Miro J, Bengochea M, Serra C, Perez N. Laparoscopic biliopancreatic diversion with duodenal switch: technique and initial experience. Obes Surg. 2002;12(2):245–8.
11. Sudan R, Puri V, Sudan D. Robotically assisted biliary pancreatic diversion with a duodenal switch: a new technique. Surg Endosc. 2007;21(5):729–33.
12. Hutter MM, Schirmer BD, Jones DB, et al. First report from the American College of Surgeons Bariatric Surgery Center Network: laparoscopic sleeve gastrectomy has morbidity and effectiveness positioned between the band and the bypass. Ann Surg. 2011;254(3):410–20. discussion 420–2.
13. Himpens J, Cadiere GB, Bazi M, Vouche M, Cadiere B, Dapri G. Long-term outcomes of laparoscopic adjustable gastric banding. Arch Surg. 2011;146(7): 802–7.
14. Nguyen NT, Goldman C, Rosenquist CJ, et al. Laparoscopic versus open gastric bypass: a randomized study of outcomes, quality of life, and costs. Ann Surg. 2001;234(3):279–89. discussion 289–91.
15. Westling A, Gustavsson S. Laparoscopic vs open Roux-en-Y gastric bypass: a prospective, randomized trial. Obes Surg. 2001;11(3):284–92.
16. Perna M, Romagnuolo J, Morgan K, Byrne TK, Baker M. Preoperative hemoglobin A1c and postoperative glucose control in outcomes after gastric bypass for obesity. Surg Obes Relat Dis. 2011 Aug 10. [Epub ahead of print].
17. Nguyen NT, Masoomi H, Magno CP, Nguyen XM, Laugenour K, Lane J. Trends in use of bariatric surgery, 2003-2008. J Am Coll Surg. 2011;213(2):261–6.
18. Taylor K. Metabolic and bariatric surgery: fact sheet. 2011. http://s3.amazonaws.com/publicASMBS/MediaPressKit/MetabolicBariatricSurgeryOverviewJuly2011.pdf. Accessed 26 Oct 2011.
19. Gould J, Ellsmere J, Fanelli R, et al. Panel report: best practices for the surgical treatment of obesity. Surg Endosc. 2011;25(6):1730–40.
20. Kapur VK. Obstructive sleep apnea: diagnosis, epidemiology, and economics. Respir Care. 2010;55(9): 1155–67.
21. Howden LM, Meyer JA. Age and sex composition: 2010 US Census Briefs. U.S. Census Bureau. 2011.
22. Young T, Shahar E, Nieto FJ, et al. Predictors of sleep-disordered breathing in community-dwelling adults: the Sleep Heart Health Study. Arch Intern Med. 2002;162(8):893–900.
23. Daltro C, Araujo L, Cotrim HP. Obstructive sleep apnoea and nonalcoholic fatty liver disease: risk factor or just coincidence? Liver Int. 2008;28(8):1047–9.
24. Mishra P, Nugent C, Afendy A, et al. Apnoeic-hypopnoeic episodes during obstructive sleep apnoea are associated with histological nonalcoholic steatohepatitis. Liver Int. 2008;28(8):1080–6.
25. Norman D, Bardwell WA, Arosemena F, et al. Serum aminotransferase levels are associated with markers of hypoxia in patients with obstructive sleep apnea. Sleep. 2008;31(1):121–6.
26. Polotsky VY, Patil SP, Savransky V, et al. Obstructive sleep apnea, insulin resistance, and steatohepatitis in severe obesity. Am J Respir Crit Care Med. 2009; 179(3):228–34.
27. Verhulst SL, Jacobs S, Aerts L, et al. Sleep-disordered breathing: a new risk factor of suspected fatty liver disease in overweight children and adolescents? Sleep Breath. 2009;13(2):207–10.
28. Kapur V, Blough DK, Sandblom RE, et al. The medical cost of undiagnosed sleep apnea. Sleep. 1999;22(6): 749–55.
29. Omachi TA, Claman DM, Blanc PD, Eisner MD. Obstructive sleep apnea: a risk factor for work disability. Sleep. 2009;32(6):791–8.

30. Sassani A, Findley LJ, Kryger M, Goldlust E, George C, Davidson TM. Reducing motor-vehicle collisions, costs, and fatalities by treating obstructive sleep apnea syndrome. Sleep. 2004;27(3):453–8.
31. Epstein LJ, Kristo D, Strollo Jr PJ, et al. Clinical guideline for the evaluation, management and long-term care of obstructive sleep apnea in adults. J Clin Sleep Med. 2009;5(3):263–76.
32. O'Keeffe T, Patterson EJ. Evidence supporting routine polysomnography before bariatric surgery. Obes Surg. 2004;14(1):23–6.
33. Lopez PP, Stefan B, Schulman CI, Byers PM. Prevalence of sleep apnea in morbidly obese patients who presented for weight loss surgery evaluation: more evidence for routine screening for obstructive sleep apnea before weight loss surgery. Am Surg. 2008;74(9):834–8.
34. Sareli AE, Cantor CR, Williams NN, et al. Obstructive sleep apnea in patients undergoing bariatric surgery–a tertiary center experience. Obes Surg. 2011;21(3):316–27.
35. Newman AB, Foster G, Givelber R, Nieto FJ, Redline S, Young T. Progression and regression of sleep-disordered breathing with changes in weight: the Sleep Heart Health Study. Arch Intern Med. 2005;165(20):2408–13.
36. Peppard PE, Young T, Palta M, Dempsey J, Skatrud J. Longitudinal study of moderate weight change and sleep-disordered breathing. JAMA. 2000;284(23):3015–21.
37. Buchwald H, Avidor Y, Braunwald E, et al. Bariatric surgery: a systematic review and meta-analysis. JAMA. 2004;292(14):1724–37.
38. Perry CD, Hutter MM, Smith DB, Newhouse JP, McNeil BJ. Survival and changes in comorbidities after bariatric surgery. Ann Surg. 2008;247(1):21–7.
39. Buchwald H, Avidor Y, Braunwald E, et al. Bariatric surgery: a systematic review and meta-analysis. JAMA. 2004;292(14):1724–37.
40. Sugerman HJ, Fairman RP, Lindeman AK, Mathers JA, Greenfield LJ. Gastroplasty for respiratory insufficiency of obesity. Ann Surg. 1981;193(6):677–85.
41. Charuzi I, Ovnat A, Peiser J, Saltz H, Weitzman S, Lavie P. The effect of surgical weight reduction on sleep quality in obesity-related sleep apnea syndrome. Surgery. 1985;97(5):535–8.
42. Sugerman HJ, Fairman RP, Baron PL, Kwentus JA. Gastric surgery for respiratory insufficiency of obesity. Chest. 1986;90(1):81–6.
43. Charuzi I, Fraser D, Peiser J, Ovnat A, Lavie P. Sleep apnea syndrome in the morbidly obese undergoing bariatric surgery. Gastroenterol Clin North Am. 1987;16(3):517–9.
44. Rajala R, Partinen M, Sane T, Pelkonen R, Huikuri K, Seppalainen AM. Obstructive sleep apnoea syndrome in morbidly obese patients. J Intern Med. 1991;230(2):125–9.
45. Charuzi I, Lavie P, Peiser J, Peled R. Bariatric surgery in morbidly obese sleep-apnea patients: short- and long-term follow-up. Am J Clin Nutr. 1992;55(2 Suppl):594S–6S.
46. Sugerman HJ, Fairman RP, Sood RK, Engle K, Wolfe L, Kellum JM. Long-term effects of gastric surgery for treating respiratory insufficiency of obesity. Am J Clin Nutr. 1992;55(2 Suppl):597S–601S.
47. Pillar G, Peled R, Lavie P. Recurrence of sleep apnea without concomitant weight increase 7.5 years after weight reduction surgery. Chest. 1994;106(6):1702–4.
48. Grunstein RR, Stenlof K, Hedner JA, Peltonen M, Karason K, Sjostrom L. Two year reduction in sleep apnea symptoms and associated diabetes incidence after weight loss in severe obesity. Sleep. 2007;30(6):703–10.
49. Schouten R, Wiryasaputra DC, van Dielen FM, van Gemert WG, Greve JW. Long-term results of bariatric restrictive procedures: a prospective study. Obes Surg. 2010;20(12):1617–26.
50. Tevis S, Garren MJ, Gould JC. Revisional surgery for failed vertical-banded gastroplasty. Obes Surg. 2011;21(8):1220–4.
51. Dhabuwala A, Cannan RJ, Stubbs RS. Improvement in co-morbidities following weight loss from gastric bypass surgery. Obes Surg. 2000;10(5):428–35.
52. Guardiano SA, Scott JA, Ware JC, Schechner SA. The long-term results of gastric bypass on indexes of sleep apnea. Chest. 2003;124(4):1615–9.
53. Kalra M, Inge T, Garcia V, et al. Obstructive sleep apnea in extremely overweight adolescents undergoing bariatric surgery. Obes Res. 2005;13(7):1175–9.
54. Rasheid S, Banasiak M, Gallagher SF, et al. Gastric bypass is an effective treatment for obstructive sleep apnea in patients with clinically significant obesity. Obes Surg. 2003;13(1):58–61.
55. Haines KL, Nelson LG, Gonzalez R, et al. Objective evidence that bariatric surgery improves obesity-related obstructive sleep apnea. Surgery. 2007;141(3):354–8.
56. Varela JE, Hinojosa MW, Nguyen NT. Resolution of obstructive sleep apnea after laparoscopic gastric bypass. Obes Surg. 2007;17(10):1279–82.
57. Marti-Valeri C, Sabate A, Masdevall C, Dalmau A. Improvement of associated respiratory problems in morbidly obese patients after open Roux-en-Y gastric bypass. Obes Surg. 2007;17(8):1102–10.
58. Dixon JB, Schachter LM, O'Brien PE. Polysomnography before and after weight loss in obese patients with severe sleep apnea. Int J Obes (Lond). 2005;29(9):1048–54.
59. Lettieri CJ, Eliasson AH, Greenburg DL. Persistence of obstructive sleep apnea after surgical weight loss. J Clin Sleep Med. 2008;4(4):333–8.
60. Rao A, Tey BH, Ramalingam G, Poh AG. Obstructive sleep apnoea (OSA) patterns in bariatric surgical practice and response of OSA to weight loss after laparoscopic adjustable gastric banding (LAGB). Ann Acad Med Singapore. 2009;38(7):587.
61. Campos GM, Rabl C, Roll GR, et al. Better weight loss, resolution of diabetes, and quality of life for

laparoscopic gastric bypass vs banding: results of a 2-cohort pair-matched study. Arch Surg. 2011;146(2):149–55.
62. Boza C, Gamboa C, Awruch D, Perez G, Escalona A, Ibanez L. Laparoscopic Roux-en-Y gastric bypass versus laparoscopic adjustable gastric banding: five years of follow-up. Surg Obes Relat Dis. 2010;6(5):470–5.
63. Woodard GA, Peraza J, Bravo S, Toplosky L, Hernandez-Boussard T, Morton JM. One year improvements in cardiovascular risk factors: a comparative trial of laparoscopic Roux-en-Y gastric bypass vs. adjustable gastric banding. Obes Surg. 2010;20(5):578–82.
64. Angrisani L, Lorenzo M, Borrelli V. Laparoscopic adjustable gastric banding versus Roux-en-Y gastric bypass: 5-year results of a prospective randomized trial. Surg Obes Relat Dis. 2007;3(2):127–32. discussion 132–3.
65. Greenburg DL, Lettieri CJ, Eliasson AH. Effects of surgical weight loss on measures of obstructive sleep apnea: a meta-analysis. Am J Med. 2009;122(6):535–42.
66. Higa K, Ho T, Tercero F, Yunus T, Boone KB. Laparoscopic Roux-en-Y gastric bypass: 10-year follow-up. Surg Obes Relat Dis. 2011;7:516–25.
67. Al Harakeh AB, Burkhamer KJ, Kallies KJ, Mathiason MA, Kothari SN. Natural history and metabolic consequences of morbid obesity for patients denied coverage for bariatric surgery. Surg Obes Relat Dis. 2010;6(6):591–6.
68. Adams TD, Pendleton RC, Strong MB, et al. Health outcomes of gastric bypass patients compared to nonsurgical, nonintervened severely obese. Obesity (Silver Spring). 2010;18(1):121–30.

Editor Profiles

Priyattam J. Shiromani, Ph.D. is a Professor in the Department of Psychiatry and Behavioral Science at the Medical University of South Carolina (MUSC) and the Ralph H. Johnson Veterans Affairs Medical Center in Charleston, SC. At MUSC he is the director of the Sleep Research Program of the Neuroscience Institute. Dr. Shiromani's research identifies the neural circuitry underlying sleep and wakefulness and how the brain shifts between states of consciousness. He has pioneered the use of molecular tools to identify key neurons regulating sleep and created orexin-saporin as a tool to dissect wake active neurons. He published the first proof-of-principle studies in animal models that gene therapy can be successfully applied to correct sleep disorders. In recognition of his research, he has received the Young Investigator Award from the Sleep Research Society. He is the author of 100+ research papers in leading scientific journals, and edited a book on post-traumatic stress disorder. His research has been supported by the National Institutes of Health and by the Veterans Administration. He has served as chair of an NIH review group and as a regular member of several NIH review groups. He was the first editor of the Sleep Research Society Bulletin.

Tamas L. Horvath is the Jean and David W. Wallace Professor of Biomedical Research at Yale University. He is Chair of the Section of Comparative Medicine and Funding Director of the Program in Integrative Cell Signaling and Neurobiology of Metabolism at Yale University School of Medicine, New Haven, Connecticut. He is a native of Hungary where he received a Doctor of Veterinary Medicine (D.V.M.) degree from the Faculty of Veterinary Sciences in Budapest, Hungary, and a doctor of philosophy (Ph.D.) degree from the University of Szeged in Hungary. He has been at Yale University since 1990.

His research has been focusing on neuronal circuitries that support physiological and pathological homeostatic conditions, including processes associated with reproduction, energy metabolism, and neurodegeneration. Experiments in his laboratory aim to decipher the intra- and intercellular processes that bring about synaptic plasticity and mitochondrial adaptations.

Susan Redline, M.D., M.P.H. is the Peter C. Farrell Professor of Sleep Medicine, Harvard Medical School and a Senior Physician at Brigham and Women's Hospital and Beth Israel Deaconess Medical Center. She directs

P.J. Shiromani et al. (eds.), *Sleep Loss and Obesity: Intersecting Epidemics*,
DOI 10.1007/978-1-4614-3492-4, © Springer Science+Business Media, LLC 2012

programs in Sleep and Cardiovascular Medicine and Sleep Medicine Epidemiology. For the past 20 years, she has developed a research program addressing the epidemiology of sleep disorders, with a focus on understanding the genetic etiology for sleep apnea and the cardiovascular and neurocognitive consequences of sleep disorders. She has participated in the Institute of Medicines Committee on sleep disorders, is a Board of Director for the American Academy of Sleep Medicine, is Deputy Editor for the journal Sleep, and has played leadership roles in other national initiatives, including the Sleep Research Network.

Eve Van Cauter, Ph.D. is the Frederick H. Rawson Professor of Medicine at the University of Chicago, where she directs the Sleep, Metabolism and Health Center (SMAHC). A native of Belgium, she obtained her M.S. in Physics (1970), M.S. in Actuarial Sciences (1972), and Ph.D. in Biophysics (1977) from the Université Libre de Bruxelles. Her initial research focus was the analysis and modeling of biological rhythms, including development of computer algorithms quantifying circadian and pulsatile variations of hormones, and their relationship to sleep stages and other physiological variables. She joined the Section of Endocrinology in the Department of Medicine at the University of Chicago in 1982.

During the past decade, research in her laboratory has focused on the impact of decreases in sleep duration and quality on markers of health and the interaction of sleep loss with the aging process. Her group identified sleep loss and poor sleep quality as novel risk factors for obesity and diabetes, opening up a new field of enquiry of high relevance to the current epidemic of obesity and diabetes and the increased prevalence of age-related chronic diseases. In recent years, her efforts have focused on the impact of obstructive sleep apnea on the risk and severity of type 2 diabetes. Van Cauter has been the recipient of numerous awards and has authored more than 250 publications.

Index

P.J. Shiromani et al. (eds.), *Sleep Loss and Obesity: Intersecting Epidemics*,
DOI 10.1007/978-1-4614-3492-4, © Springer Science+Business Media, LLC 2012

D

E